Praise for
Worldviews

"Quite simply, this is one of the most accessible – and teachable – introductions to the history and philosophy of science I've seen in over two decades of teaching. DeWitt's exposition and discussion – manifestly honed by extensive classroom teaching experience – are exceptionally clear, and helpfully complemented by some of the best diagrams I've seen. DeWitt thus makes complex ideas and developments cogent and straightforward, especially for undergraduates and those approaching the history and philosophy of science for the first time." *Charles Ess, Drury University*

"Richard DeWitt's *Worldviews* is a splendid introductory text. It is organized around themes – traditions and their overthrow – geared to engage undergraduates. It is historically informed and philosophically sensible. Best of all, it abounds in examples skillfully drawn from the physical sciences and made accessible to the non-specialist. The philosophy of science students encounter through *Worldviews* will strike them as the philosophy of *real* science – the science of Newton, Einstein, Copernicus, and Aristotle – and not some denatured surrogate for science concocted by philosophers so that it might succumb to the tools of their trade." *Laura Ruetsche, University of Pittsburgh*

"This is a brilliantly clear introduction (and indeed reframing) of the history and philosophy of science in terms of world-views and their elements ... In addition, the book is incredibly well-informed from both a scientific and philosophical angle. Highly recommended." *Scientific and Medical Network*

Worldviews
An Introduction to the History and Philosophy of Science

Second Edition

Richard DeWitt

A John Wiley & Sons, Ltd., Publication

This second edition first published 2010
© 2010 Richard DeWitt

Edition history: Blackwell Publishing Ltd (1e, 2003)

Blackwell Publishing was acquired by John Wiley & Sons in February 2007. Blackwell's publishing program has been merged with Wiley's global Scientific, Technical, and Medical business to form Wiley-Blackwell.

Registered Office
John Wiley & Sons Ltd, The Atrium, Southern Gate, Chichester, West Sussex, PO19 8SQ, United Kingdom

Editorial Offices
350 Main Street, Malden, MA 02148-5020, USA
9600 Garsington Road, Oxford, OX4 2DQ, UK
The Atrium, Southern Gate, Chichester, West Sussex, PO19 8SQ, UK

For details of our global editorial offices, for customer services, and for information about how to apply for permission to reuse the copyright material in this book please see our website at www.wiley.com/wiley-blackwell.

The right of Richard DeWitt to be identified as the author of this work has been asserted in accordance with the UK Copyright, Designs and Patents Act 1988.

Wiley also publishes its books in a variety of electronic formats. Some content that appears in print may not be available in electronic books.

Designations used by companies to distinguish their products are often claimed as trademarks. All brand names and product names used in this book are trade names, service marks, trademarks or registered trademarks of their respective owners. The publisher is not associated with any product or vendor mentioned in this book. This publication is designed to provide accurate and authoritative information in regard to the subject matter covered. It is sold on the understanding that the publisher is not engaged in rendering professional services. If professional advice or other expert assistance is required, the services of a competent professional should be sought.

Library of Congress Cataloging-in-Publication Data

DeWitt, Richard.
    Worldviews : an introduction to the history and philosophy of science / Richard DeWitt. – 2nd ed.
        p. cm.
        Includes bibliographical references and index.
    ISBN 978-1-4051-9563-8 (pbk.)
    1. Science–History. 2. Science–Philosophy. I. Title.
    Q125.D38 2010
    509–dc22

A catalogue record for this book is available from the British Library.

Set in 11 on 13 pt Dante by Toppan Best-set Premedia Limited
Printed in Singapore by C.O.S. Printers Pte Ltd

10   2016

For Susie

Contents

List of Figures

Acknowledgments

Countless people made contributions to both the first and second editions of this work. Some contributions were large, some small, but all of them were important. In the first edition, and for this edition as well, numerous anonymous reviewers provided helpful feedback, sometimes catching outright mistakes and sometimes providing good suggestions for clarifying discussions. I would like to acknowledge their contributions. Over the years my philosophy of science students have read various earlier drafts of this book, and lately drafts of additional chapters, and they provided terrific feedback on which ideas worked and which did not, and which explanations were clear and which not so clear. There are too many of them to name, but I would like to thank them all for their help. Likewise I cannot begin to name, but I do appreciate, all the colleagues who have discussed these issues with me, read portions of the manuscript, and helped me clarify and oftentimes correct my thinking on various issues. I would again like to note the contributions of Charles Ess of Drury University, and Marc Lange of the University of Washington, both of whom read the drafts of the entire first edition, most of which is included in this edition, and provided lengthy, detailed, and helpful comments and suggestions (not to mention saving me from several embarrassing mistakes). In addition I'd like to thank Todd Disotell and Shara Bailey of the Center for the Study of Human Origins, New York University, for their help in clarifying the material on evolution in an invigorating 2009 seminar on evolution, and the Faculty Resource Network for the financial support for that seminar. Finally, I would again like to thank my editor, Jeff Dean, for once again helping with the content of this work as well as smoothly facilitating the process of bringing it to press.

RD

Introduction

This book is intended primarily for those coming to the history and philosophy of science for the first time. If this description fits you, welcome to a fascinating territory to explore. This field involves some of the most deep, difficult, and fundamental questions there are. But at the same time, the "lens of science," so to speak, focuses these questions more sharply than they are often otherwise focused. I hope you enjoy this field as much as I do, and I especially hope your appetite is whetted to the point where you will want to return to explore these subjects in more depth.

This sort of introductory work provides special challenges. On the one hand, I want to be accurate with the history, the philosophy, and the interconnections between the two. On the other hand, I want to avoid the level of detail and minutiae that might swamp someone approaching this subject for the first time. Those of us who do history and philosophy of science full-time – most of us are academics – tend to get caught up in the details of our disciplines, and I think we often lose sight of what such detail must look like to those new to the subject. When faced with the minutiae, newcomers often come away with the sense "Why would anyone care about *that*?"

The question is an understandable one. The details and minutiae are important, but their importance can be understood only in the context of a broader picture. So I hope, in this text, to paint one such broader picture. But although this text provides a rather broad-brushstroke picture, to the best of my knowledge what I say is accurate, though it admittedly leaves out a good deal of detail.

The connections between history, science, and philosophy are endlessly complex and fascinating. As mentioned, I hope to whet your appetite, to make you want to explore these issues in more detail, and perhaps even come to appreciate and enjoy the minutiae. Nothing would please me more than if, at the

Worldviews: An Introduction to the History and Philosophy of Science.   Richard DeWitt
© 2010 Richard DeWitt

end of this book, you visit your bookstore, or fire up your web browser, and order works that will enable you to explore these topics further.

A Note on the Structure of the Book

In the barest of outlines, my approach is (a) to introduce some fundamental issues in the history and philosophy of science; (b) to explore the transition from the Aristotelian worldview to the Newtonian worldview; and (c) to explore challenges to our own western worldview brought on by recent developments, most notably relativity theory, quantum theory, and evolutionary theory.

To accomplish these goals, the book is divided into three parts. Part I provides an introduction to some fundamental issues in the history and philosophy of science. Such issues include the notion of worldviews, scientific method and reasoning, truth, evidence, the contrast between empirical facts and philosophical/conceptual facts, falsifiability, and instrumentalism and realism. The relevance of and interconnections between these topics are illustrated throughout Parts II and III.

In Part II, we explore the change from the Aristotelian worldview to the Newtonian worldview, noting the role played by some of the philosophical/conceptual issues involved in this change. Of particular interest is the role played by certain philosophical/conceptual "facts" that are central to the Aristotelian worldview. Discussion of these beliefs serves to illustrate many of the issues from Part I, and also sets the stage for the discussion, in Part III, of some of our own philosophical/conceptual "facts" that we must abandon in light of recent discoveries.

Part III provides an introduction to recent discoveries and developments, most notably relativity theory, quantum theory, and evolutionary theory. As we explore these, we will see that these new discoveries and developments require substantial changes in some of the key beliefs that almost everyone in the western world was raised with. And having emphasized, in Part II, the role played by philosophical/conceptual beliefs in the Aristotelian worldview, we now see that some of the beliefs we have long taken as obvious empirical facts turn out, in light of recent developments, to be mistaken philosophical/conceptual "facts."

At this point in time it is clear that changes in our overall view of the world will be required as recognition of these mistaken philosophical/conceptual beliefs becomes more widespread. It is difficult to say at this point just what shape these changes will take, but it is becoming increasingly likely that our grandchildren will inherit a view of the world substantially different from our own. I hope you enjoy exploring and thinking about not only the changes that have taken place in the past, but also the changes we find ourselves in the midst of.

At the end of the book, in the Chapter Notes and Suggested Reading, I provide further information on some of the topics discussed, as well as suggestions for where to find additional information on these topics. As mentioned, nothing would please me more than if, at the end of this work, you find yourself interested in further investigating these issues.

A final note on the structure of the book: although this book is intended to be read as a whole, and its three main parts are connected in the ways described above, it is possible to read Parts I, II, and III more or less independently of each other. For example, those more interested in the scientific revolution of the 1600s and the development of Newtonian science and the Newtonian worldview, and less interested in related issues in the philosophy of science, could largely start with Chapter 9, at the beginning of Part II. I would, however, encourage such readers to take at least a quick pass through Chapters 1, 3, 4, and 8. Likewise, readers interested primarily in more recent developments in science, especially relativity theory, quantum theory, and evolutionary theory, could jump immediately to Chapter 23, at the beginning of Part III. I would encourage such readers to take at least a quick look at Chapters 3 and 8.

Once again: I hope you enjoy your exploration.

Part I
Fundamental Issues

In Part I, we explore some preliminary and basic issues involved in the history and philosophy of science. In particular, we will discuss the notion of worldviews, truth, evidence, empirical facts versus philosophical/conceptual facts, common types of reasoning, falsifiability, and instrumentalism and realism. These topics provide the necessary background for our exploration, in Part II, of the transition from the Aristotelian worldview to the Newtonian worldview, and also for our exploration, in Part III, of recent developments that challenge our own view of the world.

Chapter One

Worldviews

The main goal of this chapter is to introduce the notion of a *worldview*. As with most of the topics we will explore in this book, the notion of a worldview turns out to be substantially more complex than it first appears. We will begin, though, with a relatively straightforward characterization of this notion. Then as the book progresses, and we come to appreciate more about the Aristotelian worldview and about our own worldview, we will come to a better appreciation of some of the complexities involved.

Although the term "worldview" has been used fairly widely for over 100 years, it is not a term that carries a standard definition. So it is worth taking a moment to clarify how I will be using the term. In the shortest of descriptions, I will use "worldview" to refer to a system of beliefs that are interconnected in something like the way the pieces of a jigsaw puzzle are interconnected. That is, a worldview is not merely a collection of separate, independent, unrelated beliefs, but is instead an intertwined, interrelated, interconnected *system* of beliefs.

Often, the best way to understand a new concept is by way of an example. With this in mind, let's begin with a look at the Aristotelian worldview.

Aristotle's Beliefs and the Aristotelian Worldview

In the western world, what I am calling the Aristotelian worldview was the dominant system of beliefs from about 300 BC to about AD 1600. This worldview was based on a set of beliefs articulated most clearly and thoroughly by Aristotle (384–322 BC). It is worth noting that the term "Aristotelian worldview" refers not so much to the collection of beliefs held specifically by Aristotle himself, but rather to a set of beliefs shared by a large segment of western culture after his death and that were, as noted, largely based on his beliefs.

Worldviews: An Introduction to the History and Philosophy of Science.   Richard DeWitt
© 2010 Richard DeWitt

To understand the Aristotelian worldview, it will be easier to begin with Aristotle's own beliefs. Following this, we will discuss some of the ways these beliefs evolved in the centuries after the death of Aristotle.

Aristotle's beliefs

Aristotle held a large number of beliefs that are radically different from the beliefs we hold. Here are a few examples:

(a)　The Earth is located at the center of the universe.

(b)　The Earth is stationary, that is, it neither orbits any other body such as the sun, nor spins on its axis.

(c)　The moon, the planets, and the sun revolve around the Earth, completing a revolution about every 24 hours.

(d)　In the sublunar region, that is, the region between the Earth and the moon (including the Earth itself) there are four basic elements, these being earth, water, air, and fire.

(e)　Objects in the superlunar region, that is, the region beyond the moon including the moon, sun, planets, and stars, are composed of a fifth basic element, ether.

(f)　Each of the basic elements has an essential nature, and this essential nature is the reason why the element behaves as it does.

(g)　The essential nature of each of the basic elements is reflected in the way that element tends to move.

(h)　The element earth has a natural tendency to move toward the center of the universe. (That's why rocks fall straight down, since the center of the Earth is the center of the universe.)

(i)　The element water also has a natural tendency to move toward the center of the universe, but its tendency is not as strong as that of the earth element. (That's why, when dirt and water are mixed, both tend to move downward, but the water will eventually end up above the dirt.)

(j)　The element air naturally moves toward a region that is above earth and water, but below fire. (That's why air, when blown into water, bubbles up through the water.)

(k)　The element fire has a natural tendency to move away from the center of the universe. (That's why fire burns upward, through air.)

(l)　The element ether, which composes objects such as the planets and stars, has a natural tendency toward perfectly circular movement. (That's why the planets and stars continuously move in circles about the Earth, that is, about the center of the universe.)

(m)　In the sublunar region, an object in motion will naturally tend to come to a halt, either because the elements composing it have reached their natural place in the universe, or far more often because something (for example, the surface of the Earth) prevents them from continuing toward their natural place.

(n) An object that is stationary will remain stationary, unless there is some source of motion (either self-motion, as when an object moves toward its natural place in the universe, or an external source of motion, as when I push my pen across my desk).

These beliefs are only a small, small handful of Aristotle's views. He also had extensive views on ethics, politics, biology, psychology, the proper method for conducting scientific investigations, and so on. Like most of us, Aristotle held thousands of beliefs, most of which were quite different from ours.

Importantly, Aristotle's beliefs were anything but a random collection of beliefs. When I say that the beliefs were not random, part of what I mean is that he had good reason to believe most of them, and the beliefs were far from naive. Every single one of the beliefs listed above turned out to be wrong, but given the data available at the time, every one of them was quite justified. To take just one example, the best scientific data of Aristotle's time strongly indicated that the Earth was at the center of the universe. The belief turned out to be wrong, but naive it was not.

By saying the beliefs were not random, I also mean that they form an interrelated, interlocking *system* of beliefs. To illustrate the ways in which Aristotle's beliefs were interrelated and interlocking, consider a wrong way and a right way of picturing them.

First, the wrong picture, which I will illustrate by an analogy with grocery lists. When most of us make grocery lists, we end up with a haphazard collection of items related only by the fact that we can, we hope, find them when we get to the grocery store. We could organize our grocery lists – with the dairy items in one part of the list, the bakery items in another part, and so on – but most of us simply do not bother. And the result is a haphazard list with no particular relation between the items on it.

When you think of Aristotle's beliefs, do not think of them as like a grocery list of unrelated items. That is, do not picture the collection of beliefs as like the somewhat haphazard list in Figure 1.1. Instead, here is a better picture. Think of the collection of beliefs as like a jigsaw puzzle. Each piece of the puzzle is a particular belief, with the pieces fitting together in a coherent, consistent, interrelated, interlocking fashion, as the pieces of a jigsaw puzzle fit together. That is, picture Aristotle's system of beliefs more as it appears in Figure 1.2.

The jigsaw puzzle metaphor illustrates the key features of the way I am using the notion of a worldview. First, pieces of a jigsaw puzzle are not independent and isolated; rather, puzzle pieces are interconnected. Each piece of a puzzle fits with the piece next to it, and that piece fits with the pieces next to it, and so on. All the pieces are interconnected and interrelated, and the overall result is a system in which the individual pieces fit together into an interlocking, interconnected, coherent, and consistent whole.

(a)	The Earth is at the center of the universe.
(b)	The Earth is stationary.
(c)	The moon, planets, and sun revolve around the Earth about every 24 hours.
(d)	Objects in the sublunar region are composed of the four basic elements: earth, water, air and fire.
(e)	Objects in the superlunar region are composed of the basic element ether.
(f)	Each element behaves as it does because of its essential nature.
(g)	The essential nature of each of the basic elements is reflected in the way that element tends to move.
(h)	The element earth has a natural tendency to move in a straight line toward the center of the universe.
(i)	...
(j)	...

Figure 1.1 A "grocery list" of Aristotle's beliefs

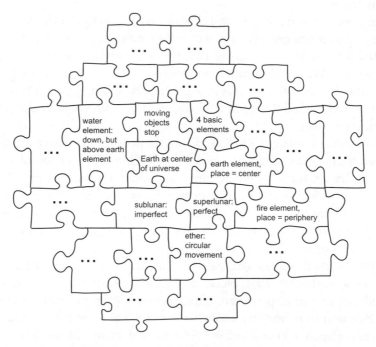

Figure 1.2 Aristotle's "jigsaw puzzle" of beliefs

Likewise, Aristotle's beliefs fit together, forming an interlocking, consistent system. Each belief is closely tied to the beliefs around it, and those beliefs in turn are closely tied to their surrounding beliefs, and so on.

To take just one example of how Aristotle's beliefs fit together, consider the belief that the Earth is the center of the universe. This belief is closely interconnected with the belief that the element earth has a natural tendency to move toward the center of the universe. After all, the Earth itself is composed primarily of the earthy element, so the belief that the earthy element naturally goes toward the center of the universe, and the belief that the Earth itself is at the center of the universe, fit together nicely. Likewise, both of these beliefs are closely tied to the belief that an object will move only if there is a source of motion. Just as my pen will remain stationary unless something moves it, so too with the Earth. Having long ago moved to the center of the universe, or as close to the center as they could, the heavy elements comprising the Earth will now remain stationary, because there is nothing powerful enough to move an object as massive as the Earth. All of these beliefs are, in turn, closely connected to the belief that the basic elements have essential natures, and the belief that objects behave as they do largely because of the essential natures of the elements out of which they are composed. Again, the general point is that Aristotle's beliefs are interconnected in the way the pieces of a jigsaw puzzle are interconnected.

In addition, notice that in a jigsaw puzzle there are differences between the core pieces of the puzzle and the peripheral pieces. Because of the interconnections, a central core piece cannot be replaced with a different-shaped piece without replacing almost the entire puzzle. A piece near the periphery, however, can be replaced with relatively little alteration in the rest of the puzzle.

In a similar vein, among Aristotle's beliefs we can distinguish between core and peripheral beliefs. Peripheral beliefs can be replaced without much alteration in the overall worldview. For example, Aristotle believed there were five planets (not counting the sun, moon, and Earth). Five planets are all that can be distinguished without the technology of recent years. But had there arisen evidence, say, of a sixth planet, Aristotle could easily have accommodated this new belief without much alteration in his overall system of beliefs. This ability of a belief to change without substantially altering the overall system of beliefs is typical of a peripheral belief.

In contrast, consider the belief that the Earth was stationary and at the center of the universe. In Aristotle's system of beliefs, this is a core belief. Importantly, this is a core belief *not* because of the depth of conviction Aristotle had in it, but rather because, like a puzzle piece near the center, it cannot be removed and replaced without dramatically altering the beliefs to which it was connected, which in turn would require altering almost his entire system of beliefs.

To illustrate this, suppose Aristotle tried to replace his belief that the Earth was the center of the universe, and replace it with, say, the belief that the sun was the center. Could Aristotle simply remove this belief, this piece of the puzzle, and replace it with a new belief that the sun is the center, and do so while still keeping most of the rest of the jigsaw puzzle intact?

The answer is no, because the new belief, that the sun is the center of the universe, would not fit into the rest of the jigsaw puzzle. For example, heavy objects clearly fall toward the center of the Earth. If the center of the Earth is not the center of the universe, then Aristotle's belief that heavy objects (those composed mainly of the heavy elements earth and water) have a natural tendency to move toward the center of the universe has to be replaced as well. This in turn requires replacing a multitude of other interconnected beliefs, such as the belief that objects have essential natures that cause them to behave as they do. In short, trying to replace just the one belief requires replacement of all the beliefs with which it is interconnected, and in general, it would require building an entirely new jigsaw puzzle of beliefs.

Again, this is all to reinforce the idea that Aristotle's beliefs were not a random, haphazard collection of beliefs, but were rather an interconnected, jigsaw puzzle-like system of beliefs. This notion that individual beliefs fit together to form an interlocking, consistent system of beliefs is the key idea behind the way I will use the notion of a worldview. In short, when I speak of a worldview, think of the jigsaw puzzle analogy.

The Aristotelian worldview

Thus far, we have primarily discussed Aristotle's own beliefs, and one might get the impression that a worldview involves a particular individual's jigsaw puzzle of beliefs. People do sometimes speak this way. There is a sense in which each of us has a somewhat different system of beliefs, a slightly different worldview, from everyone else. And our individual systems of beliefs, of course, are part of what makes us the individuals we are.

But a more important sense of "worldview," for this book, is a more generalized notion. For example, much of the western world, from the death of Aristotle to the 1600s, shared a more or less Aristotelian way of looking at the world. This certainly does not mean that everyone believed exactly what Aristotle did, or that the system of beliefs was not added to or modified during this period.

For example, at various times during this period, Judaic, Christian, and Islamic philosopher-theologians mixed Aristotelian beliefs with religious beliefs, and these sorts of mixtures illustrate some of the ways in which Aristotelian beliefs were modified in the centuries after his death. There were also groups who took a distinctly non-Aristotelian view of the universe. For example, there were groups whose beliefs were based more closely on the ideas of Plato (428–348 BC) rather than Aristotle, and such Platonic-based belief systems provided an alternative to the Aristotelian worldview. (Plato, incidentally, was Aristotle's teacher, though Aristotle's views would eventually diverge substantially from those of Plato.)

In spite of such modifications to Aristotle's beliefs, and in spite of the existence of groups taking a non-Aristotelian view of the world, the belief systems of large segments of the western world, from about 300 BC to the 1600s, were very much in the Aristotelian spirit. The belief that the Earth was the center of the universe, that objects had essential natures and natural tendencies, that the sublunar region was a place of imperfection and the superlunar region a place of perfection, and so on, were part of the consensus of most of the western world. And these group beliefs fit together much like the beliefs of an individual fit together – into an interlocking, consistent, coherent system of beliefs. And it is this group jigsaw puzzle of beliefs, very much in the spirit of Aristotle's beliefs, that I will have in mind when I speak of the Aristotelian worldview.

The Newtonian Worldview

As an example to contrast with the Aristotelian worldview, let's look briefly at a different system of beliefs. Early in the 1600s, new evidence (largely from the newly invented telescope) arose that indicated the Earth moved around the sun. As discussed above, one cannot simply replace the Earth-centered piece of the Aristotelian jigsaw puzzle without replacing virtually all of the pieces of that puzzle. The discovery meant that the Aristotelian worldview was no longer viable. The story is fascinating and complex, and we will explore it more later in the book, but for now, suffice it to say that eventually a new system of beliefs emerged. In particular, the new system included a belief in a moving Earth.

Call the worldview that eventually replaced the Aristotelian worldview the *Newtonian worldview*. This worldview has as its foundation the work of Isaac Newton (1642–1727) and his contemporaries, but it has been added to considerably over the years. As with the Aristotelian view, the Newtonian worldview has associated with it a large number of beliefs. Here are some examples:

1 The Earth revolves on its axis, completing a revolution approximately every 24 hours.
2 The Earth and planets move in elliptical orbits around the sun.
3 There are slightly more than 100 basic elements in the universe.
4 Objects behave as they do largely because of the influence of external forces. (For example, gravity, which is why rocks fall.)
5 Objects such as planets and stars are composed of the same basic elements as objects on Earth.
6 The same laws that describe the behavior of objects on Earth (for example, an object in motion tends to remain in motion) also apply to objects such as planets and stars.

And so on for the other thousands of beliefs that compose the Newtonian worldview.

This is the worldview that most of us in the western world have been raised on. And the same story applies to the beliefs that compose the Newtonian worldview as applies to the Aristotelian worldview. In particular, the Newtonian worldview comprises a system of beliefs that tie together as the pieces of a jigsaw puzzle tie together, forming a coherent, consistent, interlocking system of beliefs. While both the Aristotelian and Newtonian systems of beliefs are coherent and consistent, they are very different jigsaw puzzles, with quite different core beliefs.

The change from the Aristotelian to the Newtonian worldview was a dramatic change, and much of the story of Part II of this book involves this transition. As we will see, this transition was spurred, in large part, by new discoveries in the early 1600s. Later, in Part III, we will explore some rather surprising recent discoveries. In something like the way the new discoveries in the 1600s required a change in the existing jigsaw puzzle of beliefs, so too the discoveries of recent decades require a change in our jigsaw puzzle of beliefs.

Concluding Remarks

Before concluding this introduction to the notion of worldviews, I want to make two quick observations. The first deals with the evidence we have for the beliefs that comprise our worldview, and the second concerns the apparent common-sense nature of many of the beliefs comprising our worldview.

Evidence

We have been speaking a great deal about beliefs and, presumably, people have reasons for holding the beliefs they do. That is, we would seem to have some sort of *evidence* for the beliefs we hold.

For example, presumably you believe Aristotle was wrong, and that the Earth is not the center of the universe. Instead, you most likely believe that the sun is the center of our solar system, and the Earth and other planets move around the sun. I suspect you have good evidence for this belief. But I also suspect that your evidence is not what you think it is. Pause for a few seconds and ask "Why do I believe the Earth moves around the sun? What is the evidence I have?" Seriously, put this book down for a few seconds and ponder these questions.

Ready? First, consider whether you have any direct evidence for your belief that the Earth moves around the sun. When I say "direct evidence," this is what I have in mind: when I ride my bicycle, I have direct evidence that I am moving. I feel the movement of the bike, I feel the wind in my face, I see myself moving past other objects, and so on. Do you have any direct evidence of this sort that the Earth is moving around the sun? It seems not. We do not feel like we are moving, nor do we feel constant high winds in our face. In fact, when you look out the window, it looks for all the world as if the Earth is stationary.

If you think about your reasons for your belief in a moving Earth, I think you will find you have no direct evidence – none at all – that the Earth is moving around the sun. Yet your belief is certainly a reasonable belief, and you certainly have some sort of evidence for it. But rather than direct evidence, the evidence you have is more like this: try for a moment to believe that the Earth does *not* move around the sun. Do you see that that belief does not fit in with your other beliefs? For example, the belief does not fit with your belief that your teachers, for the most part, have told you the truth. It does not fit with your belief that, for the most part, what you read in authoritative books is accurate. It does not fit with your belief that the experts in our society could not possibly be that wrong about something so basic. And so on.

The general point is that you believe the Earth moves around the sun largely because that belief fits in with the other pieces in your jigsaw puzzle of beliefs, and the opposite belief does not fit into that jigsaw puzzle. In other words, your evidence for that belief is closely tied with your jigsaw puzzle of beliefs, that is, with your worldview.

Incidentally, it would not be unreasonable to think that even if we ourselves do not have direct evidence that the Earth moves about the sun, surely experts in astronomy and related fields have such evidence. But as we will see in later chapters, even our experts do not have such direct evidence. This is not by any means to suggest that there is not good evidence that the Earth moves about the sun. There is good evidence. But that evidence is much more indirect than I think it is often assumed to be. And this is typical of many (probably most) of our beliefs.

In summary, we have direct evidence for a surprisingly small number of the beliefs we hold. For most of our beliefs (maybe almost all of them), we believe them largely because of the way they fit in with a large package of interconnecting beliefs. In other words, we believe what we do largely because of the way our beliefs fit into our worldview.

Common sense

Most of us were raised with the Newtonian worldview, and most of the beliefs mentioned in connection with the Newtonian worldview seem almost like common sense. But think about it a minute – such beliefs are anything but common sense. For example, it does not look as if the Earth moves around the sun. As mentioned above, if you look out the window, you will see that the Earth appears to be perfectly stationary. It also appears that the sun, stars, and planets move around the Earth approximately every 24 hours. And consider the belief that you likely learned at an earlier stage in your education, that objects in motion tend to remain in motion. Most people I know take this to be an obvious truth.

But in our everyday experience, objects in motion do nothing of the sort. For example, thrown frisbees do not remain in motion. They soon hit the ground and stop. Thrown baseballs do not remain in motion. Even if they are not caught by someone else, they soon roll to a halt. In our everyday experience, *nothing* remains in motion.

My point is that, in general, although most of us share those beliefs, the beliefs mentioned above as part of the Newtonian worldview are *not* beliefs we arrive at by common sense or by common experience. But most of us were raised with the Newtonian worldview, and since these beliefs were taught to us from an early age, they now look to us to be the obviously correct beliefs. But think about it: if we had been raised with the Aristotelian worldview, then the Aristotelian beliefs would have seemed equally like common sense.

In short, from within the perspective of any worldview, the beliefs of that worldview will appear to be the obviously correct ones. So the fact that our basic beliefs seem to be correct, seem to be common sense, seem to be obviously right, is not particularly good evidence that those beliefs are correct.

This raises the following interesting issue: there is no doubt that the Aristotelian worldview turned out to be badly wrong. The Earth is not the center of the universe, objects do not behave the way they do because of internal "essential natures," and so on. Importantly, it is not just that the individual beliefs were wrong; rather, the jigsaw puzzle formed by that system of beliefs turned out to be the wrong *sort* of jigsaw puzzle. The universe, we now think, is not anything like the way it was conceptualized from within the Aristotelian worldview. Nonetheless, although wrong, those beliefs formed a consistent system of beliefs, and a system whose beliefs seemed, for almost 2,000 years, to be obviously right and commonsensical.

Might our jigsaw puzzle, our worldview, turn out to be equally incorrect, even though our system of beliefs is consistent and seems to us to be obviously correct and commonsensical? There is no doubt that some of our individual beliefs will turn out to be wrong. But the question I am asking is whether our entire way of looking at the world might turn out to be the wrong way of looking at the world, in something like the way the Aristotelian worldview turned out to be the wrong sort of jigsaw puzzle.

Or to put the same question another way: when we look at the Aristotelian worldview, many of the beliefs of that worldview strike us as quaint and curious. If we think about our descendants, say hundreds of years in the future – or even if we think about our grandchildren or great grandchildren – might our own beliefs, those that seem to you and me to be so obviously correct and commonsensical, look to them to be equally quaint and curious?

These are interesting questions. Toward the end of the book, we will explore some recent discoveries that suggest that some parts of our worldview might indeed turn out to be the wrong sort of way of looking at the world. But for now, we will leave these as questions to ponder, and move on to our next topic.

Chapter Two

Truth

This chapter and the next focus on two related topics, truth on the one hand, and facts on the other. These topics are somewhat unusual for a book on the history and philosophy of science, but I think they are worth considering early on, largely to dispel some common misconceptions and oversimplifications.

It seems to be a fairly widespread belief that the accumulation of facts is a relatively straightforward process, and that science is, in large part at least, geared toward generating true theories that account for such facts. Both of these are largely misconceptions about facts, truth, and their relations to science. One of the goals of the next two chapters is to show that these issues are much more complex than is often appreciated. In addition, as we will begin to see in this chapter and the next, and as will become increasingly clear as the book progresses, the relationship between facts, truth, and science is much more complex and controversial than the simple view suggested above – that is, of science as a process of generating true theories to account for straightforward facts.

Preliminary Issues

We think the belief that the Earth moves around the sun, which is part of our worldview, is true, and the belief that the Earth is stationary while the sun moves around it, which is common in the Aristotelian worldview, is false. Within our system of beliefs, it seems to us obviously true that the Earth moves about the sun, and it seems to us there are innumerable facts that prove this belief to be true. But within the Aristotelian worldview, it seemed equally obvious that the Earth was stationary, and within that system of beliefs there seemed to be equally many facts that proved the Earth did not move. What is the difference between our beliefs and those beliefs? If our belief about the Earth is really true, and their belief is really false, what makes the one belief true and the other false? More generally, what is truth?

Worldviews: An Introduction to the History and Philosophy of Science.   Richard DeWitt
© 2010 Richard DeWitt

A common reaction to this question is to say that facts are what make a belief true. For example, one commonly hears that there are facts that prove the Earth moves about the sun, and these facts are what make the belief true. Interestingly, facts and truth are often defined in terms of one another. People often answer the question "What is truth?" by saying that true beliefs are those supported by facts. And the question "What is a fact?" is often answered by saying that facts are the things that are true. In fact (no pun intended), my dictionary defines truth as "a verified or indisputable fact," and then turns around and defines a fact as "that which is known to be true."

But this sort of circularity – defining truth in terms of facts and facts in terms of truth – sheds no light on our questions. What is truth? What is a fact? What is the difference between true/factual beliefs and false/nonfactual beliefs? What makes some beliefs true/factual and others false/nonfactual?

Before tackling these questions directly, take a moment to reflect on how much we take the subject of truth for granted. We all have a large stock of beliefs, and we think our beliefs are true. After all, why else would we believe them? Chances are, you would not have bought this book if you did not believe that most of what is going to be said in the book is true. If you are reading this book as part of a college course, chances are you are devoting a huge share of resources, both time and money, to attend that college, and you certainly would not do so if you did not think that you would learn a sizable share of true things during your college tenure. And think about history or, for that matter, current events, both of which are filled with various incidents (wars, assassinations, religious persecutions, and so on) motivated in large part by the conviction that certain sets of beliefs are true and others false. So even if you have not thought about it explicitly, the issue of truth is likely to be of substantial interest to you. Truth is something we take for granted every minute of every day, and often with consequences that are far from trivial.

But rarely do we reflect on the subject of truth. As mentioned, one of the main goals of this chapter is to shed some light on the subject of truth, and to appreciate some of the complexities involved. We are not going to answer our questions about truth definitively – such questions have been debated at least since the beginning of philosophy and science. Since no consensus has emerged over the past 2,000 years, it is unlikely that a consensus will emerge by the end of this chapter. But some standard views of truth have emerged over the years, and we can at least get an outline of these standard views, and in doing so, come to appreciate some of the complexities.

Clarifying the Question

In inquiries such as this, it is usually a good idea to be clear on, and to keep in mind, the question that is being addressed. It is also worth distinguishing the question being addressed from other, perhaps related, questions.

When I ask the question "What is truth?" the central question I have in mind is this: What makes true statements or true beliefs true? And what is it that makes false statements (or beliefs) false? In other words, what do true statements (or beliefs) have in common that makes them true, and what do false statements (or beliefs) have in common that makes them false?

This central question about truth is often confused with an *epistemological* question about truth. Generally speaking, epistemology is the study of knowledge, and epistemology is an important branch of philosophy. A key epistemological question about truth – how do we come to know which statements and beliefs are true? – is an important question, but again, not the key question with which we are here concerned.

Consider an analogy. Suppose we have a tract of forest, and we are interested in knowing which of the trees in this tract are oak trees. In this case, our main question would be an epistemological one – how might we come to know which trees are oaks? Employing the services of a forestry expert would be an excellent way to answer this question – we can come to know which of the trees are oaks by paying attention to what the forestry expert tells us. That the forestry expert identifies a tree as an oak is not what makes that tree an oak. In other words, the question "How do we come to know which trees are oaks?" is a different question from "What makes a tree an oak?"

And just as there is presumably something that oak trees have in common that makes them oak trees, so too presumably there is something that true statements (or true beliefs) have in common that makes them true. And that is the key question in which we are interested: What do true statements (or beliefs) have in common that makes them true?

Over the years, there have been a large number of theories of truth that have been offered as possible answers to our central question. Most such theories fall into one of two categories. We will call theories that fall into the first of these categories *correspondence theories of truth*, while theories that fall into the second category we will call *coherence theories of truth*. These are not the only types of theories of truth that have been proposed, but these two categories cover much of the territory, and will serve to illustrate many of the complexities surrounding truth. Also worth noting is that, at this point, we will not be concerned with all of the specific versions of correspondence and coherence theories. Where appropriate, we will mention some of the more notable varieties. Let us begin with correspondence theories of truth.

Correspondence Theories of Truth

In a nutshell, according to correspondence theories of truth, what makes a true belief true is that the belief corresponds to reality. What makes a false belief false is that the belief fails to correspond to reality.

For example, if "The Earth moves about the sun" is true (as most of us think it is), what makes it true is that, in reality, the Earth really does move about the sun. That is, what makes this belief true is that the belief corresponds to the way things really are. Likewise, if "The Earth is stationary with the sun moving around it" is false, it is false because it fails to correspond to reality.

"Reality" is a term used in a variety of ways, so to understand correspondence theories of truth, it is crucial to appreciate how this term is being used. In this context, "reality" most definitely does *not* refer to what you or I *believe* reality to be like. Generally speaking, what you and I believe reality to be like has no effect on what reality really is like. Likewise, what our best scientists believe reality to be, or what the majority of the population believes reality to be, or what a yoga master in an enlightened state of mind believes reality to be, has little effect on what reality really is. As used in the correspondence theory of truth, "reality" is not "your reality," "my reality," "Timothy Leary's reality," the reality of an acquaintance under the influence of strong hallucinogens, or any such thing. Instead, "reality" refers to "real" reality: a reality that is completely objective, generally independent of us, and generally speaking in no way depends on what people believe that reality to be like.

There are, of course, some uninteresting ways in which some of our beliefs might affect certain aspects of reality. For example, I might believe it is too warm in my living room, and so turn down the thermostat. In this way, certain of my beliefs might lead to a change in a certain aspect of reality, such as the air temperature in my living room. But a proponent of a correspondence theory would maintain that, in general, our beliefs do not affect reality.

So to summarize: According to correspondence theories of truth, what makes a belief true is that it corresponds to an independent, objective reality. What makes a false belief false is that it fails to correspond to that reality.

Coherence Theories of Truth

According to coherence theories of truth, what makes a belief true is that the belief coheres, or ties in, with other beliefs. For example, consider my belief that the Earth moves about the sun. I tend to believe what I read in authoritative astronomy books, and these books assure me that the Earth does indeed move about the sun. I tend to believe what experts in this area say, and such experts likewise say the Earth moves about the sun. In general, my belief that the Earth moves about the sun coheres with other beliefs, and according to coherence theories of truth, this sort of coherence is what makes a true belief true.

Think back to the jigsaw puzzle analogy used in the discussion of worldviews from the first chapter. Recall that worldviews are systems of beliefs that interlock in something like the way the pieces of a jigsaw puzzle interlock. The same analogy can be used to illustrate coherence theories of truth. A true belief is like a piece of a jigsaw puzzle. That is, in something like the way a particular piece of a jigsaw puzzle fits into the overall puzzle, so likewise a particular belief is true if it fits into the overall jigsaw puzzle of beliefs. A false belief would be like a puzzle piece that does not fit.

In summary, according to coherence theories of truth, what makes a belief true is that it coheres with some overall collection of beliefs, and what makes a belief false is that it fails to cohere with that overall collection of beliefs.

Different versions of coherence theories

Thus far, we have spoken of coherence theories only in a very generic fashion. We need to take a moment to understand how many different types of coherence theories are possible. In something like the way that a Ford is a *type* of automobile, with there being a wide variety of particular versions of Fords, so likewise coherence theories are a type of theory, with a wide variety of particular versions.

The different versions of coherence theories differ primarily with respect to whose beliefs are being counted within the jigsaw puzzle of beliefs. Are we concerned only with an individual's beliefs, so that, to be true for a particular individual, "The Earth moves about the sun" must merely cohere with that individual's other beliefs? Or are we talking about the beliefs of a group, so that to be true, "The Earth moves about the sun" must cohere with the collective beliefs of that group? And if we are speaking of the beliefs of a group, then who counts as a member of that group? Is it all those who live in a certain geographic region? Is it those who share a particular worldview? Is it the community of scientists or other experts?

Depending on how such questions are answered, one arrives at various more specific versions of coherence theories. For example, if the beliefs are the beliefs of the individual in question, then we have what might be called an *individualistic* coherence theory. On such a theory, a belief is true for Sara if it fits in with Sara's other beliefs; a belief is true for Fred if it fits in with Fred's other beliefs; and so on. It should be clear that, on an individualistic coherence theory, truth is relative to the individual in question. That is, what is true for Sara may not be true for Fred.

If we opt instead to make the collection of beliefs those of a particular group, we arrive at quite different versions of coherence theories. These might be called *group* versions of coherence theories. Just for the sake of illustration, suppose we hold that, say, a belief having to do with science is true if it fits in with the collective beliefs of the group of western scientists. For convenience, let us call such a view a *science-based* version of a coherence theory.

Note that, although the individualistic version and the science-based version are both types of coherence theories of truth, they are quite different theories. To see this, consider an acquaintance of mine, whose name is Steve. Steve quite sincerely and with deep conviction believes that the moon is further from the

Earth than the sun is, that the moon is inhabited, and that the moon is a place of frequent parties and other sorts of revelry. (Steve's beliefs stem largely from a strict literal interpretation of certain religious scriptures. Whether his beliefs are any more or less reasonable than those that stem from a literal interpretation of other religious scriptures is a topic beyond the scope of this chapter. But it is worth mentioning that literal interpretations of religious scriptures often lead to unusual collections of beliefs, such as those of the Flat Earth Society or those of the Geocentric Society, whose members believe the Earth is the center of the universe.)

Steve's jigsaw puzzle of beliefs, although quite different from my jigsaw puzzle and probably quite different from yours, forms a system of beliefs that tie together perfectly well. In particular, Steve's belief that the moon is inhabited by intelligent beings coheres with the rest of his beliefs. Thus, on the individualistic version of the coherence theory, Steve's beliefs about the moon are true. Importantly, Steve's beliefs are just as true for him as your beliefs about the moon are for you and mine are for me.

On the other hand, according to a science-based version of the coherence theory, Steve's beliefs about the moon are false, since those beliefs do not cohere with the overall set of beliefs of western scientists. In short, the individualistic version and the science-based version are two different theories of truth, although both of them are types of coherence theories.

The individualistic and science-based versions were presented mainly to illustrate that there are different versions of coherence theories possible. Since different versions of coherence theories differ mainly on whose set of beliefs counts, and since there is a wide variety of different ways of specifying whose beliefs count, it should be clear that there is a large variety of very different coherence theories possible.

Problems/Puzzles about Correspondence Theories of Truth

At first glance, some sort of correspondence theory seems like the right idea. After all, what could be more natural than saying that true beliefs are those that reflect the ways things really are? Some thought on the matter, though, suggests that there are some puzzles about correspondence theories of truth.

By far the major puzzle concerns the appeal to reality. In describing this puzzle, let's digress for a moment in order to describe what is generally referred to as the *representational theory of perception*. To call it a "theory of perception" is perhaps a bit grandiose, given that it is what most people take as the common-sense view of how perception works. Nonetheless, the "representational theory of perception" is what it has come to be called, and so it is the way we will refer to it.

To understand this theory of perception, an illustration might help. Let's consider an acquaintance, whom we will call Sara, and let's suppose we can peek into

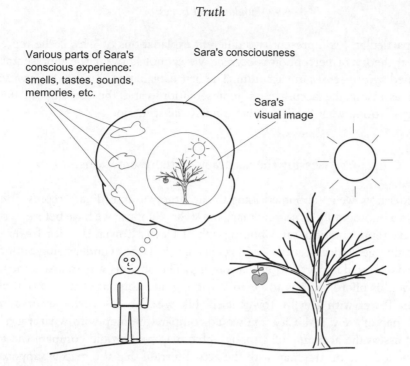

Figure 2.1   A peek into Sara's consciousness

Sara's consciousness. Borrowing the technique cartoonists commonly use to let us look into the minds of their characters, this would look as it appears in Figure 2.1.

The representational theory of perception is a general theory about perception involving all our senses, including sight, sound, taste, and so on. However, it is most easily illustrated by focusing on vision, so in what follows, most of our examples concern visual perception. It should be kept in mind, however, that similar considerations hold for the other senses.

Roughly speaking, when Sara looks at the tree, she receives a visual image of the tree, the sun, the apple, and so on. These visual images are representations of the tree. Likewise, if you or I were looking at the tree, we would have similar visual representations of the tree, the sun, and so on.

At bottom, this is all there is to a representational theory of perception: our senses provide representations (in the case of vision, representations that are in a very rough sense like pictures) of things in the external world. Again, this is a view that almost everyone takes for granted. But it is also a view that has interesting implications, and some of these implications directly affect correspondence theories of truth.

The most important of these implications is that it entails that we are all, in a sense, isolated from the world. In particular, there is no way for us to know if the representations provided us by our senses are accurate. This claim – that we cannot know whether our representations are accurate – is a strong claim, and so I will take some time to defend it.

In particular, I will present two different explanations of why, if the representational theory of perception is correct, we cannot know if the representations provided by our senses are accurate. The first explanation focuses on how we go about assessing the accuracy of representations, and the second explanation revolves around what I will call the *"Total Recall* scenario."

Assessing the accuracy of representations

Consider how we go about assessing the accuracy of an ordinary representation, such as a photograph, or a street map, and so on. Suppose we have before us some ordinary representation, say, a photograph of Devil's Tower. (Devil's Tower is an interesting geological feature – it looks like a very large cylinder rising out of the ground – located in northeastern Wyoming.) The obvious way to assess the accuracy of this photograph is to go to Wyoming and compare the photograph of Devil's Tower with Devil's Tower itself. Likewise, to assess the accuracy of a street map of New York City, you would compare the map with what it is a map of. To assess the accuracy of a topographic map, you would compare the topographic features on the map with the actual terrain that the map is supposed to be representing.

The bottom line is this: To assess the accuracy of a representation, we need to compare (a) the representation, for example, the photograph of Devil's Tower, with (b) the thing represented, for example, Devil's Tower itself.

If our senses provide us with representations of the external world, then a reasonable question to ask is whether those representations are accurate. And as we have just seen, to assess the accuracy of the representations provided by our senses, we would need to compare those representations with the things being represented.

But look again at the diagram of Sara in Figure 2.1. Suppose Sara wants to assess the accuracy of her visual representation of the apple. To do this, she would need to compare her visual representation of the apple with the apple itself. *But there is no way for Sara to do this*. The reason Sara cannot compare her visual representation of the apple with the apple itself is because she cannot step outside of her own consciousness. From Sara's point of view, all she has available is what is in her consciousness. To illustrate, consider Figure 2.2 illustrating Sara's point of view – that is all Sara has. She cannot step outside of her own conscious experience in order to compare what is in that experience with what is presumably causing that experience. In short, it seems that Sara has no way of comparing her visual representation of the apple with the apple itself, and hence has no way of assessing whether her visual representation of the apple is accurate.

Could Sara compare her visual image of the apple with, say, the sense of touch she receives when she feels the apple, or with the smell of the apple, and thereby conclude that her visual representation of the apple is accurate?

Sara could certainly compare her visual image with her tactile sensations, as well as with the olfactory sensations she receives when she smells the apple. But

Figure 2.2   Sara's conscious experience

notice that her sense of touch is itself a representation, and her sense of smell likewise is a representation. So when Sara compares the visual image of the apple with her sense of touch when she touches the apple, or with her sense of smell when she smells the apple, she is comparing representations with other representations. To assess the accuracy of her visual representation she needs to compare the representation with the thing represented, not with other representations.

This situation is very much like trying to assess the accuracy of a photograph of Devil's Tower by comparing the photograph with, say, a topographic map of Devil's Tower, or with a street map of the roads around Devil's Tower. In such a case, what is being compared is one representation with another representation. The needed comparison – of the representation with the thing represented – is not being made.

The implication is that there is no way for us to assess the accuracy of the representations our senses provide us. Or to say the same thing in different words, there is no way for us to know for sure what reality is really like.

The *Total Recall* scenario

As a second way to explain why, if the representational theory of perception is correct, there is no way for us to know whether our representations of the world are accurate, consider the *Total Recall* scenario. *Total Recall* is a science fiction movie. The movie is set in the future – say, the late twenty-fourth century – and in this period, if one wishes to take a vacation but cannot afford it, there is the less expensive option of having the experiences of the vacation implanted into one's mind. That is, there are companies who specialize in such virtual vacations. You pay the fee, and the company hooks you up to a device that implants totally realistic experiences of the vacation of your choice directly into your mind. The experiences are of a virtual reality so realistic that it is impossible to tell them from the real thing. (It is not crucial for our discussion, but the plot of the movie involves the main character's inability to tell whether his conscious experience is of reality or of realistic but unreal images being implanted into his mind. Another

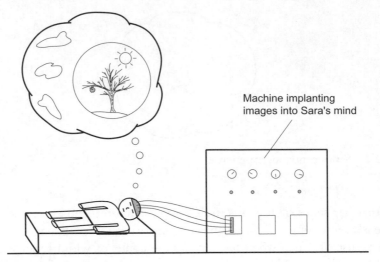

Figure 2.3   The *Total Recall* scenario

popular movie with a similar theme is *The Matrix*. This idea was not by any means invented by Hollywood – as we will discuss shortly, Descartes considered it in some detail back in the 1600s.)

With this in mind, refer again to Figure 2.1, and consider Sara's conscious experience. Sara believes her visual image of the apple, the feel of the apple, the taste and smell of the apple, and so on, are caused by there really being a tree, an apple, and so on. But if Sara were in a *Total Recall* scenario, with these sensations being implanted into her mind, her conscious experience would be exactly the same. Pictorially, the situation would be as shown in Figure 2.3. Notice that Sara's conscious experience, in both the "normal" situation depicted in Figure 2.1 as well as the *Total Recall* scenario depicted in Figure 2.3, is exactly the same. She has no way of knowing for sure that she is not in a *Total Recall* scenario. That is, she has no way of knowing for sure whether her conscious experience is caused by an external world like that in Figure 2.1, or by an external world like that in Figure 2.3. In short, Sara has no way of knowing for sure what reality is really like.

Of course, the same situation applies for you as for Sara. Suppose you are living in the twenty-fourth century, and you are a historian specializing in the early twenty-first century. Suppose you have decided to experience, via a *Total Recall* scenario, what life in the early twenty-first century was like. Part of that *Total Recall* scenario might involve reading (or having the experience as if you were reading) a book on the history and philosophy of science from that period. Your current experience – these words, this page, this book, your current surroundings – might be part of a *Total Recall* scenario. And if it were, you would have no way of knowing.

The bottom line is that, although we all believe our experiences are caused by a "normal" reality, we have no way of knowing for sure that they are not caused by the sort of reality envisioned in the *Total Recall* scenario. In short, we have no way of knowing for sure what reality is really like.

A word of caution

Be careful not to misunderstand the point of the discussions above. The proper conclusion to draw is not that reality is completely unlike what we believe it to be. Rather, the proper conclusion is that we cannot know for sure what reality is like. And if we cannot know for sure what reality is like, it follows that, if the correspondence theory of truth is correct, we can never know for sure whether any belief – or at least, any belief about the external world – is true.

This does not show that correspondence theories are wrong, or unacceptable, or incoherent. Recall that the correspondence theory of truth is a theory about what makes a belief true or false, whereas the accuracy discussion and the *Total Recall* discussion make an epistemological point about what we can know. And, as we discussed earlier, the question of what makes beliefs true or false is a different question than such epistemological questions about knowledge. But the accuracy discussion and the *Total Recall* scenario do illustrate a rather interesting aspect of correspondence theories. This aspect of correspondence theories is one of the reasons many people find them unappealing.

Problems/Puzzles for Coherence Theories of Truth

Let us begin by focusing on an individualistic version of a coherence theory. Remember that on this theory, a belief is true for an individual if it fits into the overall collection of beliefs for that individual, and false if it does not fit into that overall collection of beliefs. So what is true for my acquaintance Steve (mentioned above), and what is true for me, are two very different things. For example, it is true for Steve that the moon is inhabited, whereas it is true for me that the moon is not inhabited. It is true for Steve that the moon is further from the Earth than is the sun, whereas the opposite is true for me. In short, there are no independent truths; rather, truth is relative to individuals.

Importantly, there is no distinguishing, on an individualistic version, between "better" and "worse" truths. Steve's beliefs that the moon is inhabited is just as true (for him) as my belief that the moon is uninhabited is true (for me). All beliefs are equally true for the individual holding those beliefs. There is no way to say, on the individualistic version of the coherence theory, that my beliefs are any more true than Steve's beliefs.

In short, the individualistic version turns out to be an extreme "anything goes" sort of relativism. While this does not prove conclusively that the individualistic version is incorrect, it is worth noting that most people find a view that is this relativistic to be unacceptable.

Consider now group versions of coherence theories. Recall that on these versions of coherence theories, a belief is true if it fits in with the overall collection of some group (which group counts depends on the particular theory). The main problems for such theories are:

(a)   they do not allow for the possibility that a group might hold a mistaken belief;

(b)   there is no way to specify who exactly counts as a member of the group in question; and

(c)   with any group, there is no shared set of beliefs that are consistent.

Let's look at each of these problems more carefully.

With respect to (a), suppose Sara has been successfully framed for a crime she did not commit. When I say that Sara was successfully framed, I mean that the members of the group in question (American society, say) have come to be convinced that Sara is indeed guilty. Presumably, then, "Sara is guilty" fits in with the rest of the group's beliefs, and so, according to the group version of the coherence theory, "Sara is guilty" is true. But Sara was framed, and we want to be able to say that the group is simply mistaken in its belief about Sara's guilt. But note that, on the group version of the coherence theory, the group is not mistaken – "Sara is guilty" is true. In fact, it is *Sara* who holds the false belief. On this version of truth, when Sara thinks "I am not guilty," her belief, because it does not fit in with the overall collection of beliefs of the group, is false. In other words, this version of truth seems to get this case exactly backwards. And in general, on a group version of the coherence theory, it is difficult to see how a group could be mistaken in the beliefs its members hold. This is a very odd consequence of this version of truth.

With respect to (b), groups are not well-defined collections. For example, consider a group version of the coherence theory in which the group in question is the group of western scientists. According to this theory, what makes a belief true is whether or not it fits in with the overall collection of beliefs of the group of western scientists. But who counts as a western scientist? Consider Jim, another acquaintance of mine with unusual beliefs. Jim believes, quite sincerely, that the Earth is the center of the universe. (Incidentally, I and most of my acquaintances hold fairly mainstream beliefs, but I find it helpful to keep in contact with a number of individuals outside the mainstream.) Notably, Jim is also a practicing physicist, with a doctorate in physics from a respected institution and with publications in mainstream physics journals. Yet he holds quite unusual beliefs about the structure of the universe. So should we consider him as a member of the group of western scientists? Similar questions arise for any number of other individuals, and in general it simply is not clear whether large numbers of individuals should or should not be counted as members of the group in question. Groups have very fuzzy edges, and it is difficult if not impossible to precisely specify the members of any group.

Recall that, according to a group version of the coherence theory, a belief is true if it fits in with the overall beliefs of the group. But if the group itself is not well defined, then the theory of truth will not be well defined. In short, it is not clear that a group version of the coherence theory is itself a coherent theory.

Finally, with respect to (c), even if we could overcome the problem of specifying who counts as a member of the group in question, note that groups simply do not have consistent collections of beliefs. One member of the group may believe thus and such, and another member may believe the opposite. This will be common in any collection, any group, of people. But if there is no consistent set of beliefs among the members of the group in question, then there is no consistent jigsaw puzzle of beliefs for the group. And if there is no consistent jigsaw puzzle of beliefs, then the group version of the coherence theory, which assumes there is a consistent jigsaw puzzle of beliefs, is again not well defined.

In summary, individualistic versions of coherence theories seem to degenerate into an unacceptable sort of relativism. Group versions of coherence theories, on the other hand, seem to avoid the relativism problem, but in doing so they introduce new and substantial problems. So neither coherence theories of truth nor correspondence theories of truth provide a fully satisfying answer to our central question about truth.

Philosophical Reflections: Descartes and the *Cogito*

Before closing this chapter, it might be worth taking a moment to consider a more general philosophical question, and one raised by some of the issues we have discussed in this chapter. Earlier in this chapter, we saw that if the usual view of perception is correct – that is, the representational theory of perception that most people take as a commonsensical account of how perception works – then there is an important sense in which we cannot be certain of what reality is like. This is an extraordinarily broad-reaching claim, and given this conclusion one might reasonably wonder whether there is *anything* of which we can be certain.

Probably the best-known exploration of this question is that of René Descartes (1596–1650). Descartes considered this issue in a number of contexts, the most widely known being the discussion in his *Meditations on First Philosophy* (usually referred to as the *Meditations*). In the *Meditations*, one of Descartes' early goals is to find an absolutely certain foundation on which to build knowledge. That is, he wants to find one or more beliefs of which he can be absolutely certain, and then carefully and logically build remaining knowledge on this certain foundation.

In what we might think of as a sort of litmus test for certainty, Descartes employs a scenario closely analogous to the *Total Recall* scenario discussed above. As in the *Total Recall* scenario, Descartes considers the possibility that reality is nothing like it appears to be in his conscious experience. Descartes uses the idea of a powerful "evil deceiver" that is able to plant ideas and perceptions directly into his mind. If he can find a belief of which he can be certain even if there were such an evil deceiver, such a belief would be the sort of certain belief Descartes wants as a foundation from which to build. (Descartes' evil deceiver plays a role closely analogous to the machine in Figure 2.3 that is planting ideas and perceptions in Sara's mind, and to the devices responsible for creating the virtual realities in movies such as *Total Recall* and *The Matrix* discussed earlier.)

So Descartes is looking for a belief that can pass the evil deceiver test, that is, a belief he could be certain of even if there were an evil deceiver. Clearly, most of our beliefs will not pass such a test. For example, the belief that there is a desk in front of me would not pass such a test – if there were such an evil deceiver, it could easily fool me into thinking I am seeing a desk when there really isn't one there. Even my belief that I have a body would not pass such a test, since the deceiver could be planting images into my disembodied brain or mind.

Is there any belief that can pass such a test, that is, any belief of which we can be absolutely certain? Descartes thought he found at least one, and this belief is captured in his well-known "Cogito, ergo sum," that is, "I think, therefore I am." This belief, Descartes claims, is one in which he can be absolutely certain.

Incidentally, strictly speaking, the phrase "I think, therefore I am" does not appear in the *Meditations*, though it does appear in some of Descartes' other writings. What he does say in the *Meditations* is that "I am, I exist" is necessarily true every time he thinks it. In other words, that he exists, at least as a thinking thing, is a belief of which he can be absolutely certain. Note that he is not saying that his body necessarily exists (again, the machine in the *Total Recall* scenario, or Descartes' evil deceiver, could fool us into mistakenly thinking we had a body). Rather, what Descartes can be certain of is that every time he thinks "I am, I exist," he must exist at least as a thinking thing. Presumably, when thinking "I am, I exist," he must be thinking in order to have such a thought, and this is why he must exist at least as a thinking thing. Incidentally, to give credit where credit is due, St. Augustine (354–430) had expressed similar views, even though these views are now more commonly associated with Descartes.

There is a reasonable case to be made that Descartes' "I am, I exist" is indeed a belief of which we can be absolutely certain. So perhaps we can at least be sure of our own existence. Perhaps, contrary to the way it was beginning to appear, there is at least something of which we can be absolutely certain.

Let's return now to Descartes' foundational strategy. Recall that the idea was to find some certain beliefs from which to carefully deduce other beliefs, and thus build a structure of knowledge based on an absolutely certain foundation. At this point, you can probably guess the general problem Descartes is going to face: the foundation is just too small. A good case can be made that we can be certain of our own existence (at least of our existence as a thinking thing), and perhaps a case can be made that we can be certain of a relatively small handful of other beliefs (for example, perhaps we can be certain of highly qualified beliefs, such that there *seems* to be, say, a desk in front of me). But it is safe to say that Descartes found quite a small number of beliefs (and perhaps only one) of which he could be absolutely certain, and as it will turn out, this foundation is just too small to build on.

Descartes' foundational project was certainly one that was worth trying. And although the overall project did not succeed, it is notable that Descartes did find at least one belief of which we can be certain.

Concluding Remarks

Although we took a brief digression above to discuss the question of whether there are any beliefs of which we can be certain, the main topic of this chapter is the issue of truth. As we have seen, truth is a puzzling notion. As mentioned at the beginning of this chapter, theories of truth have been discussed for the past 2,000 years, with no consensus emerging. Our goal in this chapter was to sketch some of the main theories of truth, while providing an illustration of why those views, as well as issues surrounding truth in general, are puzzling and problematic.

As noted at the beginning of the chapter, it seems to be a reasonably common view that science is geared toward generating true theories to account for reasonably straightforward facts. It should be clear at this point that one cannot view science itself, or the history and philosophy of science, as a simple story about science generating an ever larger collection of true beliefs and true theories. As we have seen in this chapter, and as we will continue to see in Part II when we begin to look in more detail at cases from the history of science, the issues are much more complex. In the next chapter, we explore another related, and again complex, topic, involving issues surrounding the notion of facts.

Chapter Three

Empirical Facts and Philosophical/Conceptual Facts

In the previous chapter, we saw that issues surrounding truth are more complex than is generally appreciated. In this brief chapter we explore the related topic of facts.

There is no question that facts and science are closely tied to one another. Whatever else one may wish from scientific theories, there is a general consensus that such theories ought to account for the relevant facts. But the notion of facts is more complex than is often appreciated, and in this chapter we explore some of these complexities. Let's begin with some preliminary observations on the reasons we have for some beliefs we take to be obvious facts.

Preliminary Observations

I am going to go slowly through an example involving pencils and desks and drawers. Although the example may seem trivial at first, bear with me. The point is subtle, and an important one for appreciating issues involved in the history and philosophy of science.

Consider a case involving what is probably the most straightforward sort of fact we can have. For example, let's assume you are sitting at a desk, and that you put a pencil on the desk in front of you. That there is a pencil on the desk in front of you is as clear an example of a fact as you can find. You can see and feel the pencil, hear the sound the pencil makes as you tap it on the desk, and even taste and smell the pencil if you are so inclined. You have straightforward, direct, observational evidence that there is a pencil on the desk.

Facts of this sort, based on observation, are often referred to as empirical facts. As we will see shortly, what counts as an empirical fact is not as clear-cut as it

Worldviews: An Introduction to the History and Philosophy of Science. Richard DeWitt
© 2010 Richard DeWitt

might initially seem. Moreover, as we discussed in the previous chapter, there is a sense in which you cannot absolutely know for sure that reality is as you perceive it to be. Given this, you cannot be absolutely sure that there is a pencil on the desk in front of you. But nonetheless, this is a case in which you have the most direct, straightforward, "in your face" sort of observational evidence, and if anything counts as an empirical fact, it is that there is a pencil on the desk in front of you. In general, facts of this sort, which are supported by direct, straightforward, observational evidence, provide the clearest examples of empirical facts.

Now consider another case. Suppose you put another pencil on the desk in front of you. Again, you can see, feel, hear, and (if you wish) smell and taste the two pencils. Again, that there are two pencils on the desk in front of you would be as straightforward an empirical fact as you can find.

Now take one of the pencils and put it in a desk drawer, closing the drawer so that you cannot see, feel, or otherwise perceive the pencil. Chances are you believe the pencil continues to exist even when you are not perceiving it. That is, you believe it is a fact that there is a pencil in the drawer.

But now reflect on your reasons for these beliefs. Notice in particular that your reasons for believing there is a pencil in the drawer *cannot* be the same as your reasons for believing there is a pencil on the desk. Your belief about the pencil on the desk is based on direct observational evidence, whereas your belief that there is a pencil in the drawer cannot be based on any such direct observational evidence. After all, you cannot see, touch, or otherwise observe the pencil in the drawer, so you cannot have direct observational evidence for this belief. So why do you believe so strongly that there is a pencil in the drawer?

I suspect you believe this because of the way you view the world. Most of us cannot imagine that objects go out of existence when they are no longer being observed. Our conviction about the sort of world we live in – our belief that the world consists largely of stable objects that remain in existence even when they are not being observed – is the root of our belief that there is a pencil in the drawer.

So notice the substantial difference between the reasons for believing there is a pencil on the desk and the reasons for believing there is a pencil in the drawer. The one belief is based on direct observational evidence, while the other belief stems largely from our views on the sort of world we live in. Although we tend to have the same depth of conviction in each of these beliefs – that there is a pencil on the desk and that there is a pencil in the drawer – there is a substantial difference in the reasons we have for believing each of them.

What does this have to do with the history and philosophy of science? As noted, a scientific theory has to respect the relevant facts. But in looking at theories from the history of science, and in looking at the facts those theories were required to respect, we can in hindsight clearly see that some of those facts – although people believed they were reasonably clear empirical facts – were actually based more on philosophical/conceptual convictions about the sort of world the people involved inhabited.

An example may help to illustrate this point. From the time of the ancient Greeks, and until the early 1600s, it was widely believed that planets (and other objects in the heavens) moved with perfectly circular and uniform motion. For example, all motion associated with a planet such as Mars was thought to be perfectly circular. The motion was also believed to be uniform, that is, always at the same rate, never speeding up or slowing down.

In contrast, on our current theories (for which there is strong support), a planet such as Mars moves in an elliptical (not circular) orbit about the sun, and moves at varying speeds in different parts of its orbit. So both of the beliefs mentioned above – let's call them the "perfect circle fact" and the "uniform motion fact" – turned out to be mistaken.

The perfect circle and uniform motion facts sound quite alien to modern ears. A typical reaction, on first learning of the belief in these "facts," is to wonder, "Why would anyone ever have believed that?" Yet it is important to realize that, for a long period in our history, the perfect circle and uniform motion facts seemed to be obvious facts about the world in which we live. As mentioned in the first chapter, objects in the heavens were composed of the element ether, and the essential nature of this element was to move in perfectly circular, uniform motion. Obviously, then, all motion of the sun, stars, and planets must be perfectly circular and uniform. In something like the way it seems like an obvious fact to us, given the sort of universe we think we inhabit, that the pencil continues to exist when out of sight in the drawer, so too did it seem like an obvious fact to our predecessors that heavenly bodies move with perfectly circular, uniform motion.

These sorts of facts, that is, strongly held beliefs that turn out to be based heavily on philosophical/conceptual views as to the sort of world we live in, are what I will generally refer to as "philosophical/conceptual facts." However, we need to be careful here.

Importantly, empirical facts, on the one hand, and philosophical/conceptual facts, on the other, are not absolute categories, that is, most beliefs do not fall cleanly into just one or the other. Rather, most beliefs are based on a mixture of empirical, observational evidence, together with more general views about the sort of world we inhabit. For example, consider again the perfect circle and uniform motion facts discussed above. Although these beliefs were tied closely to other beliefs, such as the nature of the element ether, the heavens being a region of perfection, and so on, there was also an observational, empirical component to these beliefs as well. For example, going back at least to the beginnings of recorded history, people had observed that stars move across the sky in what appears to be perfectly circular, uniform motion. And that fact – that the points of light we call stars seem to move in perfectly circular, uniform motion – is based heavily on empirical observation. So even the perfect circle and uniform motion facts turn out to have at least some empirical component to them.

In light of such considerations, it is better to think in terms of a continuum. At one end of the continuum are the most straightforward examples of empirical facts, such as the fact that there is a pencil on the table. At the other end of the continuum are the clearest examples of philosophical/conceptual facts, for example, beliefs such as the perfect circle and uniform motion facts.

Most of our beliefs – most of what we take to be facts – lie somewhere on the continuum between the clearest examples of empirical facts and the clearest examples of philosophical/conceptual facts. That is, the reasons we have for most beliefs are tied partly to observational, empirically based evidence, and partly to the way those beliefs fit in with our overall jigsaw puzzle of beliefs.

As we will see, certain philosophical/conceptual facts, including the perfect circle and uniform motion facts, turn out to play substantial roles in the history and philosophy of science. And as we will explore in Part III, certain beliefs that most people raised in the western world take to be obvious empirical facts turn out, in light of recent discoveries, to be mistaken philosophical/conceptual "facts."

A Note on Terminology

In the discussion above, you may have noticed that I used the word "fact" when speaking of beliefs that we are now sure are incorrect. For example, I characterized the beliefs that the heavenly bodies move with perfectly uniform and perfectly circular motion as facts (albeit as philosophical/conceptual facts). Given that we generally do not use the term "fact" in this way – that is, when we discover that a previously held belief is mistaken, we usually cease to speak of it as a fact – a brief discussion of my use of the term is in order.

We simply do not have the right word to properly characterize deeply held and (at least in the context of the time) justified beliefs, such as the beliefs in perfectly uniform and in perfectly circular motion, that turn out to be mistaken. Of the two alternatives that come to mind first – characterizing such views as "assumptions," on the one hand, or as "beliefs" on the other – neither is quite right.

The views in question are much more than mere assumptions. For example, as we discussed to some extent above, and as we will explore more fully in Chapter 9, our predecessors' beliefs in perfectly circular, perfectly uniform motion were, in the context of the time, well-justified beliefs. They turned out to be mistaken, but it would be very misleading to characterize these beliefs as mere assumptions.

To illustrate this point, consider again your belief that the pencil in the drawer continues to exist. Is this belief a mere assumption? That does not seem the right way to characterize this belief. Yet, as we discussed above, our belief in the continued existence of the pencil is based largely on our overall view of the sort of universe we inhabit. But our predecessors' belief in the perfectly circular and uniform motion of the heavenly bodies was likewise based largely on their overall view of the sort of universe they inhabited. So it would not be any more proper to characterize these beliefs of our predecessors as assumptions than it would be to characterize our belief in the continued existence of the pencil as an assumption.

Likewise for the term "belief." Distinguishing facts from beliefs suggests that there is a reasonably clear distinction between the two – it suggests that facts are one thing, mere beliefs another. But there is not such a clear distinction between the two – at least, not from within one's own lifetime or from within one's own worldview (again, consider the case of the pencil on the desk versus the pencil in the drawer). Strongly held and well-supported beliefs appear, from within one's worldview, to be facts.

In short, none of the available terms is quite right. I think the best option is the one taken above – that is, with respect to strongly held and well-justified beliefs, to characterize those based more on reasonably direct observational evidence as empirical facts, and those more closely tied to one's overall worldview as philosophical/conceptual facts. And even in the cases, which again exist both for our predecessors and for us, in which some of these strongly held beliefs turn out to be incorrect, I will continue to refer to them as philosophical/conceptual facts, to remind ourselves that, from within the relevant worldview, these were much more than mere assumptions, beliefs, or opinions.

Concluding Remarks

Before closing this chapter, it is worth taking a moment to make a few final observations about empirical and philosophical/conceptual facts.

To re-emphasize a point made above, do not think of empirical facts and philosophical/conceptual facts as absolute categories. Most beliefs are based on a combination of empirical evidence and more general views about the sort of world we inhabit. As noted, it is better to think of the distinction between empirical facts and philosophical/conceptual facts more in terms of a continuum, with the clearest examples of empirically based beliefs (such as the belief involving the pencil on the table) at one end, and the clearest examples of beliefs more closely tied to general philosophical/conceptual views (such as the beliefs in perfectly circular and uniform motion of the heavenly bodies) at the other end.

Also, be careful not to make the error of viewing philosophical/conceptual facts as the sort of facts one would find only as part of an old, naive way of thinking. Our predecessors' beliefs in the perfect circle and uniform motion facts turned out to be wrong, but the beliefs were not naive. As is typical with philosophical/conceptual facts, the perfect circle and uniform motion facts fit in well with the overall system of beliefs, that is, with their overall jigsaw puzzle of beliefs. The beliefs turned out to be wrong, but they were not naive.

As a corollary, calling a belief a philosophical/conceptual fact is not meant to suggest that the people holding that belief do not, or did not, have good reasons for believing it. As noted above, the perfect circle and uniform motion facts turned out to be wrong, but the people of the time had good reasons for believing these facts.

Likewise, do not make the mistake of thinking that we, in this period of modern science, have managed to avoid the pitfalls of believing in philosophical/conceptual facts. Such facts are still a part of the modern story, and as noted above, much of the focus of Part III of this book will be to look into developments in twentieth-century science, with an eye toward identifying some facts that we have long taken as straightforward empirical facts, but that turn out, in light of recent discoveries, to be mistaken philosophical/conceptual facts.

Also, calling a fact a philosophical/conceptual fact is not to suggest that it is incorrect. Many philosophical/conceptual facts of the past did indeed turn out to be wrong. No doubt some of our own philosophical/conceptual facts will turn out to be wrong. But most such facts will, we hope, withstand the test of time, and turn out to be at least more or less correct. To put this same point another way, the distinction between empirical facts and philosophical/conceptual facts is not based on whether the facts turn out to be correct; rather, the distinction is based on the type of reasons we have for believing the fact.

Finally, it is worth noting that, in everyday life, we typically do not make much of a distinction between empirical facts and philosophical/conceptual facts. With the benefit of hindsight, especially when looking at past cultures, it is relatively easy to see which beliefs were more empirical and which were more philosophical/conceptual. But within our own time frame, facts just look like facts to us, and they all look pretty much on a par. It is only on careful reflection, and often with great difficulty, that we can see that some beliefs we hold are more empirically based, and some are more philosophical/conceptual.

Chapter Four

Confirming and Disconfirming Evidence and Reasoning

The main goal of this chapter is to explore issues surrounding some of the most common sorts of reasoning found in science. In particular, we will look at one of the most common types of evidence and reasoning used to support theories in science. Of course, we will also explore the flip side of this, that is, issues involved in evidence and reasoning indicating that theories are incorrect. In keeping with a recurring theme of this text, we will see that the issues involved are more complex than they might at first seem.

Discoveries, evidence, and reasoning in science (and in everyday life as well) are often quite complicated. Our strategy will be to begin by focusing on some of the more straightforward evidence and reasoning, and to show that even these simpler cases are surprisingly complex. In particular, we will begin by looking at two general types of evidence and logical reasoning commonly found in science (and again, also commonly found in everyday life). For convenience, I will refer to these two types as *confirmation reasoning* and *disconfirmation reasoning*. We will begin with a brief sketch of each, and then explore some of the subtleties involved.

Confirmation Reasoning

About 100 years ago, Albert Einstein proposed the general theory of relativity. It was a controversial theory, and in certain ways it conflicted with other accepted theories. Notably, using relativity theory one could make unusual predictions – unusual in the sense that other theories did not make the same predictions. For example, Einstein's theory predicted that the gravitational effect of a large body, such as the sun, would bend starlight. It would be possible to observe such

Worldviews: An Introduction to the History and Philosophy of Science.   Richard DeWitt
© 2010 Richard DeWitt

bending of starlight during a total solar eclipse, so the solar eclipse that was to occur in May of 1919 provided an opportunity to test the prediction. As it turned out, these predictions were correct, and this was taken as evidence supporting (that is, helping to confirm) Einstein's relativity theory. In other words, the fact that Einstein's theory made correct predictions, and, notably, predictions that were not made by competing theories, was taken as evidence that the theory was correct.

Notice there is nothing unique to science about this sort of reasoning. We use the same sort of reasoning all the time. In general, when we base predictions on a certain theory, and those predictions turn out to be correct, this provides at least some evidence that the theory is correct. If we use T to represent a theory, and O to represent one or more observations predicted by the theory T, we can schematically represent this reasoning as follows:

> If T, then O
> <u>O</u>
> so (probably) T

It is worth noting that the Einstein example above, and the schema just outlined, are quite simplified accounts of confirmation reasoning. Again, at this point we are just interested in a brief introduction to this type of reasoning. We will next consider a brief sketch of disconfirmation reasoning, and then move on to look at some of the factors that make such reasoning more complicated than it might at first appear.

Disconfirmation Reasoning

To understand disconfirmation reasoning, it is again easiest to use an example. In the late 1980s, two well-established scientists claimed to have discovered a way to achieve nuclear fusion at low temperatures (so-called cold fusion). This claim was very exciting but also quite controversial, since the general consensus is that nuclear fusion requires extremely high temperatures. Suppose we call their claim (essentially the claim that fusion is possible at low temperatures, and that they had the key ideas on how to accomplish such fusion) "cold fusion theory."

As is usually the case, certain predictions could be made on the basis of cold fusion theory. For example, if cold fusion theory was correct, one would expect a very high number of neutrons to be emitted during the process. Yet the expected level of neutrons was not detected, and this was taken as evidence against the cold fusion theory. Again, there is nothing at all unusual about this type of reasoning. In general, when we make predictions based on a particular theory, and those predictions turn out not to be correct, we take this as evidence against the theory. Again using T to represent the theory, and O to represent one or more observations predicted by T, we can schematically represent this reasoning as follows:

>   If T, then O
>   <u>Not O</u>
> so   Not T

It is again worth emphasizing that this reasoning scheme is overly simplified, and should be considered a first approximation to disconfirmation reasoning. We will now move on to consider some of the complicating factors involved in confirmation and disconfirmation reasoning, beginning with the distinction between inductive and deductive reasoning.

Inductive and Deductive Reasoning

Confirmation reasoning is a type of inductive reasoning, whereas disconfirmation reasoning is a type of deductive reasoning. The inductive nature of confirmation reasoning and the deductive nature of disconfirmation reasoning have certain important implications. To get at these implications, we first need to be clear about the difference between inductive and deductive reasoning.

You may have heard that inductive reasoning moves from the specific to the general, whereas deductive reasoning moves from the general to the specific. While this holds for some cases of inductive and deductive reasoning, overall it is not accurate, and so it is not a good way to characterize inductive and deductive reasoning.

There is a more straightforward, more accurate, and more insightful way to characterize inductive and deductive reasoning. Consider the following as a typical example of inductive reasoning:

> The local college men's basketball team has never won the NCAA championship. For that matter, on the few occasions that the team has been in the NCAA tournament, it has never made it past the first round of competition. This year's team is not that much different from the teams of the past, nor has anything else changed dramatically about the men's basketball program. In light of all these factors, it is extremely unlikely that the men's team will win the NCAA tournament this year.

This provides a nice example of a convincing inductive argument. Given the premises of the argument, the conclusion is very likely. However – and this is the defining characteristic of inductive reasoning – even if all the premises and the evidence are correct, it is still *possible* that the conclusion is wrong. However unlikely, it is possible that the men's basketball team will win this year's NCAA tournament. This, then, is what characterizes inductive reasoning: in a good inductive argument, even if all the premises are true, it is still possible for the conclusion to be wrong.

In contrast, in a good deductive argument, true premises guarantee a true conclusion. That is, in a good deductive argument, if all the premises are true, then the conclusion must also be true. Consider the following example borrowed from the movie *No Way Out*:

> The man who was in Linda's apartment that night killed Linda. And whoever killed Linda is Uri. Commander Farrell was the man in Linda's apartment that night. Therefore, Commander Farrell is Uri.

This argument is interestingly different from the inductive example. In particular, the premises of this argument, if true, guarantee that the conclusion is true. And this is what characterizes deductive arguments: in a good deductive argument, true premises guarantee a true conclusion.

With this in mind, let us return to our discussion of confirmation and disconfirmation reasoning. Remember that confirmation reasoning is a type of inductive reasoning. Simply because confirmation reasoning is inductive, instances of such reasoning will not guarantee the conclusion. That is, confirmation reasoning can at best provide support for a theory, but no matter how many instances of confirmed predictions there are, it will always remain possible that the theory is in fact mistaken. This is simply due to the inductive nature of such reasoning.

The inductive nature of confirmation reasoning is part of the reason why you sometimes hear the claim that scientific theories can never be proven (at least, not in the strong sense of "proven"). Most scientific theories are supported in large part by inductive evidence. As such, no matter how much confirming evidence exists for a theory, it is always possible, simply because of the inductive nature of the reasoning involved, that the theory will turn out to be wrong. The fact that theories in science cannot be shown without a doubt to be correct is not a flaw of such theories, nor is it a defect in science itself. Rather, it is simply a consequence of the fact that confirmation reasoning is a widely used type of reasoning supporting theories, and the fact that confirmation reasoning is a form of inductive reasoning.

It is also worth noting that the factors and reasoning involved in actual theories are usually much more complex and intertwined than might be suggested by the discussion so far. To illustrate this with just one example, consider again the case of the bending of starlight predicted by Einstein's theory. This would appear to be a pretty simple prediction and observation. Everyone agrees that Einstein's theory predicts the bending of starlight, and that a solar eclipse would provide an opportunity to observe such bending. So go out during the next solar eclipse, and see whether or not starlight is bent. The observation may not be trivial, but it does sound reasonably straightforward.

But in fact the case is not at all straightforward. For example, in order to do the calculations necessary to predict the position of bent versus unbent starlight, a

good number of simplifying – and, strictly speaking, incorrect – assumptions had
to be made. In the actual observations in May 1919, in order to make the calculations manageable, the sun was treated as a perfectly spherical, nonrotating body,
with no outside influences acting on it (such as the gravitational effects of bodies
such as the Earth, moon, and other planets). Of course, the sun is not in fact
spherical, and it rotates, and there are any number of outside influences acting on
it. In short, everyone knew these assumptions were wrong, but everyone also
knew that, without making such simplifying assumptions, the necessary calculations could not be done.

Most (not all, but most) of those familiar with the bent starlight observations
of 1919 agree that these simplifying assumptions did not change the overall implication of the observation, namely, that the observation provided confirming evidence for Einstein's theory. Nonetheless, the point I am trying to drive home is
that actual cases of confirming evidence tend to involve factors that are much
more complex than are generally recognized.

This situation is not unusual. More often than not, checking to see if a prediction is or is not observed involves layers of nontrivial theories and data. In short,
actual cases of confirming evidence are usually very complex. So not only does
the inductive nature of confirmation reasoning mean that such reasoning cannot
prove (in a certain strong sense of "prove") that a theory is correct, but, in addition, the actual evidence and reasoning tends to be intertwined in complex ways,
such that cases of confirming evidence are usually far less straightforward than
they might at first appear to be.

If it is not possible to prove (again, in the strong sense of the word) that a theory
is correct, is it at least possible to prove that some theories are incorrect? At first
glance, the answer would appear to be yes. After all, disconfirmation reasoning is
a type of deductive reasoning, and, as noted above, in good deductive reasoning
the premises guarantee the conclusion. So, at first glance, one would think that
disconfirmation reasoning could be used to prove a theory is incorrect. But as is
so often the case, first impressions are misleading.

Consider this example to illustrate why using disconfirmation reasoning to
show a theory is wrong is not as straightforward as it seems. Anyone who
has taken a lab course (for example, in chemistry or biology) will have had an
experience similar to the following. Suppose in a chemistry lab your professor
gives you a beaker of ethanol, and instructs you to find its boiling point. Now
suppose (when the professor is not looking, of course) that you sneak a peek
into any of a number of standard reference texts, and find that the boiling point
of ethanol is 78.5 degrees Celsius. Now you perform your experiment, confident
that the boiling point will turn out to be 78.5 degrees Celsius. Unfortunately, it
turns out that the sample does not seem to boil at 78.5 degrees Celsius. What
do you do?

This appears to be a case where disconfirmation reasoning would apply. The
reasoning scheme for disconfirmation reasoning would lead you to reason as
follows:

> If the sample in the beaker is ethanol, then I should observe the sample boiling at 78.5 degrees Celsius.
> <u>I do not observe the sample boiling at 78.5 degrees Celsius.</u>
>
> so   The sample in the beaker is not ethanol.

At this point, do you conclude that the professor was mistaken, and that the beaker does not contain ethanol? Probably not. Instead, you will likely consider alternative explanations of why the boiling point did not turn out to be 78.5 degrees Celsius. For example, a broken thermometer, or dirty glassware, or a contaminated sample, or unusual air pressure in the lab, or any other of a number of explanations are possible. In short, it would be bad practice to jump to a conclusion based on the small amount of evidence you have.

Your reasoning in this case is more accurately characterized as follows:

> If the sample in the beaker is ethanol, and the thermometer is working properly, and my glassware is clean, and the sample is not contaminated, and the air pressure in the lab is normal, and (any of a number of other alternatives), then I should observe the sample boiling at 78.5 degrees Celsius.
> <u>I do not observe the sample boiling at 78.5 degrees Celsius.</u>
>
> so   The sample in the beaker is not ethanol, or my thermometer is not working properly, or my glassware is not clean, or the sample is contaminated, or the air pressure in the lab is unusual, or (any of a number of other alternatives).

The moral is that disconfirmation reasoning, as schematically represented above, was vastly oversimplified. As we are seeing, disconfirmation reasoning is more accurately schematically represented as follows:

> If T, and A1, and A2, and A3, … , and An, then O
> <u>Not O</u>
>
> so   Not T, or not A1, or not A2, or not A3, … , or not An

This is a more accurate representation, and hereafter is what I will have in mind when I speak of disconfirmation reasoning.

In the scheme above, A1, A2, and so on, represent what are commonly called *auxiliary hypotheses*. Auxiliary hypotheses are crucial, but usually unstated, parts of any instance of disconfirmation reasoning. Auxiliary hypotheses are crucial simply because, without them, we would not expect to get the observation in question. To put the point slightly differently, auxiliary hypotheses are what, in a sense, are needed to get the *then* part of the statement from the *if* part. That is, *if* such and such is the case, *and* all the auxiliary hypotheses are correct, *then* we would expect to observe so and so.

And as the case of the beaker of ethanol illustrates, in any situation where a theory is used to make a prediction that turns out to be incorrect, it is always possible (indeed, in many cases it is likely) that the theory is fine and that one or more of the auxiliary hypotheses are mistaken.

The same situation with auxiliary hypotheses arose (and is still present) in the cold fusion example. It is true, for example, that the large number of neutrons one would expect to observe from cold fusion were in fact not observed. But the expected large number of neutrons depends on the auxiliary hypothesis that the processes involved in cold fusion are more or less similar to the processes involved in usual (hot) fusion. The proponents of the theory had the option – and indeed, they took this option – of retaining their belief in cold fusion and instead rejecting the auxiliary hypothesis that cold fusion is like usual fusion.

In the case of cold fusion theory, eventually the quantity of disconfirming evidence increased to the point where there are now relatively few who still accept cold fusion theory (though, notably, there are still those who continue to adhere to cold fusion theory and reject the ever available auxiliary hypotheses). But, in general, the question of when it is more reasonable to reject a theory in the face of disconfirming evidence, and when instead it is more reasonable to reject one or more of the auxiliary hypotheses, is an extraordinarily difficult one. And, importantly, there is no recipe for answering it.

In short, then, here are the two most important points about disconfirming evidence and reasoning. First, when one is faced with evidence that seems to disconfirm a theory, it is not only an option, but indeed it is often more reasonable, to maintain one's belief in the theory and instead reject one of the auxiliary hypotheses. And second, the question of when it is more reasonable to reject a theory, and when instead it is more reasonable to reject one or more of the auxiliary hypotheses, is not a question that can be answered by applying any cut-and-dried recipe.

Concluding Remarks

To summarize the main points of this chapter: confirmation and disconfirmation reasoning are two common types of reasoning, both within and outside of science. Confirmation reasoning, simply in light of the fact that it is a type of inductive reasoning, can never show beyond a doubt that a theory is correct. Thus, no matter how much confirming evidence there is for a scientific theory, it will always remain possible that the theory is wrong. In addition, the inductive evidence and reasoning at play in actual cases is generally complex and intertwined. Confirming evidence and reasoning tends to be far less straightforward than it appears at first sight.

On the other hand, disconfirmation reasoning is a type of deductive reasoning. However, actual instances of disconfirming evidence also tend to be complex. In particular, there are usually a substantial number of auxiliary hypotheses involved in disconfirmation reasoning. Thus, disconfirming evidence shows only that either the theory in question is wrong, or (as is often the case) one or more of the auxiliary hypotheses are incorrect. Thus, disconfirming evidence and reasoning are likewise less straightforward than they might at first appear.

Confirmation and disconfirmation reasoning are used every day, both within and outside of science. And as we will see in later chapters, the points just noted play substantial roles in the history of science. So in conclusion, and to restate the main point of this chapter, the evidence and reasoning found in science (and in everyday life as well) is surprisingly complex. In the next chapter, we explore two issues that are closely tied to the topics discussed above, these being the Quine–Duhem thesis and issues surrounding scientific method.

Chapter Five

The Quine–Duhem Thesis and Implications for Scientific Method

In the previous chapters, we looked at worldviews, truth, facts, and reasoning, as well as a number of issues related to these topics. As we will see in this chapter, many of those issues have close ties to what is often referred to as the Quine–Duhem thesis (or, sometimes, the Duhem–Quine thesis, to reflect the fact that Duhem preceded Quine). The Quine–Duhem thesis is one of the better-known views in modern philosophy of science, and for this reason alone is worth looking into. But in addition, it will give us an opportunity to better see the ways in which the issues discussed in previous chapters are intertwined, and moreover, this discussion will help set the stage for later chapters, where we will see how these intertwined issues play out in examples from the history of science.

Another point worth noting is that these issues have implications for various views on scientific method. Toward the end of the chapter, then, we will consider various proposals that have been suggested concerning scientific method. This section on scientific method will do double duty: first, it will allow us to see, historically, some of the views presented on the proper way to conduct science. So, for example, we will see how the Aristotelian approach to science differed substantially from what is typically taken now as the proper sort of approach. And second, discussing scientific method will give us a chance to see issues surrounding methodology in science, and this in turn will be useful for later chapters, where we will see the (often surprising) methods used in cases from the history of science.

The Quine–Duhem Thesis

The Quine–Duhem thesis is a well-known view in the philosophy of science, and one involving a number of intertwined and controversial issues. By way of a brief

Worldviews: An Introduction to the History and Philosophy of Science. Richard DeWitt
© 2010 Richard DeWitt

introduction to the principal players, Pierre Duhem (1861–1916) was a well-respected French physicist who had substantial interests in broader questions, including questions involved in the testing of scientific hypotheses and theories, while Williard Quine (1908–2000) was one of the most influential philosophers of the 1900s, who had a lifelong interest in issues surrounding the philosophy of science.

In this section, we will look at three of the key ideas associated with the Quine–Duhem thesis, namely, the idea that (to borrow a phrase from Quine) our beliefs face the "tribunal of experience" not singly, but in a body; the claim that there can typically be no "crucial experiments" to decide which of two competing theories is correct; and the notion of *underdetermination*, that is, the idea that the available data typically does not pick out a unique theory as being correct.

Bodies of beliefs and the tribunal of experience

Recall from our discussion in the previous chapter that, when faced with disconfirming evidence, there are almost always crucial (but usually unstated) auxiliary hypotheses involved. As we saw in the previous chapter, it is always possible to reject an auxiliary hypothesis rather than rejecting the main view.

Given the role played by auxiliary hypotheses, when we perform an experiment, presumably to test a particular hypothesis, we are not really testing just the individual hypothesis. Rather, in an important sense the test is more of a test of the main hypothesis plus the accompanying auxiliary hypotheses. So what we are typically testing is really a *body* of claims, any one of which can be rejected or modified in the face of disconfirming evidence. And this is one of the key elements of the Quine–Duhem thesis – that is, the key idea is that a hypothesis typically cannot be tested in isolation. Rather, what is tested is an entire group of claims, any of which is available for rejection or modification should the experimental results not be as expected. And this is the key idea behind Quine's phrase noted above, that our beliefs face the tribunal of experience not individually, but as a body.

The emphasis here on bodies of claims brings to mind our discussion of worldviews from Chapter 1. And indeed, this aspect of the Quine–Duhem thesis is closely tied to the notion of worldviews. To see this, recall our discussion of interconnected systems of beliefs from Chapter 1, in which we discussed such collections of beliefs by an analogy with a jigsaw puzzle. Quine tended to speak of such collections of beliefs as "webs of beliefs," suggesting an analogy with a spider's web. In a spider's web, changes in the outer regions of a web affect the more central regions in only minor ways. Likewise, beliefs toward the outer edge of a "web of beliefs" can be modified with only minor alterations of more central beliefs (such beliefs would be the peripheral beliefs we discussed in Chapter 1). In contrast, changes in the central regions of a web will cause changes throughout the web, and, in a similar way, modifications to central beliefs (the core beliefs) will cause changes throughout one's web of beliefs.

We noted above that, according to the Quine–Duhem thesis, tests of a hypothesis are typically not tests of the individual hypothesis, but, rather, are tests of groups or collections of beliefs. How large a group of beliefs are we speaking of here? For example, if we design an experiment to test a hypothesis, how large a collection of beliefs are we really testing? Are we testing just a relatively small subset of our overall collection (or jigsaw puzzle) of beliefs, or much more radically, is every experiment and test we perform in some sense really a test of our entire jigsaw puzzle (or web of beliefs, or worldview)?

There is no consensus on the answers to these questions. Quine at times defended the more radical view, maintaining that it is one's entire web of beliefs – that is, our entire interconnected collection of beliefs – that face the tribunal of experience as a whole. And, faced with evidence that runs counter to views we hold, no belief, not even a core belief, is immune from revision. We would, of course, typically be more willing to modify beliefs that are more toward the periphery, but Quine's point is that in principle any belief is subject to revision. Tests are tests of the whole package. Duhem, on the other hand, was a bit more reserved on this point in that, on his view, although tests might involve large collections of beliefs, it is not typically our entire collection of beliefs – our entire worldview – that is put to the test.

In spite of the differences in the details between Quine's and Duhem's views, there is a general agreement that tests are not typically tests of a hypothesis in isolation, but rather, such tests are typically tests of large bodies of beliefs. And, as noted, this is generally taken as a key component of the Quine–Duhem thesis.

Crucial experiments

Another aspect of the Quine–Duhem thesis, and one closely related to what we have just been discussing, involves the notion of crucial experiments in science. The idea of a crucial experiment goes back at least to Francis Bacon (1561–1626), the idea being that, when faced with two competing theories, it should be possible to design a crucial experiment for which the two theories give conflicting predictions. Ideally, since the predictions of the competing theories conflict, such an experiment should show at least one of the theories to be mistaken. Because of the issues involved in confirmation reasoning discussed in the previous chapter (mainly that confirming evidence can at best support a theory, but not show definitely that the theory is correct), such an experiment would not show that the theory making the correct prediction was definitely the correct theory. Nonetheless, the key idea is that a crucial experiment, even if it could not show one of the competing theories to be definitely correct, could at least serve to rule out one of the competing theories.

However, if tests are typically tests of collections of beliefs, and if when faced with disconfirming evidence it is always an option to reject an auxiliary hypothesis rather than the main theory, then it seems that crucial experiments will typically not be possible. The reason should be clear: in any such experiment designed to show that at least one of the two competing theories gives a wrong prediction, whichever theory seems to make the wrong prediction can still be kept, simply by rejecting one of the auxiliary hypotheses. And again, as we noted in the previous chapter, it is often perfectly reasonable to reject an auxiliary hypothesis rather than the main theory.

It is worth noting that this skepticism about the possibility of crucial experiments can be understood in a variety of ways, some much stronger and more controversial than others. There is little question that, in some cases, results of experiments for which competing theories give conflicting predictions can be accommodated within both of the conflicting theories. For example, the lack of observed neutrons during early cold fusion experiments was clearly compatible with the usual theories about fusion, but as we saw in the previous chapter, this result could also be accommodated within cold fusion theory by rejecting one of the auxiliary hypotheses involved. If we take the Quine–Duhem skepticism about crucial experiments in a relatively weak sense, as claiming only that competing theories can often both accommodate the results of an alleged crucial experiment, then the claim is reasonably uncontroversial. There are numerous examples from the history of science (the cold fusion example above being just one) that support this weaker version of the claim.

Another reading of this aspect of the Quine–Duhem thesis, in which this part of the thesis is construed as claiming that *any* experimental result whatsoever can be accommodated within *any* theory whatsoever, is a much stronger and much more controversial version of the thesis. It is much more difficult to find clear examples from the history of science to support this stronger claim. However, Quine did sometimes speak this way, and, not surprisingly, there is far less consensus on this stronger claim. So, to summarize this brief subsection, while there is general consensus that a key part of the Quine–Duhem thesis involves a certain skepticism about the idea of crucial experiments, there is less consensus on how strongly this claim should be interpreted.

The underdetermination of theories

One other often discussed issue in the philosophy of science is generally termed the "underdetermination" of theories. Recall again from the discussion above that theories can generally be preserved in the face of disconfirming evidence, and that generally it will be difficult if not impossible to design a crucial experiment to decide between competing theories. Add to this our discussion of confirming evidence from the preceding chapter, especially where we noted that, given the inductive nature of confirming evidence, such evidence can at best support a theory, but never demonstrate conclusively that a theory is correct.

Putting all these factors together, we arrive at the view that the available data, including the outcomes of all relevant experiments, can never fully determine that a particular theory is the correct theory. Nor can all the data and experimental results ever show definitively that any competing theory is incorrect. In short, a variety of competing theories will often be compatible with all the available evidence. This is often summarized by saying that theories are *underdetermined* by the available data.

It is worth noting that, as with the aspects of the Quine–Duhem thesis we discussed above, the notion of underdetermination can be read in a variety of ways, some stronger and more controversial than others. There is no question that, at times, the available data does not uniquely point toward one of two or more competing theories. To use the cold fusion example again, in the late 1980s the data simply did not point cleanly toward either the cold fusion theory or the existing hot fusion theory (that is, the usual view of fusion as requiring extremely high temperatures). Both the cold fusion and hot fusion theories were compatible with the available data. Understood in this relatively mild way, there is little question that theories are underdetermined in this sense.

At the other end of the spectrum, in contrast to this mild sense of underdetermination, it is not uncommon to see discussions involving a much more radical notion of underdetermination. On this much more radical view of underdetermination, scientific theories and scientific knowledge are viewed as "social constructs," more or less inventions of the relevant community. According to this view, scientific theories are seen as more closely tied to, and reflections of, social conditions rather than being tied to, and reflections of, the physical world. On this much more radical and controversial notion of underdetermination, there is no more a uniquely determined and objectively correct scientific theory than there is a uniquely determined and objectively correct set of table manners. On this view, table manners and scientific theories are both reflections of society, and one theory cannot be said to be the uniquely correct theory in any deep or objective sense of the word "correct."

In short, although there is agreement that the underdetermination of theories is a key aspect of the Quine–Duhem thesis, the notion of underdetermination is construed in a variety of ways. And as noted, some of these ways are substantially stronger and more controversial than others.

In summary, consider again the key issues associated with the Quine–Duhem thesis – the underdetermination of theories, the idea that hypotheses are typically not tested in isolation, and the notion that crucial experiments are typically not possible. All of these are reasonably uncontroversial, at least when construed in a somewhat mild way. How broadly such claims should be construed, and whether such broader claims can be supported by actual cases, is much more controversial. Watch for these sorts of issues as we discuss, in Part II, historical cases such as that involving the dispute over the Earth-centered and sun-centered views of the universe. As we will see, such disputes involve a surprisingly wide range of issues, including those central to the Quine–Duhem thesis.

Implications for Scientific Method

As mentioned earlier, the sorts of issues we have been discussing have some interesting implications for views on scientific method. Before closing this chapter, then, we will look briefly at a variety of proposals that have been made concerning the proper way to conduct science. This will allow us a chance to see how scientific method was viewed within the Aristotelian worldview (and especially note how much it differs from the way scientific method is more typically viewed today). These discussions will also help set the stage for our discussion, in Part II, of cases from the history of science.

At some point in your education, you may well have been taught what is commonly termed "the scientific method." Although the exact formulation of this method varies somewhat from book to book, and from school to school, in general terms this method is generally presented as one that involves (a) gathering the relevant facts, (b) generating hypotheses to explain those facts, and (c) testing the hypotheses, typically by performing experiments that either confirm or disconfirm (using something like the patterns of confirmation and disconfirmation reasoning discussed earlier) the hypotheses.

Given our earlier discussions, especially concerning the nature of facts from Chapter 3, of confirmation and disconfirmation reasoning in the previous chapter, and of issues associated with the Quine–Duhem thesis discussed above, we might reasonably wonder whether the method just outlined could possibly be as straightforward as it is often presented to be. In what follows, we will look at some of the methods that have been proposed for doing science, and we will also explore some of the issues surrounding such methods. We will by no means be surveying every proposed scientific method, but we will look at enough to get a good idea of some of the factors complicating any attempt to give a single, definitive method for doing science. Let's begin with a look at some of Aristotle's ideas on the subject.

Aristotle's axiomatic approach

Within the Aristotelian worldview, science was generally viewed as geared toward generating knowledge that was certain. That is, it was generally thought that scientific knowledge had to be necessarily true, and not merely probable. If we ask how we might arrive at such necessarily true knowledge, there seems only one possible approach, and that is to use deductive reasoning that is based on necessarily true basic principles. If such necessarily true basic principles could be found, and if the reasoning used is deductive, then the conclusions (that is, the scientific knowledge) would "inherit" the certainty of the basic principles, so to speak, and we would thus arrive at necessarily true scientific knowledge.

Such approaches are often termed *axiomatic* approaches – that is, these are approaches based on deductive reasoning from basic principles that are in some sense certain or necessarily true. Aristotle was an advocate of such an approach, and during the period in which the Aristotelian worldview was dominant, the Aristotelian approach to scientific knowledge was generally viewed as the correct approach. Looking at Aristotle's approach, then, will give us an idea of the sort of scientific method dominant for much of (at least recorded) western history, and also give a good sense of the fundamental problems that will be faced by any attempt to generate necessarily true scientific knowledge.

Aristotle viewed logic as a tool to be used in investigations, including (but not limited to) scientific investigations. In fact, for Aristotle, providing a scientific explanation was essentially a matter of providing a certain sort of logical argument. We typically do not view scientific explanations and logical arguments as all that similar, but in fact the two are closely related. To illustrate this, consider the following example. (The example is chosen for ease of explanation, and since it uses notions discovered well after the time of Aristotle, it is not one that Aristotle himself would have or could have given.)

Suppose you are curious as to why copper conducts electricity. Suppose someone explains that copper contains free electrons, and that things with free electrons conduct electricity, and that's why copper conducts electricity. Note how closely related this explanation is to the following argument:

> All copper contains free electrons.
> <u>All things containing free electrons conduct electricity.</u>
>
> so   All copper conducts electricity.

In fact, aside from the style of presentation, there is little difference between the explanation given in the preceding paragraph, and the argument given immediately above.

Arguments of the sort just given, consisting of two premises and a conclusion, are termed *syllogisms*. For Aristotle, a proper scientific explanation consisted of a *demonstration*, which in essence was a chain of syllogisms in which the conclusion of the final syllogism is the item that is being explained. (I should note that, strictly speaking, Aristotelian syllogisms are two-premise arguments meeting certain conditions as to the form and arrangement of the statements involved. Likewise, strictly speaking there are more conditions on a demonstration than those just mentioned. However, these additional details need not concern us here.)

As noted, for Aristotle, scientific knowledge had to be certain knowledge. Or to put it another way, the conclusion of the final syllogism in the chain had to be necessarily true. Note how this differs importantly from the modern conception of scientific knowledge. Science is now generally viewed as producing theories that are probably correct, but we do not expect science to guarantee that theories are correct (nor do we think that this is possible). Not so for Aristotle, nor for the typical view of scientific knowledge up until the 1600s. Scientific knowledge was to be certain knowledge, and the certainty was tied importantly to its being derived deductively.

But how might such a deduction guarantee that the conclusion is not merely true, but necessarily true? As noted, there is only one way, and that is to use premises that are themselves necessarily true, so that the conclusion inherits, so to speak, the certainty of the premises.

But this raises the question as to where the necessity of the premises comes from. One solution would be to derive such premises, via other syllogisms higher in the chain of syllogisms, from other premises that were themselves necessarily true. And indeed, this is the way Aristotle envisions a full scientific explanation as proceeding. That is, in the chain of syllogisms, the final syllogism will consist of a conclusion that is necessarily true because it has been derived from premises that are necessarily true. And these premises typically will themselves be conclusions of syllogisms earlier in the chain, in which the premises of these earlier syllogisms are necessarily true.

Of course, the chain of syllogisms cannot go on forever, so at some point there must be some premises that are necessarily true, but that have not themselves been derived from earlier syllogisms. These starting points, the premises that are necessarily true in and of themselves, are typically termed *first principles*. First principles are envisioned as basic, necessarily true facts about the world. But how does one recognize first principles, and, in particular, how does one know a first principle is necessarily true? An analogy with geometry might be helpful.

Consider the axiom of Euclidean geometry that, given a line on a plane, and a point on the plane not on that line, one and only one line can be drawn on the plane through the point parallel to the given line. This axiom is illustrated in Figure 5.1. In the figure, the paper represents the plane, the solid line at the top represents the given line, the point is the point on the plane, and the dotted line represents the one line that can be drawn through the point parallel to the given line. This axiom cannot be proven in Euclidean geometry, and thus it (or an axiom equivalent to the way it is phrased here) is taken as a basic, unproved starting point (that is, an axiom or postulate) of Euclidean geometry. Although

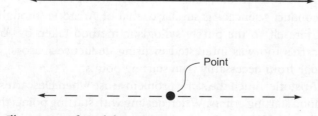

Figure 5.1 Illustration of Euclidean axiom

the claim cannot be proven, it seems that if one has adequate education and intelligence, and an understanding of the terms involved, then one can simply "see" that this axiom must be true. (Incidentally, the discovery, in the 1800s, of non-Euclidean geometries casts serious doubt on whether it makes sense to talk about such axioms being "true" in any meaningful sense.)

In a way somewhat similar to the way we presumably "see" the truth of axioms such as the one illustrated above, if one has adequate education, intelligence, and training, and a certain amount of scientific savvy, so to speak, then according to Aristotle one will simply "see" that certain basic facts about the world are not only true, but necessarily true. And that, in outline, is how one comes to see first principles.

What you can probably see more clearly at this point is that this approach is simply not going to work. The basic problem lies with the first principles. Consider again our discussions from our earlier chapters on worldviews, truth, and empirical facts and philosophical/conceptual facts. Given what we saw in those chapters, it is highly unlikely that there will ever be any such agreement on what constitute basic facts, much less what might constitute basic facts that have to be necessarily true. And so the basic problem with the sort of deductive approach envisioned within Aristotle's method lies with the very starting points of that approach.

As noted, Aristotle envisioned science as resulting in theories and claims that were not merely probable, but rather were certain. This sort of axiomatic approach, based on necessarily true first principles, seems the only way such certain scientific knowledge might be achieved. And as you might guess, the problem noted above, that is, the problem of finding agreed-upon, necessarily true starting points, is going to be a general problem for all such approaches. Largely for this reason, there is now general consensus that scientific claims and theories cannot be guaranteed to be correct. And as we discussed in the previous chapter, this is not a defect of science, but rather simply a consequence of the inductive nature of much of scientific reasoning. However, before moving on to consider other approaches, one other axiomatic approach, that of Descartes, is worth considering briefly.

Descartes' axiomatic approach

We discussed Descartes at the end of Chapter 2, where we saw that Descartes was interested in finding necessarily true beliefs to serve as a foundation on which to build a structure of certain knowledge. In many ways, Descartes' view of the proper way to conduct science was similar to that of Aristotle (though Descartes did not restrict himself to the purely syllogistic method taken by Aristotle). In particular, Descartes too was interested in using deductive reasoning to derive certain knowledge from necessarily true starting points.

Also as with Aristotle, much the same problem arose when Descartes attempted to find agreed-upon starting points. When dealing with starting points that concern matters about the world, there simply do not seem to be any agreed-upon basic principles about the world that we can know with certainty. And so with respect to basic starting points concerning the world, Descartes' approach is going to run into essentially the same problem as Aristotle's.

But as we saw in Chapter 2, at one point Descartes considered his own mind in his search for necessarily true starting points. And as we saw in that chapter, there is a case to be made that his "I am, I exist" is necessarily true. So Descartes may have found at least one (generally) agreed-upon, necessarily true belief to serve as a starting point.

However, as we also discussed at the end of Chapter 2, the basic problem with this is simply that it is not enough of a foundation. In short, when searching for necessarily true starting points concerning the world, Descartes had the same problem as Aristotle, namely, there seem to be no such agreed-upon, necessarily true starting points. And although there is probably more agreement that one can have certainty in the proposition that one exists (at least as a thinking being), it is too slim a foundation on which to build.

Popper's falsificationism

Karl Popper (1902–94) is the best-known advocate of what is generally termed the *falsificationist approach*. Popper himself did not take falsificationism as being a definitive scientific method; in fact, he did not think there was a single definitive scientific method. He did, though, view falsification as a key element of science, and a key criterion by which to distinguish scientific theories from nonscientific theories. In what follows, we will look at an outline of Popper's views.

In general, Popper argued that science should emphasize the attempted refutation of theories rather than emphasizing the confirmation of theories. According to Popper, it is too easy to find confirming evidence for many theories. To use one of Popper's examples, consider Freudian psychoanalysis. According to Popper, "predictions" made by this theory are general enough that almost any events could be interpreted as instances of confirmation. And thus, for Popper, confirming evidence for such theories is simply not of much interest.

In contrast, consider Einstein's theory of relativity. As we noted early in Chapter 4, Einstein's relativity theory predicted that starlight should be bent when passing near a massive body such as the sun. And since such bending of starlight, if it indeed occurred, could be observed during a solar eclipse, Einstein's theory made a specific, dramatic prediction not made by any competing theory. In that Einstein's theory made such a dramatic prediction, and one that could fairly easily be shown to be incorrect, Einstein's theory was, in this way, taking a substantial risk.

In a certain sense, for Popper, the riskier a theory is, the more scientific it is. For example, for the reasons just mentioned (that is, that Einstein's theory made a specific and dramatic prediction and hence was at risk of quickly being shown to be wrong), Einstein's theory is a much better example of a scientific theory than, say, Freudian psychoanalysis. And in general, for Popper this was the hallmark of good science, that is, science should emphasize falsification over confirmation, and should strive for at-risk theories.

As noted, Popper did not place great emphasis on confirming evidence. For Popper, what characterized a successful scientific theory was not that there was a great deal of confirming evidence for it; rather, what characterized a successful scientific theory was that it had survived repeated attempts to refute it via the testing of specific dramatic predictions. And this sort of falsificationist approach, that is, the emphasis on trying to falsify theories rather than confirm them, is central to Popper's view.

This is a fairly brief outline of Popper's views, but it should be sufficient to give a sense of the approach he favored. And as you might guess, the issues we discussed earlier concerning disconfirmation reasoning, as well as the issues noted in the discussion of the Quine–Duhem thesis, are especially relevant here. As we noted earlier, seeming cases of disconfirmation are rarely, if ever, as simple as they appear. Rather, where a prediction made by a theory does not come out as expected, it is always an option, and indeed it is often more reasonable, to reject an auxiliary hypothesis rather than the main theory. In short, although there is no question that cases of disconfirming evidence play important roles in science, the issues surrounding such evidence are sufficiently complex to make it unlikely that disconfirmation – that is, falsification – can serve as the central feature of science.

The hypothetico-deductive method

One often sees reference to what has come to be called the *hypothetico-deductive method*, and given how prominent such references are, the hypothetico-deductive method is certainly worth discussing here. But our discussion can be brief, since at bottom the hypothetico-deductive method does not involve much beyond issues we have already discussed.

The basic idea behind the hypothetico-deductive method is that from a hypothesis or set of hypotheses (or theories, broadly speaking) one deduces observational consequences, and then tests to see if those consequences are observed. If so, then for the reasons discussed earlier in relation to confirmation reasoning, this is taken as support for the hypothesis. If the consequences are not observed, then again for the reasons discussed earlier in the context of disconfirmation reasoning, this is taken as evidence against the hypothesis.

A quick note: the hypothetico-deductive method is not generally concerned with how the hypotheses themselves are generated, but rather with the justification or confirmation of hypotheses. In the philosophy of science, this distinction (between how hypotheses are generated versus how hypotheses are justified or confirmed) is usually described as the difference between the *context of discovery* versus the *context of justification*. The context of discovery is generally regarded as the more complex of the two, and as we will see in later chapters, the ways in which actual hypotheses or theories are developed are surprisingly varied and complex. However, as we are seeing, even the context of justification – that is, roughly speaking, how we go about justifying or confirming hypotheses or theories – is plenty complex by itself.

There is no question that confirmation and disconfirmation reasoning play important roles in science. Given the close ties between these patterns of reasoning and the hypothetico-deductive method, it is safe to say that this method plays an important role in science. However, consider again the issues we have discussed – the inductive nature of confirmation reasoning, the possibility of rejecting auxiliary hypotheses in the face of disconfirming evidence, the underdetermination of theories, the difficulty if not impossibility of designing crucial experiments, the notion that hypotheses are tested in groups rather than singly, and so on. The view that science proceeds by a relatively simple process of generating predictions from hypotheses and then accepting or rejecting hypotheses depending on whether the prediction is observed seems, given what we have discussed, to be at best an overly simplistic account of science.

Again, there is no question that the hypothetico-deductive method – that is, essentially confirmation and disconfirmation reasoning – plays an important role in science. But given the issues explored earlier, although the hypothetico-deductive method is *a* method used in science, it would be misleading to call it *the* scientific method.

Concluding Remarks

The Quine–Duhem thesis, and issues surrounding the topic of scientific method, illustrate some of the ways in which issues in science and the philosophy of science are intertwined, and intertwined in complex ways. As noted at the outset of this chapter, our main goal was to get these issues on the table, so that we will be in a position to appreciate the way such issues come into play in cases from the history of science. We will be turning to such cases in Part II; however, before doing so we need to consider a few more fundamental issues. We will turn next to some puzzling issues involving inductive reasoning.

Chapter Six

Philosophical Interlude: Problems and Puzzles of Induction

The issues discussed in Part I of this book are, by and large, issues involving basic topics in the history and philosophy of science, and ones that provide background material for topics explored in Parts II and III. In this chapter, we take a sort of philosophical interlude. The problems and puzzles we explore here are primarily philosophical, in the sense that they arose from and are mainly discussed by philosophers, rather than being issues that have a practical effect on the everyday workings of science. These topics also provide something of an interlude in the sense that, unlike the other topics of this part of the book, these issues are not necessarily needed as background for material in later chapters. Nonetheless, the problems and puzzles we discuss are of general interest, in that they illustrate some deeply puzzling aspects of some of the reasoning that is most basic to science.

I should mention that these issues do not tend to strike people, on first hearing of them, as deep or puzzling or profound. I recall that my first reaction to these problems, when I first learned about them years ago, was that they seemed to consist mainly of philosophical nonsense. And they did not initially strike me as deep or difficult at all – my initial sense was that they could all be resolved without a great deal of hard thought.

But after a while, one realizes that these problems do not admit of easy answers, and that they raise issues that are deeply puzzling. My main goal in this chapter is to introduce you to a few of these philosophical problems, all of which concern inductive reasoning. I encourage you to let these problems simmer for a while, and in so doing (one hopes) come to appreciate how puzzling the issues are. We will look, in particular, at Hume's problem of induction, Hempel's raven paradox, and Goodman's new riddle of induction. We will begin with Hume's problem.

Worldviews: An Introduction to the History and Philosophy of Science.   Richard DeWitt

Hume's Problem of Induction

David Hume (1711–76) was apparently the first to notice a puzzling aspect of inductive reasoning, and his observation now is typically referred to as Hume's problem of induction. Understanding Hume's point requires reaching one of those "Aha ..." moments. If you really grasp Hume's point, you will see that it is an extraordinarily puzzling point about our most common types of everyday reasoning, in particular, about inferences concerning the future. Let's begin with a quick point about reasoning in general.

When we reason, for example, when we present or consider arguments, our arguments almost always contain implied premises. Implied premises, as the name suggests, are premises that are necessary in order for the reasoning to be plausible, but that are implied rather than explicitly stated. For example, suppose we agree to meet for lunch downtown this Sunday, but your car is being repaired and you are not sure how you can get to the restaurant. Suppose I tell you that there is a local bus that runs from your house to the restaurant, and so you can take the bus to our lunch meeting. Implied in this informal bit of reasoning, but not explicitly stated, is the premise that the buses run on Sundays. If we use brackets to indicate implied premises, the reasoning can be summarized as follows:

> There is a bus that runs from your house to the restaurant.
> [The buses run on Sundays.]
> so   You can take the bus to our lunch meeting this Sunday.

Again, almost all reasoning contains implied premises, and there is nothing particularly surprising or unusual about this fact.

As noted, Hume's problem of induction concerns inferences involving the future, so let's now consider a typical inference about the future. Consider, for example, the following perfectly ordinary piece of inductive reasoning:

> In our past experience, the sun has always risen in the east.
> so   In the future, the sun will probably continue to rise in the east.

Note the logical form of this piece of reasoning, which is:

> In our past experience, ❏ has always (or at least regularly) occurred.
> so   In the future, ❏ will probably continue to occur.

So far, there is nothing particularly unusual about this reasoning. We simply have a typical inductive inference with a quite common logical form, of the sort we use all the time. But Hume was apparently the first to notice something interesting about reasoning of this sort. In particular, Hume noticed that this sort of reasoning contains the following implied, but crucial, premise:

The future will continue to be like the past.

Given this, and again using brackets to indicate implied premises, the reasoning above is more accurately captured as:

> In our past experience, the sun has always risen in the east.
> [The future will continue to be like the past.]
>
> so   In the future, the sun will probably continue to rise in the east.

And more generally speaking, the form of reasoning noted above is better captured as:

> In our past experience, ❏ has always (or at least regularly) occurred.
> [The future will continue to be like the past.]
>
> so   In the future, ❏ will probably continue to occur.

The first important point to note is why this implied premise is necessary. This implied premise is necessary for any inference about the future simply because, if the future does not continue to be like the past, then there is no reason to think that past experience will be any guide to what future experience will be like. In other words, if the statement noted above, that the future will continue to be like the past, is not correct, then past experience is no guide to the future. And so inferences about the future would not be reliable.

This is a crucial point to understand, so we should pause for a moment to get clear on this issue. To help illustrate, consider Robert Heinlein's novel *Job*. In this novel, the two key characters continually wake up to find themselves in worlds that are slightly different from the world they inhabited the day before. For example, one day they might wake up in a world in which the monetary system is slightly different from the world of the day before (and hence any money they have on them from the previous day is no longer worth anything). One day they might inhabit a world in which everyone obeys traffic laws, and the next day they wake up in a world in which violating traffic laws is the norm. And in general, each day their world is different from the day before. Since the world they inhabit is constantly changing, they don't know what to expect from day to day. For them, the future is *not* like the past. As a result, they are unable to make the sorts of inductive inferences about the future that we take for granted. (About the only inductive inference they can make about the future is that the future will *not* continue to be like the past. And this, of course, is not a particularly helpful inference.)

So the first key point to recognize in understanding Hume's problem of induction is this: the statement noted above, that the future will continue to be like the past, is a necessary, though generally unrecognized, implied premise in every piece of reasoning we do about the future.

Now, if the statement "The future will continue to be like the past" is a necessary implied premise in any reasoning about the future, then it is clear that the degree of confidence we have in inferences about the future depends crucially on the degree of confidence we have in this statement. The obvious next question, then, is: What reason do we have to think the future will continue to be like the past?

Our main (perhaps only) reason for believing the future will continue to be like the past seems to boil down to the fact that today was pretty much like yesterday (heavy things still fall down today, the sun again rose in the east, day was followed by night, and so on). Yesterday was pretty much like the day before it, and that day was pretty much like the day before it, and so on. In short, in our past experience, each day has tended to be more or less like the day before. And this seems to be the basis for our believing that, in the future, things will be more or less like they have always been. In short, if we ask the question "Why believe the future will continue to be like the past?" the best reason we can give is summed up in the following inference:

<u>In our past experience, the future was like the past.</u>
so   The future will probably continue to be like the past.

But note that this inference is an inference about the future. And again, any inference about the future, including the inference immediately above, depends on the implied premise that the future will continue to be like the past. When this implied premise is made clear, the inference above is better represented as:

In our past experience, the future was like the past.
<u>[The future will continue to be like the past.]</u>
so   The future will probably continue to be like the past.

But this reasoning is blatantly circular, that is, it assumes, as one of its premises, the very conclusion it is trying to establish. In other words, the inference summarized above depends on the assumption that its own conclusion is true. And that is blatantly circular and hence provides no justification for accepting the conclusion.

In summary, Hume's point is that every instance of inductive reasoning depends upon the implied premise that the future will continue to be like the past. But the main (and seemingly only) way to justify this implied premise is circular, and hence it appears that this crucial implied premise cannot be justified. So inferences about the future depend on an assumption that cannot be justified, and thus they cannot themselves be logically justified.

Before closing this section, a few final comments are in order. First, note how general Hume's point is. It covers *all* inferences about the future, whether they are garden-variety inferences (such as the sun rising in the east), or inferences about scientific laws continuing to hold in the future, or beliefs about mathematics being the same in the future as it has been in the past, and so on.

Second – and this is an important point in understanding him – Hume was *not* trying to convince us that we ought not to make inferences about the future. He thought that making inferences about the future is part of our nature – we can no more stop making inferences about the future than we can voluntarily stop breathing. His question was whether or not we can *logically justify* our inferences about the future, and his answer was that we cannot.

Hempel's Raven Paradox

Carl Hempel (1905–97) was an influential twentieth-century philosopher who worked mainly in the philosophy of science. As you might guess, his raven paradox was originally presented using ravens as an example, though it might be easier to see the relevance of the paradox if we use a somewhat different example. As an example to illustrate Hempel's raven paradox, suppose you and I are astronomers, and our main project involves gathering information on quasars. As a brief background, quasars are a relatively recent discovery, with the first being discovered about 40 years ago. Even after 40 years of research not a great deal is known about quasars (though there are some interesting and reasonably plausible theories about quasars that have been developed recently). At any rate, some of the basic facts about quasars are that they seem to emit an enormous amount of energy, and that they all seem to be located a great distance from Earth.

Now, suppose we are working in the early years of research on quasars. Suppose we notice that the first few quasars detected are all located a great distance from Earth, and one question we become interested in is whether all quasars are located a great distance from Earth. As the years go on, we (and other astronomers) continue to observe more quasars, and to note that every one we observe is located far away. So far, so good. We seem to be dealing with a fairly common situation, in which our observations are providing inductive support for the statement "All quasars are located a great distance from Earth."

There is nothing particularly puzzling about the situation described so far. When we are considering a general statement such as the one about quasars, and we observe a good number of instances that are in agreement with the statement, and none that run contrary to it, we tend to take this as inductive support for the statement.

The puzzle arises, as Hempel noted, when we consider the logical structure of general statements such as "All quasars are located a great distance from Earth." A general statement such as this is logically equivalent to its contrapositive, in this case, to the statement "All objects not located a great distance from Earth are not quasars." In other words, the statement

(a)   All quasars are located a great distance from Earth.

and the statement

(b)   All objects not located a great distance from Earth are not quasars.

are logically equivalent statements.

We noted above that each time we observed a quasar that was located a great distance from Earth (and again assuming that we observe no instances running contrary to the statement), each observation helps support the statement that all quasars are located a great distance from Earth. To be consistent, then, each time we observe an object not located a great distance from Earth that is not a quasar, we have to accept that this observation supports (b), that is, that all objects not located a great distance from Earth are not quasars.

This, too, by itself is not necessarily a problem or a puzzle. But now recall what we noted above, namely, that (a) and (b) are equivalent. If (a) and (b) are equivalent, then any support for (a) would have to count equally as support for (b), and likewise, any support for (b) would have to count equally as support for (a). And this gets right to the heart of the puzzle: whenever we have an observation that supports (b), it seems that the observation must equally support (a).

So, for example, the book in your hand is an object not located a great distance from the Earth and is also not a quasar, so observing this book supports (b). And for the reasons just noted, this observation should equally support (a). But this seems just crazy – surely you cannot help confirm a substantive scientific claim about quasars by making a trivial observation about the book in your hand.

As with Hume's problem, do not misconstrue Hempel's point. He is certainly not claiming that a trivial observation about a book in front of you actually helps support a substantive scientific claim about quasars. But he is pointing out that there is something odd about what seems a very basic pattern of inductive reasoning. Also, as noted above, Hempel's raven paradox does not constitute a practical problem, in that it is not generally a problem that affects the actual conduct of science. But undoubtedly, inductive reasoning supporting general statements, such as that all quasars are located a great distance from the Earth, is an important component of science. And Hempel's raven paradox suggests that there is something deeply puzzling about the nature of such reasoning.

Goodman's Gruesome Problem

Hume's problem of induction, discussed above, is now sometimes referred to as the "old" riddle of induction, in contrast to the "new" riddle of induction put forth by Nelson Goodman. Goodman (1906–98) was a broad-ranging philosopher, with interests ranging from logic to epistemology to the arts. He was apparently the first to notice another odd feature of certain types of inductive reasoning, and here we focus just on this issue.

Consider a statement such as "All emeralds are green." This statement seems to be highly supported by experience, in particular: every emerald we have observed has been green and, moreover, we have never observed an emerald that was not green. With respect to emeralds, the predicate "green" would seem to be what Goodman called a "projectible" predicate, that is, a predicate for which we can, based on our past experience that all observed emeralds have been green, project that in the future all observed emeralds will be green.

Now, define a new predicate, which Goodman named "grue." There are a number of ways to define "grue," but for our purposes (and this follows Goodman's formulation fairly closely), say an object is grue if it is green and first observed before New Year's Day 2020, or blue and first observed after that day. As noted above, all emeralds observed so far have been green, and none has failed to be green. And again, this seems to give us some reason to think that any emeralds we observe in the future will also be green.

But now note that every emerald observed so far was green and was first observed before New Year's Day 2020. In other words, every single emerald observed so far has been grue, and moreover, none has failed to be grue. In other words, at least with respect to emeralds observed so far, the inductive support for the statement that, in the future, all observed emeralds will be green, is *exactly* the same as the inductive support for the statement that, in the future, all observed emeralds will be grue.

But of course, we would never make the inference that, in the future, all observed emeralds will be grue. That is, although we feel justified in thinking that emeralds we observe in the future will continue to be green, we are sure that emeralds observed in the future (in particular, those first observed after New Year's Day 2020) will not be grue.

But if it is so obvious that emeralds observed after New Year's Day 2020 will be green, but not grue, there must be some difference between the predicate "green" and the predicate "grue." The former, to use the terminology mentioned above, is what Goodman's calls a projectible predicate (again, one for which we are justified in projecting its application to emeralds into the future), but the latter is not a projectible predicate. But in general, what is the difference between projectible and nonprojectible predicates?

This question, which at first glance seems as if it would be quite easy to address, has proved to be difficult. Of the responses that come immediately to mind – that predicates such as "grue" are constructed rather than "natural" predicates, that unlike ordinary predicates they involve references to time, and so on – none has withstood scrutiny. So, although there are numerous suggestions for distinguishing projectible from non-projectible predicates, none of the suggestions has reached anything that might be considered a consensus view.

As with Hume's problem of induction, and Hempel's raven paradox, it is important not to misunderstand Goodman's point. He is certainly not suggesting that we should believe that all emeralds observed in the future will continue to be grue. Obviously they will not. But given how obvious the difference between a predicate such as "green" and a predicate such as "grue" seems to be, one would think it would be easy to provide an account that plausibly captures the difference between projectible and nonprojectible predicates. Goodman's main question was what that difference was. And as noted, even though the problem seems at first glance an easy one to solve, after decades of proposed solutions there is no agreement that any of the proposals provide an adequate solution. So once again, although Goodman's new riddle of induction is certainly not a practical problem, in that it does not affect the everyday workings of science, the problem raises puzzling questions about inductive reasoning, and in particular about the differences between predicates whose use we are confident can be projected into the future, and those that cannot be.

Concluding Remarks

As noted at the outset of this chapter, the issues discussed above are clearly philosophical issues, rather than issues that affect working scientists. And they tend to seem, at first, to be problems that should be easy to solve. Yet the fact that these problems resist solution, even after decades or more of extensive discussion, suggests that there is something deeply puzzling about some of our most basic types of inductive reasoning.

Also, as noted at the beginning of this chapter, it generally takes some time to fully appreciate these problems. With this in mind, I would encourage you to keep these problems in the back of your mind and let them simmer a while. In the meantime, we will move on to discuss issues that will arise repeatedly in examples from the history of science, these being issues surrounding the notion of falsifiability.

Chapter Seven

Falsifiability

In this chapter, we introduce the notion of *falsifiability*. The issues surrounding falsifiability appear, at first glance, as if they could not be more simple or straightforward. But in fact, especially when applied to real-life cases, they can get quite complex. In this chapter we will begin with a somewhat simplified account of falsifiability, and then move on to consider some of the complicating factors. In later chapters, especially as we examine some cases from the history of science, we will see examples of some of the more complex issues involving this notion.

Basic Ideas

In a sense, falsifiability is perfectly straightforward. It is an attitude toward theories. In particular, it is the attitude one has when one allows for the possibility that a particular theory might be false. For example, suppose Sara is a physicist who believes that the Big Bang theory of the origin of the universe is probably correct. Suppose that, like most physicists, Sara is not dogmatic in her beliefs. That is, if a sufficient amount of new evidence came to light that provided convincing reasons to think the Big Bang theory was not correct, then Sara would be willing to give up her belief in the Big Bang theory. In short, although Sara believes the Big Bang theory is correct, she is willing to admit that it might be false, and so we say that she treats the theory as falsifiable.

In contrast, suppose Joe is a member of the Flat Earth Society. The Flat Earth Society is a group whose members believe, quite sincerely, that the Earth is flat. Suppose Joe believes the theory that the Earth is flat. Moreover, no matter what evidence is presented that suggests this theory is false, Joe finds some way around the evidence. For example, suppose we point out that almost everyone believes the Earth is spherical. Joe replies (perhaps not unreasonably) that popular opinion is no guide to the truth. So we show Joe a photograph of the Earth that was taken during a space shuttle flight. Joe replies that there is good reason to believe that the space program was a complete fraud, that the photographs and television coverage have been faked; and he expresses sympathy that we have been suckers for such fraudulent reports. We argue that history books are filled with reports from voyagers who have circumnavigated the globe, which is possible only if the Earth is a sphere. Joe tells us about an article he recently read suggesting that, on a flat Earth, compass bearings would be skewed as one approached the periphery of the Earth, and what probably happened to explorers such as Ferdinand Magellan was that they sailed in a large circle around the periphery of a flat Earth and, due to the skewed compass bearings, they mistakenly thought they were sailing in a straight line around the circumference of a sphere.

We soon realize that Joe will hold to his theory no matter how much evidence is presented that suggests the theory is false. Unlike Sara, Joe appears unwilling to admit that his theory might be false, and so it appears that Joe is treating the theory as unfalsifiable.

When writing and speaking on falsifiability, authors tend to speak of falsifiability as if it were a characteristic of theories. That is, there is a prevalent but bad habit of speaking of this or that theory as being falsifiable or unfalsifiable. But with a little reflection, it should be clear that this is not the best way of speaking. Generally, falsifiability is an attitude that a person might hold toward a particular theory, rather than being a characteristic of the theory itself. For example, consider again the flat Earth theory. There is nothing about the flat Earth theory that makes the theory inherently unfalsifiable. We can easily imagine two individuals, both believers in the flat Earth theory, yet such that one becomes convinced the theory is wrong while the other (like Joe above) refuses to abandon the theory no matter how much evidence is presented. In each case the theory is the same; what differs is the attitude of the individuals toward the theory. Thus, it is generally not accurate to speak of a theory itself as being unfalsifiable; rather, the crucial factor is one's attitude toward the theory, and it is one's attitude that determines whether one is treating a theory as falsifiable or unfalsifiable.

Complicating Factors

At this point, the notion of falsifiability might seem like a pretty simple notion, and the question of whether someone is treating a theory as falsifiable might seem to be a straightforward question. But in many cases, especially those from the history of science involving substantial changes in theories (for example, the change from the Earth-centered to the sun-centered view), it is not at all easy to say when theories are being treated as unfalsifiable. Let's consider a few reasons why this issue is difficult.

In describing Sara above, we said that she was willing to give up the Big Bang theory if there was a "sufficient amount" of new evidence providing "convincing reasons" that the theory was wrong. As discussed in Chapter 4, evidence against a theory often comes in the form of predictions that turn out to be incorrect. That is, when a theory is used to make predictions, and those predictions turn out to be wrong, that poses a problem for the theory. However, also as discussed in Chapter 4, incorrect predictions are often the result of incorrect auxiliary hypotheses rather than an incorrect theory. So when faced with incorrect predictions, it is often more reasonable to reject one or more auxiliary hypotheses than to reject the theory itself.

Since one can (and often should) reject one or more auxiliary hypotheses, an extremely difficult question arises: What counts as a "sufficient amount" of evidence to reject a theory? At what point does one have "convincing reasons" that a theory (rather than one or more auxiliary hypotheses) is wrong?

There are no clear answers to these questions. It is certainly unreasonable to abandon a theory the first time problems arise, but on the other hand, for some theories there comes a time when the evidence against the theory reaches the point where it would be unreasonable to continue to hold the theory.

The cold fusion example from Chapter 4 provides a good illustration of this. Initially, in the late 1980s, there were some interesting experimental results suggesting that fusion was indeed taking place at low temperatures. Moreover, the two scientists reporting these results were by no means oddballs or fringe scientists. These were well-respected, well-published, well-established scientists reporting (although via the press rather than via mainstream scientific journals) quite intriguing experimental results. However, in the ensuing months, problems arose for the cold fusion theory. In particular, one could use cold fusion theory to make certain predictions, and many of these predictions were not observed. Initially, the supporters of cold fusion handled these problems by rejecting various auxiliary hypotheses – the cold fusion apparatus was set up using incorrect materials, the experimenters did not give the cold fusion apparatus sufficient time to charge itself, and so on. As the years went on, disconfirming evidence continued to pile up. In addition, plausible alternative explanations of the initially interesting results were offered. By the end of the 1990s, 10 years after the initial cold fusion announcement, the dwindling number of cold fusion supporters were forced to appeal to increasingly complicated auxiliary hypotheses. Such auxiliary hypotheses included, for at least some defenders of the theory, that the problems for cold fusion were the result of a conspiracy by large oil companies to suppress new energy sources.

The point is that continuing to believe the cold fusion theory by rejecting various auxiliary hypotheses was a reasonable thing to do initially. But by the time one is appealing to conspiracy theories to save one's theory, one has crossed the line from reason to unreason. But, importantly, the line is not a precise, well-defined line. And as a consequence, it is not a precise matter to say at what point one is treating a theory as unfalsifiable.

These issues become even more difficult when we recall earlier discussions on evidence and worldviews. Consider again my acquaintance Steve, first discussed in Chapter 2. Steve takes an extremely literal interpretation of certain passages from Vedic scriptures, and as a result of his beliefs about the reliability of these scriptures, he believes that the moon is inhabited with intelligent life, that the moon is further from the Earth than the sun is, and that the Apollo moon landings were faked. My students and I have had numerous discussions with Steve on these matters, generally in the form of presenting him with evidence that his beliefs are wrong. Steve rejects all this evidence in favor of evidence from his scriptures. It looks to us, given our way of looking at the world, that Steve's views on these matters provide a clear case of his treating the views as unfalsifiable. After all, he refuses to change his views in spite of the overwhelming evidence we provide.

But now, *look at the matter from Steve's perspective*. During our discussions with Steve, he often presents us with what he takes to be convincing evidence that the scriptures in question are accurate. And if his scriptures are correct, then Steve's beliefs are justified and our beliefs are the ones that are incorrect. But notice that we do not accept his evidence, refusing to change our views in spite of what Steve takes to be overwhelming evidence supporting those views. Notably, from Steve's perspective it is *we* who are treating *our* views as unfalsifiable.

Also worth noting is that, from Steve's perspective, he *is* treating his theory as falsifiable. Steve clearly agrees that he would be willing to give up his views if faced with a sufficient amount of evidence. But what Steve takes to be the relevant evidence is very different from what I and most of my acquaintances take to be the relevant evidence. I and most of my acquaintances put the most emphasis on what we would consider empirically based evidence – evidence from physics, astronomy, cosmology, and the like. But for Steve, the most important evidence is evidence from his scriptures. So, if he were given evidence based on his scriptures (perhaps in the form of newly discovered scriptures, or a new and better translation of existing scriptures, or the like), Steve readily agrees that he would be willing to change his views. Hence, from his perspective, he is indeed willing to give up his views if he were faced with a sufficient amount of evidence, and so he is, from his perspective, treating his theories as falsifiable.

The key and difficult issue here is the question of what counts as the relevant evidence. This is a subtle but important point, and one that comes up again and again in the history and philosophy of science. Given its importance, it is worth repeating: in almost all real-life cases, the main point of disagreement is *not* whether one party or the other is willing to give up their theories if faced with sufficient evidence. Rather, the main point of disagreement is what counts as the most relevant and most important evidence.

Importantly, what one takes to be the most relevant and important evidence ties in closely with one's overall worldview. Steve's trust in his scriptures is a core piece of his jigsaw puzzle. He could not give up his trust in his scriptures without drastically altering, and really replacing, most of his overall jigsaw puzzle. And if

I am honest, my emphasis on what I take to be the appropriate sort of empirically based evidence is likewise a core piece of my jigsaw puzzle. In other words, our respective systems of beliefs heavily influence what we take to be the relevant evidence, and this in turn heavily influences our views on who is treating his or her theory as unfalsifiable.

Concluding Remarks

Before closing, I want to stress an important point. In the discussion above, I am *not* suggesting that some sort of relativism is correct, nor am I suggesting that all evidence and worldviews are equally reasonable, nor am I suggesting that Steve's views are reasonable. I think Steve's views are entirely unreasonable. I am more than happy to argue that basing evidence on literal readings of religious scriptures is a bad and outdated idea, and that individuals such as Steve are treating their views as unfalsifiable.

What I do want to suggest is that whether someone is treating a theory as unfalsifiable, and if so why, is a much more subtle issue than is often appreciated. As the case of Steve illustrates, we cannot merely claim that he refuses to accept our evidence, and thereby conclude that he is treating his theory as unfalsifiable. Again, he can say the exact same thing about us – that we refuse to accept his evidence. So if we are to make a case that Steve is treating his theory as unfalsifiable, we have to do better than this.

Likewise, it would not be reasonable to merely assert, dogmatically, that our preferred evidence is the correct type of evidence. That is, we cannot claim that Steve is treating his theory as unfalsifiable based on dogmatically asserting that our evidence is the correct type of evidence.

To build a case that Steve is treating his theory as unfalsifiable, we would need to consider a number of interrelated issues, including issues such as whether it is more reasonable to appeal to empirical evidence rather than to rely on ancient scriptures. Such a case can be made, that is, I do not think there is much question that when such factors are considered, the correct conclusion is that Steve is indeed treating his theory as unfalsifiable. The main point here, though, is that making this case is more complicated than merely saying that someone does not accept one's evidence.

So, as stated at the opening of this chapter, falsifiability turns out to be a much more subtle and complicated issue than it first appears to be. Watch for these issues as we explore, in subsequent chapters, some of the major developments from the history of science.

Chapter Eight

Instrumentalism and Realism

The goal of this chapter is to introduce two common attitudes toward scientific theories. These attitudes are generally labeled *instrumentalism* (or *operationalism*) and *realism*. We will begin with a discussion of two central issues relevant to scientific theories, namely, prediction and explanation.

Prediction and Explanation

Suppose we ask the question "What do we want from scientific theories?" Certainly, the ability to make accurate predictions is one feature we want from a scientific theory. As discussed in Chapter 4, when Einstein introduced his theory of relativity in the early 1900s, one of the points in the theory's favor was that it made accurate predictions, not made by any other theory, as to what would be observed during the solar eclipse of 1919. Clearly, the ability to make accurate predictions such as this is a feature we desire.

In addition, there is widespread agreement that the ability to explain the relevant data is another feature we desire in a theory. However, although there is agreement that explanation is an important feature, there is much less agreement on what exactly counts as an adequate explanation. For example, can there be more than one correct explanation for a single event, or does each event have only one correct explanation? Does an adequate explanation have to specify the exact series of events that produced the data? Is it sufficient for a theory to specify *that* certain data should be observed, or does an adequate explanation need to go further and specify *how* or *why* the events in question occurred? These and other questions about the nature of explanation are difficult and controversial.

Worldviews: An Introduction to the History and Philosophy of Science.   Richard DeWitt
© 2010 Richard DeWitt

To help clarify some of these issues, philosophers of science sometimes distinguish between *explanation* (or what is sometimes referred to as "formal explanation"), on the one hand, and what is often termed *understanding* on the other. In this distinction, "explanation" is used in a fairly minimalist way. More specifically, one says that a theory explains a piece of existing data or observation if one could have predicted the data from the theory. Used this way, explanation is sort of a retroactive prediction.

An example may help to illustrate this notion of explanation. In the early 1900s it had been noted, for some decades, that there were certain peculiarities about the orbit of Mercury. Einstein's theory of relativity did not come about until after these observations about the orbit of Mercury had been made, but had Einstein's theory of relativity been developed before these observations, the theory could have been used to predict that such peculiarities would be observed. In other words, when Einstein's theory was developed in the early 1900s, it could be used to explain (again, in this minimalist and sort of retroactive sense of "explanation") these peculiarities. And it was certainly a point in favor of Einstein's theory of relativity that it could explain this unusual data.

In contrast, and broadly speaking, "understanding" is used to refer to a somewhat more thorough appreciation of the data and observations. For example, consider the observation that falling bodies accelerate at about 10 meters per second squared. One can use Newton's theories and equations concerning gravity to show that falling objects should accelerate at about this rate. That is, Newton's physics can be used to explain (in the minimalist sense of "explanation" described above) this data. Now if you take gravity as a really existing force acting on objects (that is, you take what is called a *realist* attitude, which we will explore more fully in a moment, toward Newton's concept of gravity), then you might say you know not only *that* objects accelerate at about 10 meters per second squared, but you also know *why* they do so (because they are under the influence of a gravitational force). That is, you have both an explanation for the data, and also an understanding of the data.

The notion of explanation (again, in the somewhat minimalist sense described above) is a reasonably straightforward and uncontroversial notion, whereas issues surrounding understanding are quite complex and controversial. Many of the reasons for this complexity will emerge as the book progresses. But for now, to keep our discussion relatively straightforward, we will take the minimalist approach to the concept of explanation described above. That is, we will say that a theory explains an existing piece of data or observation if the theory could have been used to predict that data or observation.

As noted, there is widespread agreement that prediction and explanation are very important requirements for any adequate theory. While explanation and prediction are the most important characteristics of theories, it should be noted that they are not the only desirable characteristics. For example, characteristics such as simplicity, elegance, and beauty are regularly appealed to in arguing for and against theories. In what follows, I will usually focus just on prediction and explanation, since these are agreed to be the most important characteristics, but the other attributes just mentioned should also be kept in mind.

So, there is general agreement that accurate predictions and explanations are required from scientific theories. But are these characteristics enough? Or is it the case, as Einstein came to believe (at least, the older Einstein – the younger Einstein was not so committed to this view), that reality is the business of physics (and, of course, other sciences as well)? That is, is it important that a theory reflect, or model, reality?

This issue – whether we require theories to reflect the way things really are – is a controversial issue. It is the issue that distinguishes instrumentalists and realists. For an *instrumentalist*, an adequate theory is one that predicts and explains, and whether that theory reflects or models reality is not an important consideration. For a *realist*, on the other hand, an adequate theory must not only predict and explain, but in addition it must reflect the way things really are.

To illustrate the differences between instrumentalists and realists, it will help to look at an actual theory. In this case, let us look at certain aspects of the Ptolemaic astronomical system.

The Ptolemaic system was formulated by Claudius Ptolemy around AD 150. Ptolemy's approach is an Earth-centered one, with the sun, planets, and stars revolving around the Earth. Ptolemy considers each of the relevant objects – for example, the moon, sun, and planets – in turn, specifying the mathematics needed to predict and explain the observed positions of these objects.

One of the more interesting aspects of Ptolemy's approach is his use of *epicycles*. Ptolemy did not invent the technique of epicycles, but he put them to more extensive use than anyone before him. To understand the notion of epicycles, it might help to begin with the picture presented in Figure 8.1. It should be noted that this is an overly simplified picture of Ptolemy's approach, and we will not look at details of Ptolemy's approach until later chapters. This picture will, however, be sufficient to illustrate the points of interest to us in this chapter. In particular, and roughly speaking, a planet such as Mars moves in a circle around a point (labeled A on the diagram), while this point moves in a circle around the Earth. The circle traced out by the movement of Mars around A is called an epicycle. In short, an epicycle is a small circle around which a planet moves, with the

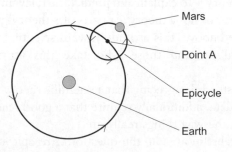

Figure 8.1   Mars' motion on the Ptolemaic system

center of the epicycle itself moving around some other point (usually, though not always, the center of the system).

On an Earth-centered view of the universe, epicycles, or something at least as complex and (to our eyes, at least) as odd-looking as epicycles, are necessary in order for the theory to predict and explain the relevant data. In this case, the relevant data consists largely of the observed position of the planets (and other heavenly bodies) in the night sky. For example, consider the point of light we call Mars. The observed position, in the night sky, of this point of light varies from night to night, week to week, year to year. A theory such as Ptolemy's needs to accurately predict and explain this data, and to do so, Ptolemy's theory (and any other Earth-centered approach) needs epicycles, or something at least as complex. We will defer discussion of why an Earth-centered system requires such complexities until later, but at this point, trust me: an Earth-centered model without epicycles (or something similar) cannot accurately predict and explain the motion of the planets.

So without epicycles, the Ptolemaic system (and any other Earth-centered system) would be unacceptable due to a failure to adequately explain and predict. On the other hand, with epicycles the Ptolemaic system is quite good at explanation and prediction. In fact, the Ptolemaic system is a marvelous mathematic model, which is able to explain and predict the motion of all the visible planets and stars with remarkable accuracy. It is not perfect with respect to prediction and explanation (few theories are perfect), but it is very good, and certainly far superior to any other approaches available at the time.

Thus, in terms of explanation and prediction, the Ptolemaic system, with its peculiar-looking epicycles, is excellent. But are epicycles real, or are they included simply because they are necessary in order to predict and explain the motion of the planets?

Suppose we were back in the second century and considering the question of whether Mars really moved in a small circle around the point A. And suppose we were of the opinion that the only important point is that Ptolemy's system, with epicycles, is good at prediction and explanation, and whether Mars really moves on an epicycle is just not important. This attitude would not have been at all unusual in Ptolemy's day, nor is it at all unusual today. A large percentage of working scientists, and philosophers of science, take the approach that the main task of a scientific theory is to explain and predict the relevant data, and it is simply not important whether or not a theory (or parts of a theory) reflect the way things "really are." As noted above, this attitude toward scientific theories is generally labeled instrumentalism, and those who take this attitude are considered instrumentalists.

Realists, in contrast, while agreeing that a scientific theory ought to explain and predict the relevant data, additionally require that a good scientific theory be real, that is, that it reflect the way things really are.

To an instrumentalist in AD 150, the question "Are epicycles real?" would not have been an important question. Ptolemy's theory accurately predicts and explains the relevant data, and that is all that is important. On the other hand, for a realist this question would be important. Even though Ptolemy's theory accurately predicts and explains, a realist would also require that it reflect the way things really are. So if Mars does not really move on an epicycle, that is, if epicycles are not real, then Ptolemy's system would be unacceptable.

Incidentally, with respect to the question "Are epicycles real?" it is not obvious how Ptolemy himself would have answered. In recent writings about the Ptolemaic system, Ptolemy is almost always portrayed as an instrumentalist. But this is not entirely accurate. It is true that Ptolemy generally is concerned with explanation and prediction, with little discussion as to whether his system reflects the way things really are. When Ptolemy speaks like this (which, again, is most of the time), he sounds like an instrumentalist. On the other hand, there are passages where Ptolemy discusses issues such as the mechanism by which planets are carried around on their epicycles. Such discussions make sense only from a realist perspective, and if Ptolemy had a purely instrumentalist attitude, it is difficult to explain why he includes such discussions. I think the most accurate view is that Ptolemy, like many of us, held something of a mixture of instrumentalist and realist attitudes.

Mixtures of these attitudes are not uncommon. It is certainly possible, and not at all contradictory, to hold a realist attitude toward certain *parts* of a theory while having an instrumentalist attitude toward other parts of the theory. For example, before the 1600s it would not have been unusual for someone to hold a realist attitude toward the Earth-centered part of the Ptolemaic system, while holding an instrumentalist attitude toward the part of the theory involving epicycles. That is, prior to the 1600s it was widely, and quite reasonably, believed that the Earth really was the center of the universe. Thus, people would generally view the Earth-centered part of Ptolemy's theory as reflecting the way things really are. Many, perhaps most, of those same people took an instrumentalist attitude toward the epicycles in the Ptolemaic system.

It is also common for someone to have a realist attitude toward theories in one branch of science while holding an instrumentalist attitude toward theories in other branches of science. For example, virtually everyone I know holds a realist attitude toward the current sun-centered model of our solar system. Many of the same people, though, take an instrumentalist attitude toward much of modern quantum theory.

It is also possible to accept two competing theories by having an instrumentalist attitude toward one and a realist attitude toward the other. For example, the Copernican system (a sun-centered view) was published in the 1550s, and in the late 1500s it was not unusual for a European university to teach both the Ptolemaic and the Copernican systems. Before the invention of the telescope (about 1600), there was good reason to think the Earth was indeed the center of the universe. Thus, people typically took a realist attitude toward the Ptolemaic system (or at least, the Earth-centered part of the system). On the other hand, in certain ways the Copernican system was somewhat easier to use, and so an instrumentalist attitude was taken toward this system. That is, the Copernican system was not viewed as reflecting the way things really are, but it was accepted and widely used as a convenient theory for making predictions and explanations. In short, between about 1550 and 1600, the Ptolemaic and Copernican systems coexisted peacefully. A realist attitude was typically taken toward the former and an instrumentalist attitude toward the latter. This relatively peaceful coexistence changed dramatically with the invention of the telescope and the discovery of evidence indicating that the Earth-centered view was wrong. But that is a story for later chapters.

In summary, instrumentalism and realism are attitudes toward theories. Both instrumentalists and realists agree that an adequate theory must accurately predict and explain the relevant data. But realists require, in addition, that an adequate theory pictures, or models, the way things really are. Finally, it is neither contradictory nor unusual to find mixtures of instrumentalist and realist attitudes, or for an individual to hold a realist attitude toward certain theories and instrumentalist attitudes toward others.

Concluding Remarks

I will close with two quick notes. As is the case with the concept of falsifiability discussed in the previous chapter, authors often speak of instrumentalism and realism as if they were features of scientific theories. But instrumentalism and realism are better thought of as *attitudes toward* scientific theories, rather than as aspects of the theories themselves. That is, just as typical theories are not inherently falsifiable or unfalsifiable, so also typical theories are not inherently instrumentalist or realist. Rather, one's attitude toward a theory is what is better classified as instrumentalist or realist.

Lastly, in Chapter 2 we discussed correspondence and coherence theories of truth. Recall that advocates of correspondence theories see truth as a matter of a belief corresponding with reality, whereas advocates of coherence theories view truth as a matter of a belief cohering, or fitting in with, some overall set of beliefs. One might wonder whether correspondence and coherence theories go hand in hand with realism and instrumentalism.

It should be noted that there is no necessary connection between theories of truth, on the one hand, and instrumentalism and realism on the other. For example, it would not be logically contradictory to be an advocate of a coherence theory of truth and at the same time be a realist about theories. And likewise, it would not be contradictory to be an advocate of a correspondence theory of truth while being an instrumentalist.

However, it should come as no surprise that there is some linkage between instrumentalist/realist attitudes and coherence/correspondence theories of truth. Recall from Chapter 2 that advocates of coherence theories of truth are often motivated by qualms about reality, or more precisely, about our knowledge of reality. It would be rather odd (though, strictly speaking, not contradictory) for someone having qualms about reality when considering theories of truth to then insist, in the context of instrumentalism and realism, that theories model or reflect they way things really are. And so it is not surprising that advocates of coherence theories of truth are more likely to hold instrumentalist attitudes.

Likewise, it is not surprising that advocates of correspondence theories are more likely to hold realist attitudes toward theories. The reasons are essentially the same – if one sees truth as a matter of corresponding to the way things really are, it would be natural to insist that scientific theories likewise model or reflect the way things really are.

This, then, ends our survey of some of the preliminary and basic issues involved in the history and philosophy of science. With an understanding of these issues, we are in a better position to explore the issues raised in the next part of the book, in which we investigate the transition from the Aristotelian to the Newtonian worldview.

Part II

The Transition from the Aristotelian Worldview to the Newtonian Worldview

In Part II, we explore the transition from the Aristotelian worldview to the Newtonian worldview. This transition was largely initiated by new discoveries in the early 1600s. In this transition, the issues explored in Part I – worldviews, empirical and philosophical/conceptual facts, confirming and disconfirming evidence, auxiliary hypotheses, falsifiability, instrumentalism, realism – are intertwined in interesting and complex ways. Discussion of this transition, and the issues involved, sets the stage for our exploration, in Part III, of challenges to our own worldview brought on by recent discoveries.

Chapter Nine

The Structure of the Universe on the Aristotelian Worldview

In this part of the book, we will be exploring the transition from the Aristotelian worldview to the Newtonian worldview. The main goal of the current chapter is to provide a general sense of how people typically viewed the universe from around 300 BC to AD 1600. This includes both views on the physical structure of the universe, as well as more conceptual beliefs as to what sort of universe we inhabit. We will begin with a sketch of the physical structure of the universe.

The Physical Structure of the Universe

As previously indicated, the Aristotelian worldview was the dominant worldview in the western world from about 300 BC to about AD 1600. When I say this was the dominant worldview, I mean that a system of beliefs, with strong roots in the views of Aristotle (though not necessarily identical to his views), was the main system of beliefs in the western world. This worldview was certainly not the only system of beliefs during that period – as with any period, there were a variety of alternative and competing systems of beliefs – but the Aristotelian system of beliefs was the most common.

On the Aristotelian worldview, the Earth was believed to be the center of the universe. Contrary to what I think is often assumed, people did not believe in an Earth-centered view merely for egotistical reasons. That is, the Earth-centered view was not based, at least not originally, on the view that humans were special, and thus should be at the center of all there is. True enough, the view that humans are special fits in well with the Earth-centered view, but the original reasons for the view were a result of solid, empirically based reasoning. We will look at some of the reasons in the next chapter.

Worldviews: An Introduction to the History and Philosophy of Science. Richard DeWitt
© 2010 Richard DeWitt

Also contrary to what is often assumed, during the Aristotelian worldview the Earth was viewed as spherical, not flat. It was clear to our predecessors, since even before the time of Aristotle, that the Earth was almost certainly spherical. Again, we will look at the reasons for this belief in the next chapter, but they largely overlap the reasons we have today.

With respect to the moon, sun, stars, and planets, the view was as follows. The moon, of course, is the closest heavenly body to the Earth. The region between the moon and the Earth – that is, the sublunar region – was thought to be importantly different from the region from the moon outward – that is, the superlunar region. We will discuss some of these differences in a moment.

Beyond the moon, the usual consensus was that the order of the planets and sun was as follows: first Mercury, then Venus, the sun, Mars, Jupiter, Saturn, and then the so-called sphere of the fixed stars. A few comments on the planets and stars are in order.

Consider, for example, the planet Mars. In our time, when we think of Mars, we picture a rocky body, something like the Earth, perhaps with a barren landscape, with more reddish soil than is found on Earth. But overall, we tend to think of Mars as basically being like the Earth – a big rocky body moving through space.

Our view of Mars is colored largely by the technology available. We have seen photographs of the surface of Mars, seen data from spacecraft that have visited Mars, perhaps observed Mars personally through a telescope, and so on. In short, our view of Mars is heavily influenced by technology.

This technology was not available during the period of the Aristotelian worldview. In fact, beliefs about the stars and planets had to be based primarily on naked-eye observations. What can one observe about stars and planets with the naked eye? Not much. In fact, without modern technology, stars and planets appear quite similar. Basically, stars and planets both appear to be points of light in the night sky. The main observational difference between stars, on the one hand, and the five points of light we call the planets (at least, the five planets visible to the naked eye) on the other hand, is that stars and planets move across the night sky in different ways. This different movement between planets and stars is the main factor distinguishing planets from stars.

As such, there was no reason whatsoever for people in the Aristotelian worldview to think of planets as being at all like the Earth. In fact, during the Aristotelian worldview the sun, stars, and planets were thought to be composed of a similar substance, quite unlike anything found on Earth. This substance, ether, was thought to be found only in the superlunar region, and was thought to have unusual properties that explained why objects in the heavens behaved as they did.

At the periphery of the universe was the sphere of the fixed stars. The thinking was that all the stars were the same distance from the Earth, embedded in a sphere. The sphere turned on its axis, completing a revolution about once every 24 hours. As the sphere turned, it carried the stars with it, and this explained the observational fact that the stars appear to rotate in a circle around the Earth every 24 hours.

Finally, a note on the size of the universe. During the Aristotelian worldview, how big did people consider the universe to be, that is, how far away was the sphere of the fixed stars? We have to be careful here. By the standards of the time, the universe was considered to be quite large. But we consider the universe to be unimaginably larger than they did, and perhaps infinite, so by modern standards their view of the universe strikes us as a concept of a relatively small universe. To put the same point another way, they considered the universe to be large, but they had no idea how huge it would turn out to be.

Conceptual Beliefs about the Universe

When we turn from beliefs about the physical structure of the universe, and consider instead more conceptual beliefs about the universe, the two most important such beliefs concern *teleology* and *essentialism*. That is, the universe was considered to be teleological and essentialistic. Importantly, the teleology and essentialism were closely intertwined, to the extent that they can be viewed as flip sides of the same coin. A few words of explanation about these concepts are in order.

To understand teleology, let's begin by understanding the notion of a *teleological explanation*. Suppose we ask the question "Why do fruit-bearing plants bear fruit; for example, why do apple trees have apples?" Clearly, the answer has to do with reproduction, that is, apples contain seeds, and seeds are the means by which apple trees reproduce, so clearly apples have something to do with reproduction. But note that most plants do not encase their seeds in fruit, so why do fruit-bearing plants do so? Incidentally, contrary to what is often thought, fruit does not provide any source of nutrients to the seeds (in this way, fruits are very different from nuts, the interiors of which are a source of nutrients). So the question is a good one – apple trees, for example, expend a lot of their resources to develop apples, and encase their seeds in these apples, yet the apples are of no direct use in providing nutrients to the seeds. So why in the world do apple trees go to so much trouble and expense to encase their seeds in apples?

One good answer is that the apples provide a means of dispersing the seeds. Speaking anthropomorphically for a moment, look at the situation from the point of view of the apple tree. Remember, plants do not move. So if you drop your seeds straight down, they will land on a piece of ground that is already taken, so to speak. You need some way to get the seeds away from yourself. Most plants have this problem, and the problem is solved in different ways. Some plants encase their seeds in light, fluffy structures that are caught and carried off by the wind, some encase their seeds in burry containers that stick to and are carried off by the fur of passing animals, some encase their seeds in helicopter-like structures that spin away from the plant, and so on. Fruit-bearing plants encase their seeds in structures that are a good food source for animals, and when the animals eat the fruit they also eat the seeds. A day or two later the animal excretes the seeds, at this time some distance from the parent plant (along with a convenient supply of fertilizer too, it might be noted).

In short, if we ask the question "Why do apple trees have apples?" a good answer is that they have apples in order to disperse seeds. This explanation is a prime example of a teleological explanation. Here are some other examples. Why does the heart beat? To pump blood. Why are you reading this book? To learn about the history and philosophy of science. Why did the stegosaur have large bony plates on its back? To regulate heat.

In general, a teleological explanation is any explanation that is given in terms of a goal, or purpose, or function to be fulfilled. In all the cases above, the goal, purpose, or function should be clear: dispersing seeds, pumping blood, learning, and regulating heat are all goals, purposes, or functions.

Now let's contrast the explanations above with *mechanistic explanations*. A mechanistic explanation is an explanation that is not given in terms of goals, or purposes, or functions. For example, suppose I drop a rock. If we ask the question "Why did the rock fall?" since the late 1600s the standard explanation is that the rock fell because of gravity. Note that in this explanation there is no hint – none at all – of any goal, or purpose, or function. The rock has no goal or purpose in falling, nor is any function involved. It is simply an object under the influence of an external force. Such goalless, purposeless, functionless explanations are mechanistic explanations. So in general, a teleological explanation is one given in terms of goals, purposes, or functions, while a mechanistic explanation is one that does not use goals, purposes, or functions.

Note that many questions admit of both teleological as well as mechanistic explanations. The example above, that apple trees have apples to disperse seeds, is a teleological explanation. But the same question can be given a perfectly good mechanistic explanation, as follows. In the evolutionary history of apple trees, those ancestors of modern apple trees that produced apples (or the predecessors of apples) survived and reproduced more successfully than those that did not, and hence apple-bearing apple trees (or again, their predecessors) came to be represented more highly in the population. In short, the answer to why apple trees have apples is simply a matter of an evolutionary account involving different rates of survival and reproduction.

Notice that this evolutionary account does not make use of any goals, purposes, or functions. And again, this is a perfectly accurate explanation of why apple trees have apples. In general, questions often admit of both teleological and mechanistic explanations.

I have been going on at some length about teleological explanations because such explanations illustrate an important difference between the way we conceptualize the universe, and the way our predecessors conceptualized the universe. Within the Aristotelian worldview, teleological explanations were considered the proper sort of scientific explanation. This is in stark contrast to modern science, in which mechanistic explanations dominate. The reason teleological explanations were considered the proper sort of scientific explanation is straightforward: within the Aristotelian worldview, the universe was considered really to be teleological. That is, the teleology was not merely a feature of the explanation; rather, the teleology was considered to be a feature of the universe.

Some examples might help to illustrate this. Suppose we return to the example of the dropped rock. Again, in the modern era the explanation for why the rock drops is given in terms of gravity. But the notion of gravity (in our modern sense of the term) did not come about until the 1600s, so whatever the explanation is for the falling rock within the Aristotelian worldview, it cannot be given in terms of our notion of gravity. (Incidentally, the term "gravity" does appear often in writings before the 1600s, but it does not refer to anything like our usual sense of the term, that is, as an attractive force. Rather, before the 1600s "gravity" generally simply referred to the tendency of heavy objects to move downwards.) Within the Aristotelian worldview, the rock falls downward because it is composed primarily of the heavy earth element, and as mentioned in Chapter 1, the earth element has a natural tendency to move toward the center of the universe. To say the same thing in slightly different terms, the earth element has a natural tendency to fulfill a certain goal, namely, of being at the center of the universe.

Each of the basic elements has a natural goal of reaching its natural place in the universe, and these natural goals explain why objects naturally move as they do. Fire burns upward, because the fire element has a natural goal to move toward the periphery, away from the center. And so on for other cases of natural movement. (Enforced movement, for example when I throw a rock upwards, is a different story, but not of direct importance here.)

Similar accounts hold for the superlunar region. The element ether has a natural goal to move in a perfectly circular manner, and this explains the circular movement of heavenly bodies such as the sun, stars and planets. In general, the universe was viewed as a teleological universe, full of natural goals and purposes.

Going hand in hand with this teleology was the other key concept mentioned above, that of essentialism. Natural objects were viewed as having essential natures, and these essential natures were why objects behave as they do. All objects are composed of matter organized in a certain way, and given the matter out of which an object is composed, and the way that matter is organized, the object will have certain natural abilities and natural tendencies, which we can summarize by saying the objects have an essential nature. The simplest objects – the basic elements – of course have the simplest essential natures. Their essential natures consist simply of their tendency to move toward their natural place.

Importantly, notice how closely the teleology and essentialism are tied together. An object's essential nature is a teleological nature, and again, the teleology and essentialism are like two sides of the same coin.

More complex objects have more complex essential natures, but the general story is still the same. For example, consider an acorn. An acorn, as with all objects, is composed of certain types of matter arranged in a certain way, and as noted above, because of the matter of which it is composed, and the organization of that matter, the acorn will have certain natural abilities and tendencies. Again, in other words, the acorn will have an essential nature, and that essential nature is why the acorn behaves as it does. An acorn has a natural goal of becoming a mature oak, and under the appropriate conditions the acorn will grow into an oak, eventually reproducing by producing further acorns. Again, all this is because of the essential nature of the acorn, which arises out of its matter and the organization of that matter.

Notice again the close tie between the essential nature of the acorn, and its goal-oriented, teleological behavior. In outline, the acorn's essential nature is closely tied to its growth, maturation, and eventual reproduction. In other words, the acorn's essential nature is a teleological one – to grow and reproduce.

Within the Aristotelian worldview, the job of a natural scientist is largely the job of coming to understand the teleological, essential natures of categories of objects. A biologist, for example, will want to understand the essential natures of species of animals. This will usually not be an easy or trivial task, but the outline of the task is clear. One needs to understand the matter out of which an object is composed, the organization of that matter, how that matter came to be organized in this way, what sorts of natural goals or functions the object is geared toward, and so on. In doing so, one will come to understand the teleological, essential nature of the object.

To sum up the main points of this section: all natural objects have essential natures; the essential natures are teleological; and the essential natures are why objects behave as they do. In short, the universe was viewed as a teleological, essentialistic universe.

Concluding Remarks

In summary, with respect to the physical structure of the universe, the universe was conceived of as having the Earth at its center, with the moon, sun, stars, and planets revolving about it. As we will see in the next chapter, these were the beliefs that were best supported by the available evidence of the time.

In more conceptual terms, the Aristotelian worldview considered the universe to be teleological and essentialistic. The universe was chock-full of natural goals and purposes, and understanding such goals and purposes was one of the key tasks of a natural scientist in coming to understand the universe.

In the western world, this general view of the universe was the usual view for a long period of time – almost 2,000 years. It goes without saying that over this long course of time, various additions and modifications were made to this view of the world. For example, the main religions of the western world – Judaism, Christianity, and Islam – each made contributions. But the contributions were still within the general Aristotelian framework – that is, within the framework of an Earth-centered, essentialistic, teleological universe.

Chapter Ten

The Preface to Ptolemy's *Almagest*: The Earth as Spherical, Stationary, and at the Center of the Universe

In the previous chapter, we looked at beliefs about the general structure of the universe in the Aristotelian worldview. In this chapter, we explore some of the reasons behind those beliefs; in particular, we look at the arguments supporting the belief that the Earth was spherical, stationary, and at the center of the universe.

One of the main goals of this chapter is to illustrate that beliefs of the Aristotelian worldview, although very different from ours, were nonetheless well-supported beliefs. There is an unfortunate tendency to view the beliefs of our predecessors as being in some way childish or naive, but in this chapter we will see that this is not at all the case. As you consider the arguments presented in this chapter, note that, in general, they are quite good arguments. Most (aside from those about the Earth being spherical) turned out to be mistaken, but mistaken in subtle ways and for reasons that are far from obvious. In fact, discovering the roots of the flaws in the arguments took the combined efforts of some of the best-known names in the history of science (including Galileo, Descartes, and Newton, to name just a few).

Most of the arguments we will consider can be found in Aristotle's *On the Heavens* and the opening section of Ptolemy's *Almagest*. Most of the arguments in these two works are similar. Ptolemy's writing, however, is generally somewhat easier to follow, and so in this chapter I focus mainly on the arguments as they are presented in his work.

As a final introductory note, it is worth mentioning that we focus here on just a small set of arguments from the Aristotelian worldview – those presented by Ptolemy supporting the belief that the Earth was spherical, stationary, and at the center of the universe. But the same moral holds for most of the other beliefs of the Aristotelian worldview: although they are different beliefs than ours, and most turned out to be wrong, the people holding those beliefs generally had good

Worldviews: An Introduction to the History and Philosophy of Science. Richard DeWitt
© 2010 Richard DeWitt

reasons for doing so. We will begin with some preliminary comments on Ptolemy's *Almagest*.

The *Almagest* was published around AD 150. It is a highly technical work and, with text and illustrations, modern printings of this work run to about 700 pages. It is a substantial and difficult work.

The arguments we will consider come from the preface to the *Almagest*. This is by far the least technical section of the book (in fact, it is not at all technical). In this preface, Ptolemy presents a number of arguments concerning the general structure and workings of the universe. In this chapter we focus only on the arguments supporting the beliefs about the structure of the universe, although in later chapters we will consider some of Ptolemy's arguments concerning the way the universe works (for example, arguments supporting the belief about what keeps the sun, stars, and planets moving). We begin with the arguments supporting the view that the Earth is spherical.

The Earth as Spherical

There is a common but mistaken belief that, before the 1500s, people tended to believe the Earth was flat. In fact, very few educated people, at least since the time of the ancient Greeks (such as Plato and Aristotle around 400 BC) have believed the Earth to be flat. It is an interesting question how this misperception about our predecessors came to be so widespread, but one that goes too far afield of our focus. Suffice it to say that our predecessors, going back at least to 400 BC, had good reasons for believing the Earth to be spherical. Consider, for example, the following passage from the preface to Ptolemy's *Almagest*. (Hereafter, all quotations, unless otherwise noted, are from Ptolemy's preface. Numbers in brackets, for example [1], are my additions for use when referring to particular passages.)

Section 4. That the Earth, Taken as a Whole, Is Sensibly Spherical

Now, that also the Earth taken as a whole is sensibly spherical, we could most likely think out in this way. [1] … it is possible to see that the sun and moon and the other stars do not rise and set at the same time for every observer on the Earth, but always earlier for those living towards the orient [toward the east] and later for those living towards the occident [toward the west]. [2] For we find that the phenomena of eclipses taking place at the same time, especially those of the moon, are not recorded at the same hours for everyone … [3] And since the differences in the hours is found to be proportional to the distances between the places, one would reasonably suppose the surface of the Earth spherical, with the result that the general uniformity of curvature would assure every parts' covering those following it proportionately. But this would not happen if the figure were any other, as can be seen from the following considerations.

[4] For, if it [the Earth] ... were flat, the stars would rise and set for all people together and at the same time ... But none of these things appears to happen. [5] It is further clear that it could not be cylindrical ... [since] the more we advance towards the north pole, the more the southern stars are hidden and the northern stars appear. So it is clear that here the curvature of the Earth covering parts uniformly in oblique directions proves its spherical form on every side. [6] Again, whenever we sail towards mountains or any high places from whatever angle and in whatever direction, we see their bulk little by little increasing as if they were arising from the sea, whereas before they seemed submerged because of the curvature of the water's surface. (Munitz 1957, pp. 108–9)

Ptolemy first notes, in the passage I have referenced as [1], that the sun, moon, and stars rise and set at different times, depending on where one is located on the Earth. For example, consider the rising sun this morning. I am sure you are aware that when the sun rose in your location this morning it had already risen for those living any substantial distance east of you, and had not yet risen for those living west of you. Ptolemy and his contemporaries were also aware of this fact, and it is most straightforwardly explained if the Earth is spherical. In passage [2], Ptolemy notes that the times that eclipses are recorded to have occurred are likewise best explained by a spherical Earth, and in [3], he notes that since the differences in times is proportional to the distance between observers, the curvature of the Earth must be pretty much uniform.

Note that Ptolemy's implicit reasoning here is just the common type of confirmation reasoning we discussed in Chapter 4. That is, Ptolemy's implicit reasoning in [1] is that, if the Earth is spherical, one should observe the sun, moon, and stars rising earlier for those living in the east and later for those living in the west, and since this is what is observed, this supports the view that the Earth is spherical. Similar considerations hold for [2] and [3], that is, these also support, via straightforward confirmation reasoning, the conclusion that the Earth is uniformly spherical.

Ptolemy next, in [4], turns to a pattern of disconfirmation reasoning, arguing that if the Earth were some shape other than spherical, then we would not observe what we in fact do observe. For example, Ptolemy notes that if the Earth were flat, we should observe the sun, moon, and stars rising at the same time everywhere on the Earth, but since we do not observe this, we have disconfirming evidence that the Earth is flat.

Notice that, thus far, Ptolemy's arguments really only show that the Earth is uniformly curved in the east–west direction. In other words, Ptolemy's observations thus far are consistent with a cylindrically shaped Earth with a north–south orientation. So to finish this section, Ptolemy considers evidence showing that the Earth cannot be shaped like a cylinder. In [5], Ptolemy notes that different stars are visible as one moves north and south. For example, those of us in the northern hemisphere can see the star we call Polaris (that is, the North Star), whereas those in the southern hemisphere cannot see this star. Likewise, those in the southern hemisphere can see the constellation we call the Southern Cross; those in the northern hemisphere cannot. This is exactly what one would expect if the Earth were spherical, and contrary to what one would expect if the Earth were any other shape, such as cylindrical. Finally, in [6], Ptolemy notes the long-known fact that as one sails toward land, the first parts of the land that are visible are the tops of mountains, and then the lower areas gradually become visible as one sails closer to the land. And again, this is disconfirming evidence for a flat Earth, but exactly what one would expect if the Earth were spherical.

In summary, it was well established that the Earth was most likely spherically shaped. We will next consider the arguments (which, although solid arguments, turned out to be mistaken) that the Earth is stationary.

The Earth as Stationary

Prior to the 1600s, there were excellent reasons for believing the Earth was stationary, that is, for believing the Earth did not move in an orbit about another body (such as the sun), or by revolving on its axis. Although these arguments turned out to be mistaken, they were mistaken for subtle reasons.

As far back as the ancient Greeks, people had considered the possibility that the Earth moves, either around the sun or on its own axis (or both). Aristotle and Ptolemy, for example, explicitly consider this possibility. They and others were well aware that the apparent daily movement of, say, the sun around the Earth, could be explained either by supposing the Earth is stationary, and the sun revolves once a day around the Earth, or by supposing the sun is stationary and the Earth revolves once a day on its axis. Both suppositions would explain the apparent daily motion of the sun around the Earth, and in the *Almagest* we see Ptolemy explicitly considering the latter possibility.

But Ptolemy concludes that the view that the Earth is moving, either on its axis or about the sun, is contrary to some solid evidence and thus the view that the Earth is stationary is the better-supported view. Ptolemy provides a number of what I will call *common-sense* arguments, as well as two somewhat difficult but quite powerful arguments. These latter arguments I will call the *argument from objects in motion* and the *argument from stellar parallax*. We will begin with the common-sense arguments.

Common-sense arguments

Notice that a stationary Earth is the view that we (and our predecessors as well) arrive at by common sense. For example, if you look out your window, it certainly appears that the Earth is stationary. After all, when I am in motion – say, in a car or train, or on my bike – I can certainly notice that I am moving. Even at relatively slow speeds, for example when bicycling, I feel the vibration caused by movement,

I feel the wind against my face, and so on. Or if you are in a convertible moving 70 miles per hour down the interstate, there is no question in your mind that you are in motion. Again, you feel the vibration, the wind, and in general there are observable consequences of being in motion.

Now suppose the Earth is in motion. First, consider the possibility that the Earth revolves once a day on its axis. The Earth is about 25,000 miles in circumference (the people of Ptolemy's time, and for that matter those going back as far as the ancient Greeks, also had a pretty good idea that the Earth was about this size). Given this circumference, if the Earth revolves once a day on its axis, then the surface of the Earth at the equator is traveling at over 1,000 miles per hour (this is the speed the surface would have to be traveling in order to go 25,000 miles in 24 hours). In short, if the Earth is revolving once a day on its axis, then you and I, on the surface of the Earth, would currently be moving at about 1,000 miles per hour. But again, when we move at relatively slow speeds, on bicycles or even in convertibles on the interstate, we clearly notice the effects of the motion. So surely, if we were currently moving at 1,000 miles per hour, we would notice the effects. Since we (and likewise, those in Ptolemy's time) do not observe such effects, this provides disconfirming evidence that the Earth revolves on its axis.

The situation is more dramatic if we consider the possibility that the Earth moves about the sun, completing an orbit once a year. We know the radius of the Earth's orbit is almost 100,000,000 miles. (Incidentally, in Ptolemy's time people did not have any good idea of the distance between the Earth and sun, but nonetheless it would have been clear that the distance would have been substantial.) Given the distance between the Earth and the sun, the Earth would have to be moving at about 70,000 miles per hour to complete a yearly orbit around the sun. But again, we notice dramatic effects when driving in a convertible at 70 miles per hour. We feel 70-mile-per-hour winds in our face, we feel the vibration caused by the movement, we would tumble off if we tried to stand up in the convertible, and so on. So, surely, if we were moving at 70,000 miles per hour we would notice *some* sort of effect of that motion. But where are the 70,000-mile-per-hour winds? Where is the vibration such dramatic motion would certainly cause? How could we possibly stand on an Earth that is moving at 70,000 miles per hour?

In short, these are some obvious effects we would expect to observe if the Earth were in motion, and given that we do not observe such effects, we have good reason to believe the Earth is not in motion.

Consider one more common-sense argument, again one which Ptolemy suggests. In my front yard is a reasonably large boulder, about 4 feet high and 3 feet wide. That boulder just sits there, unmoving. And it is not going to move unless something moves it. Furthermore, if I were to move it, say, with a garden tractor, the boulder will continue to move only so long as I continue to push on it – as soon as I stop pushing on it, it will halt.

Now consider the Earth. The Earth seems to be basically a large rock, vastly huger and heavier than the boulder in my front yard. So just as the boulder will not move unless something moves it, likewise the Earth will not move unless something moves it. And just as the boulder will continue to move only so long as something continues to move it, likewise the Earth would continue to move only if there were something continually moving it. But there is seemingly nothing substantial enough to ever move the Earth in the first place, and even if there were, there is nothing that could keep the Earth continually moving. So it is far more reasonable to believe the Earth is not moving.

In summary, even these basic, common-sense arguments provide good reasons to believe the Earth is stationary. Again, these arguments are flawed, since we now know the Earth is in motion, both on its axis and about the sun. But the flaws in even these common-sense arguments are not obvious, and it took a lot of talent, not to mention decades and even centuries of work, for our predecessors to figure out how we can be moving at the speeds noted above, yet not observe any of the expected effects. This will be part of the story of later chapters.

The argument from objects in motion

The argument from objects in motion is one of the most powerful arguments supporting a stationary Earth. Again it is based on simple observations. Ptolemy notes that dropped objects fall perpendicular to the surface of the Earth. In what follows, I will modify Ptolemy's argument somewhat, considering instead objects that are thrown straight up in the air, noting that such objects go straight up, perpendicular to the surface of the Earth, and then fall straight down, again perpendicular to the Earth. The idea behind my example and Ptolemy's example is exactly the same, although I believe the point is somewhat easier to see when considering the case of objects thrown in the air. As we will see, the fact that a dropped object falls perpendicular to the Earth, or that a thrown object goes straight up and then straight down, suggests that the Earth must be stationary.

To understand the argument, we must discuss common views of the behavior of moving objects, such as objects moving because they have been thrown straight up. I want you to consider two options involving thrown objects, and ask yourself which you think is closer to what would actually happen.

In both options, we will imagine that Sara is holding a ball and is moving, left to right, on a skateboard. As she travels along, she throws the ball straight up in the air. All the while she continues to move. The key question is this: while the ball is in the air, will Sara (moving along on the skateboard) move out from under the ball, so that the ball lands behind her? Or will instead the ball travel in an arc, landing back in (or at least close to) her hand?

Pictorially, the two options are represented in Figures 10.1 and 10.2. Now, the question is whether the ball will travel as depicted in Figure 10.1, such that, while the ball is in the air, Sara moves out from under it and thus it lands behind Sara, or whether the ball will travel in an arc, landing approximately back in her hand, as in Figure 10.2. Again, ask yourself which of these options is the way you believe

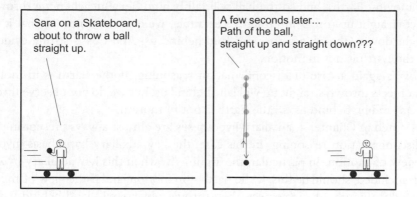

Figure 10.1   Does the ball follow this path?

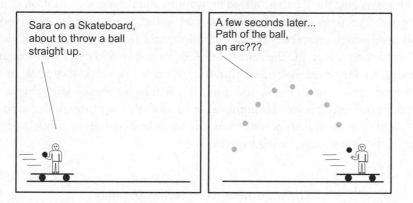

Figure 10.2   Or does the ball follow this path?

the thrown ball will behave. Generally speaking, the question is whether when we are in motion and throw an object straight up, the object will land behind us, or travel in an arc and land back in or close to our hand?

When asked this question, the majority of people chose the option depicted in Figure 10.1, and indeed, this does seem to be the common-sense view of motion. But importantly, if you believe that is the correct view of motion, then to be logically consistent *you must also believe the Earth is stationary.*

Here is why. In the scenario above, the source of motion is not relevant. That is, nothing would change if Sara's motion was due to her standing on roller skates, standing on a moving car, standing on the pedals of a bicycle, or whatever. And nothing would change if Sara's motion is due to her standing on the surface of a moving Earth. That is, if Sara is moving because she is standing on the surface of a moving Earth, and if thrown objects behave as pictured in Figure 10.1, then when Sara is standing in her front yard and throws a ball straight up, she will move out from underneath it (because of her motion due to her standing on the surface of a moving Earth), and so the ball will fall behind her. But when we throw an object straight up (or, as in Ptolemy's example, we drop an object so that it falls straight down), the object does not land behind us. And this is strong evidence that the Earth is not in motion.

This is again a case of disconfirmation reasoning. If the Earth is in motion, then objects thrown straight up will land behind us; but we do not observe thrown objects landing behind us; so the Earth is not in motion.

As noted in Chapter 4, auxiliary hypotheses are almost always present in cases of disconfirmation reasoning. In this case, the key auxiliary hypothesis involves the view of motion. In particular, the argument, when this key auxiliary hypothesis is included, should be this: If the Earth is in motion, and if the view of motion depicted in Figure 10.1 is correct, then thrown objects will land behind us; but thrown objects do not land behind us, so either the Earth is not in motion, or the view of motion depicted in Figure 10.1 is not correct.

As it turns out, the Earth is indeed in motion, and the view of motion depicted in Figure 10.1 is incorrect. But again, that view of motion is, even today, a common (though mistaken) view of motion, and it was the accepted view of motion during most of the time the Aristotelian worldview was dominant. Figuring out a more correct view of motion is one of the tasks that took an enormous amount of talent, work, and time, and will be a topic of later chapters. At this point, though, it is worth emphasizing again that, even though this argument of Ptolemy's turns out to be mistaken, the mistake is based on subtle and (even today) difficult issues involving motion.

The argument from stellar parallax

In Section 6 of the preface, Ptolemy notes that the "angular distances" of the stars always remain unchanged, and then in the next section he notes that this fact about the stars supports the view that the Earth is stationary. This argument is again one of the more compelling arguments for a stationary Earth, but will take some work to understand.

When Ptolemy notes that the angular distances of the stars appear the same everywhere, he is referring to what we would call stellar parallax. In particular, Ptolemy is noting that we cannot observe stellar parallax, and this supports the view that the Earth is stationary. To understand Ptolemy's argument, let's start by understanding parallax.

Parallax is the apparent shift in the position of objects due to your (not their) motion. For example, hold a pen vertically at arm's length in front of your eyes. Keeping the pen still, move your head to the left and right, and notice the shift in the apparent positions of the pen and background objects. The apparent shift in their positions is due, of course, to the movement of your head, and not to any movement of the pen or background objects. This is an example of parallax – again, the apparent shift in the position of objects due to your movement.

As mentioned, when Ptolemy speaks of the angular distances of the stars remaining the same everywhere on the Earth, he is referring to the fact that we cannot observe stellar parallax. Stellar parallax would be an apparent shift in the position of stars (that is, a shift in where the stars appear relative to each other) due to our movement. Ptolemy's point is that, if the Earth were moving, either on its axis or about the sun, then we should observe stellar parallax. Since we do not, the Earth must not be moving.

To see this more clearly, suppose the Earth is spinning on its axis. As noted above, the circumference of the Earth is roughly 25,000 miles, so if the Earth is revolving on its axis, then every hour we are traveling about 1,000 miles. Suppose we go out at night and carefully plot the position of several stars, then we wait several hours and again carefully plot the position of those same stars. In between these two observations we will (if the Earth is revolving on its axis) have traveled several thousand miles, so since we have moved several thousand miles, we should detect an apparent shift in the position of the stars we plotted earlier (again, the change would be in the stars' positions relative to each other). That is, we should be able to detect stellar parallax. But we do not observe any such parallax. So – and this is Ptolemy's point – the Earth must not be revolving on its axis.

Once again, the situation is even more dramatic if we consider the possibility that the Earth is orbiting the sun. Again, in Ptolemy's time there was no good estimate for the distance from the Earth to the sun, but we know it to be almost 100,000,000 miles, and they knew it had to be a substantial distance. Using our known distance for this example, if the Earth is orbiting the sun, then we are traveling almost 200,000,000 miles from one point in our orbit to the farthest point. Now, recall the example of parallax using your pen and background objects. In that case, a movement of your head of just a few inches resulted in a clearly detectable parallax. So if we are moving 200,000,000 miles, there would seem to be no way we could fail to observe stellar parallax. But again, as Ptolemy points out, we do not observe such parallax, and so we must not be in motion. In short, Ptolemy's argument from stellar parallax provides an extremely strong, logically solid, empirically based argument that the Earth is not in motion.

As should be clear, this is again a case of disconfirmation reasoning, and, as is usually the case, there are various auxiliary hypotheses lurking beneath the surface. Before reading on, you may want to pause and see if you can identify the key auxiliary hypothesis in this case.

In this case, the key auxiliary hypothesis involves distances. You may have noticed, in exploring examples of parallax (such as the pen), that the amount of the apparent shift very much depends on the distance objects are from you. In particular, the further away objects are, the less the apparent shift. So one explanation for the lack of stellar parallax would be that the stars are unbelievably far away from us. But – and this is an important point in understanding our predecessors' reasoning here – keep in mind that if the Earth is moving about the sun, the distance we travel from one part of our orbit to the farthest part is large – almost

200,000,000 miles. So for us to travel that large distance, and there still to be no detectable stellar parallax, the stars would have to be incredibly – and I do mean incredibly, unbelievably, almost inconceivably – far away.

So the reasoning involved here is really more as follows: If the Earth is in motion, and if the stars are not almost inconceivably far away, then we should detect stellar parallax; but we do not detect such parallax, so either the Earth is not in motion or the stars are almost inconceivably far away.

As an almost final note in this section, recall a point made in the previous chapter, concerning our predecessors' views on the size of the universe. They considered the universe to be large, by their standards, but nothing like the size we consider the universe to be. You and I have no problems considering the universe to be unimaginably huge, but then you and I were raised with a worldview in which a belief in a huge universe fits. But a belief in such a huge universe would not fit comfortably into the Aristotelian jigsaw puzzle of beliefs. So, given the worldview of the time, the notion of an unimaginably huge universe was not really a viable option. As such, the argument from stellar parallax provides another quite strong argument for a stationary Earth.

As a genuinely final note in this section, stellar parallax was eventually observed, although the first accurate measurement of it did not occur until 1838, almost 1,700 years after Ptolemy wrote the *Almagest*. And in fact, the detection of stellar parallax currently provides some of the strongest empirical evidence that the Earth does in fact orbit the sun.

The Earth as the Center of the Universe

If one accepts that the Earth is spherical and stationary, it would seem natural that it would be located at the center of the universe. Indeed, the view of the Earth as the center of the universe is one that fits best with the other relevant beliefs. Section 5 of the preface to the *Almagest* deals specifically with Ptolemy's reasons for viewing the Earth as the center. In this section, he refers in passing to a number of Aristotle's arguments from *On the Heavens*, and it seems clear that Ptolemy endorsed Aristotle's arguments. In what follows I will present somewhat of a mixture of Aristotle's and Ptolemy's arguments.

For the first argument, note that the Earth certainly appears to be the center of the universe. The moon, sun, stars, and planets all appear to revolve around the Earth, and it would seem natural to think that the common point around which they all revolve – that is, the Earth – is the center of the universe. In other words, an Earth-centered view would be the most straightforward option. (Incidentally, it is widely known that the moon and sun appear to revolve about the Earth, but somewhat less widely known that the stars and planets also appear to revolve about the Earth. We will discuss such motions more fully in the next chapter.)

In addition, recall that, on the Aristotelian worldview, the earth element has a natural tendency to move toward the center of the universe, and the fire element has a natural tendency to move toward the periphery, away from the center. This is, again, why heavy objects such as rocks fall and why fire burns upwards. Since the Earth itself seems to be composed primarily of the earth element, and the natural place for that element is the center of the universe, then the Earth itself would naturally be located at the center of the universe.

Recall the discussion above about the behavior of objects in motion. I noted that an object, such as the boulder in my front yard, will not move unless there is something to move it. Since the Earth itself, being primarily composed of the earth element, would naturally be located at the center of the universe, and since it (like the boulder in my front yard) will not move unless something moves it, and since there seems to be nothing that could possibly move the Earth (again, see the discussion above), the most reasonable conclusion is that the Earth is naturally located at the center of the universe, and it is not going to move from that location.

The view that heavy objects have a natural tendency to move toward the center of the universe gives rise to another consideration supporting the Earth-centered view. Since we know (see the arguments above) that the Earth is spherical, and given the observation above that dropped objects fall perpendicular to the surface of the Earth, it follows immediately that the center of the Earth must be the center of the universe. To see this, consider objects dropped from various locations on the Earth. Those objects are moving toward the center of the universe, so the line of their fall points toward the center of the universe. Since the various lines (of the objects dropped from various locations) converge at the center of the Earth, it must be that the center of the Earth is the center of the universe.

As with the arguments for a stationary Earth, notice how these arguments tend to tie in with, and depend upon, other beliefs in the Aristotelian worldview. For example, several of the arguments just mentioned depend closely on the view that objects have natural places in the universe. This again reinforces the point from Chapter 1, that the individual beliefs within a jigsaw puzzle of beliefs are closely interconnected, and one cannot change many of the beliefs without substantially altering the overall jigsaw puzzle.

Concluding Remarks

To return to a point made in the opening paragraphs of this chapter, our predecessors had good reasons for believing the Earth was spherical, stationary, and at the center of the universe. Their arguments for the Earth as spherical were, as it turns out, absolutely correct. Their arguments for the Earth being stationary and at the center of the universe turned out to be mistaken, but in ways that are subtle and not at all obvious. As mentioned, it would take the combined efforts, over many decades and even centuries, of many of the best-known names in the history of science to figure out a new belief system compatible with a moving Earth.

This completes our survey of the main arguments supporting the view of the Earth as spherical, stationary, and at the center of the universe. Eventually evidence would arise suggesting that the latter two beliefs were mistaken, and this would cause serious problems for the Aristotelian worldview. And, as noted earlier, eventually that worldview would be replaced by the Newtonian worldview. The transition from the Aristotelian to the Newtonian worldview would, importantly, involve various theories as to the structure of the universe. Given this, our next area of exploration is the data such theories needed to account for, after which we will look into the various astronomical theories themselves.

Chapter Eleven

Astronomical Data: The Empirical Facts

In a few chapters we will begin looking at the Ptolemaic, Copernican, Tychonic, and Keplerian astronomical theories, with an eye toward understanding some of the factors and issues involved in the transition from the old Aristotelian way of looking at the universe, to the newer Newtonian view of the universe. This transition importantly involved the astronomical theories just mentioned, so as background to understanding these theories, we need to look at some of the data the theories were primarily designed to handle.

As we discussed in earlier chapters, whatever else we wish from theories, they must at a minimum be able to explain and predict the relevant data. In other words, generally speaking there will be a body of facts that are relevant to a particular theory, and the theory should be able to explain and predict these facts.

Moreover, as we discussed in Chapter 3, the notion of "facts" is not as straightforward as it might at first appear. In particular, we noted that some facts are relatively straightforward empirical facts – the clearest examples of which are straightforward observations, such as the observation that, where I live, the body of light we call the sun appeared over the eastern horizon at 6:33 a.m. this morning; and we noted that there are also philosophical/conceptual facts, that is, beliefs that are generally strongly held, and often appear to be empirical facts, but that turn out to be based more on one's worldview than on straightforward empirical observation.

The primary goal of the next two chapters is to explain the major facts, both empirical and philosophical/conceptual, that are relevant to astronomical theories such as those of Ptolemy, Copernicus, Tycho, and Kepler. In this chapter we will focus on some of the more important empirical facts, and in the next chapter we will turn to the philosophical/conceptual facts.

Worldviews: An Introduction to the History and Philosophy of Science.   Richard DeWitt
© 2010 Richard DeWitt

The Ptolemaic, Copernican, Tychonic, and Keplerian theories are astronomical theories, and so the relevant facts these theories must explain and predict primarily concern astronomical events. By "astronomical events" I mean events concerning the heavenly bodies, such as the moon, sun, stars, and planets. Such events largely involve their observed movements. What follows is not an exhaustive catalog of the movement of these heavenly bodies, but it will provide a good sense of the range of empirical facts the various astronomical theories needed to explain and predict.

Importantly, this is a chapter on the relevant empirical facts, and so when we are speaking of movement, the emphasis is on the *observed* movement of the sun, moon, stars, and planets. For example, when we speak of the movement of, say, Mars, the issue is not whether Mars moves in an elliptical orbit, or a circular orbit, or any other type of orbit. Rather, the emphasis is on the observed movement of Mars. More specifically, there is a point of light, visible in the night sky, that by convention we call "Mars." That point of light moves in a particular way (described more fully below). When we speak of the movement of Mars, then, we are speaking of the straightforward, empirical, directly observable fact about how that point of light moves across the night sky.

With that in mind, let's begin with the observed movement of the stars.

The Movement of the Stars

The stars appear to move in a regular pattern, which repeats approximately every 24 hours. For example, suppose you are in the northern hemisphere, and you go out at 9:00 p.m. to observe the stars. Suppose you focus on the movement of the points of light we call the Big Dipper. During the night, you will notice the Big Dipper moving in a circular motion, counterclockwise, around the point of light we call the North Star. If you stand out there for an entire 24 hours, you will of course lose sight of the Big Dipper during the daylight hours, but when darkness returns you will see the Big Dipper apparently continuing its circular path around the North Star. After 24 hours, at 9:00 p.m. the following night, you will notice the Big Dipper is very close to the position it was in at 9:00 p.m. the previous night. In short, the Big Dipper, and other stars near the North Star, appear to follow a circular path, with the North Star at the center of the circle. Moreover, such stars seem to complete a circle about the North Star approximately every 24 hours.

Suppose the next night you go out and observe stars further away from the North Star, say, stars that early in the evening are close to the eastern horizon. As the night progresses, you will see these stars move in an arc (much like the arc the sun makes moving across the sky), eventually setting below the western horizon. Once again, if you observe for a full 24 hours, you will see the same stars in approximately the same location as you saw them at that time the previous night.

Stars in the southern part of the sky also trace an arc across the sky, rising over the southeastern horizon and setting over the southwestern horizon. And again, such stars will be found close to their original position 24 hours later.

Two final points are worth noting. First, the observations above assumed you were viewing the stars from the northern hemisphere. If instead you were in the southern hemisphere, you would be viewing different stars (for example, the North Star would not be visible), but the pattern of movement would be analogous to that described above.

Second, as each particular star moves across the sky (other than the North Star, which does not move any noticeable amount), it remains in the same position relative to the other stars. That is, the stars move as a group across the night sky. If you pick out a particular star to observe, the star whose course you are plotting will always remain in the same position relative to the other stars in the sky. This is why the stars are traditionally referred to as the "fixed stars." They are not really fixed in place – they do appear to move about the Earth every 24 hours – but they move as a group, remaining fixed in place relative to each other.

In summary, the points of light we call the stars move in a predictable pattern, and a pattern that has been recognized since well before the beginning of recorded history. Let's now move on to consider the movement of the sun.

The Movement of the Sun

The most straightforward movement of the sun is its daily movement across the sky. The sun rises in the east, moves in an arc across the sky, sets in the west, and then rises again approximately 24 hours later.

In addition, the point on the eastern horizon where the sun rises moves north and south during the course of the year. On the day of the winter solstice (this will be the first day of winter, the shortest day of the year, falling on or near December 22), the sun rises at its most southern point on the eastern horizon. Over the next few months, the point on the horizon at which the sun rises moves north, until on or about March 22 (the day of the spring equinox, marking the first day of spring) the sun rises almost due east, and day and night are close to being equal in length (for somewhat complicated reasons, and contrary to popular belief, day and night are not quite equal on the day of the equinox, but that need not concern us here). Again over the next few months the point on the horizon where the sun rises continues to move north, reaching its most northerly point on the day of the summer solstice (marking the first day of summer, the longest day of the year, on or about June 21). Then the point of sunrise begins to move south, rising almost due east on the day of the fall equinox (the first day of fall, on or near September 22). Finally, for the next few months the point of sunrise continues to move south, until on or near December 22 it again reaches its most southerly point, again marking the first day of winter. And this cycle repeats itself year after year, as it has as far back as we know. (Incidentally, note again that I am describing the situation from the point of view of the northern hemisphere. Similar movements of the sun are observed in the southern hemisphere as well, though some of the descriptions above – for example, of the seasons – would be different.)

Nor are these the only movements of the sun. The sun's position in the sky, relative to the fixed stars, changes from day to day. Although we do not ordinarily plot the position of the sun against the stars, it is not difficult to do. If you go out at sunset, and note what stars are visible on the western horizon immediately after sunset, you will notice that the stars are in a slightly different position each evening. Using the fixed stars as a reference point, the sun appears to drift eastward with respect to the stars. In other words, each day the sun is in a slightly more eastward location with respect to the fixed stars. (As we will see below, the planets drift also. This is why, in astrology, the sun and planets are described as being in different constellations at different times of the year. So, for example, as the sun drifts eastward relative to the fixed stars, it may be near the constellation Capricorn one month, and hence in astrological circles it may be said to be in Capricorn, then another month in Pisces, and so on.)

This completes the description of the more obvious motions of the sun. We will now move on to briefly consider the movement of the moon.

The Movement of the Moon

The motions of the moon are more complex, but we will just briefly describe some of the more obvious ones. On nights the moon is visible (which is most, but by no means all, nights), the moon, like the sun, rises in the east, moves in an arc similar to the sun across the sky, and sets (not necessarily while it is still dark out) in the west. Unlike the stars and sun, it does not rise again 24 hours later. Rather, each night the moon rises later than it rose the previous night (the amount later varies during the year, but on average is a bit less than an hour).

The moon also goes through a series of phases, taking a little over 29 days to complete a cycle of phases. That is, the moon will sometimes be a crescent moon, sometimes a half-moon, sometimes a three-quarters moon, sometimes full, and so on. And whatever phase the moon is in tonight, it will again be in that phase a little more than 29 days from now.

Like the sun, the moon also drifts eastward with respect to the fixed stars, but it drifts faster than the sun does. The moon returns to the same spot, relative to the fixed stars, about every 27 days. In other words, if you go out tonight and plot the position of the moon relative to the stars, in a little over 27 days it will be in the same relative position.

As mentioned, these are by no means the only motions of the moon, but they are the more obvious motions. Let's now turn to the more complicated issue of the movement of the planets.

The Movement of the Planets

We need to be careful when discussing the planets. You and I were raised in a time dominated by technology. We have been privileged to see photographs of planets, both from technological marvels such as the Hubble space telescope as well as, in some cases, from spacecraft that flew close to, or even landed on, some of the planets.

As such, the image of a planet that immediately comes to our minds is very different from the image of a planet that would have come to the mind of someone living before modern technology. But keep in mind two important points. First, our discussion is taking place as background for discussing astronomical systems such as those of Ptolemy, Copernicus, Tycho, and Kepler, none of whom had access to the sorts of technology we have. And second, we are discussing the empirical facts, which again, in the most clear cases, consist of straightforward observational data.

So the relevant question is, what sorts of direct, observational data do we have about the planets? In other words, if we restrict ourselves to just straightforward, naked-eye observation, what facts are there about the planets?

The first point to note is that, on any given night, a point of light we call a planet does not look appreciably different from a point of light we call a star. In general, stars and planets look a lot alike.

By the way, you may have heard that stars twinkle, and planets do not, and there is some truth to this. But I have never met anyone not already knowledgeable about the night sky who could distinguish stars from planets based on whether or not the points of light twinkle. It is only after one learns to distinguish stars from planets – based on other criteria – that one can start to notice the twinkling/nontwinkling aspect.

Moreover, on any given night the points of light we call stars, and those we call planets, move in a similar way. That is, all the points of light, both stars and planets, move across the night sky during a single night in the way described in the section on the movement of the stars.

In short, if one does not already know how to tell stars from planets, then one will not be able to distinguish any difference on a given night. However, going back to before the beginning of recorded time, our predecessors distinguished five points of light in the night sky as being different from the thousands of other points of light. The distinction was based primarily on the way those five points of light moved, not on a single night, but over the course of many nights. (Incidentally, we generally think of there being nine planets. But until the 1700s, with advances in telescopes, the only known planets were those visible to the naked eye, namely Mercury, Venus, Mars, Jupiter, and Saturn.)

As mentioned, on any given night a planet will typically appear (to the naked eye, at least) to be no different from a star. For example, if you observe Jupiter over the course of several hours, you will see it move with the fixed stars, and in general it will not look any different from a star. But, if you watch Jupiter carefully over the course of several days or weeks, you will notice that, like the moon and sun, it drifts with respect to the fixed stars. Generally speaking, each night Jupiter will be very slightly eastward, relative to the fixed stars, than it was the night before, and over the course of weeks or months Jupiter will have drifted noticeably eastward relative to the stars.

It is also worth noting that the planets, unlike the stars, vary noticeably in brightness from time to time. For example, when Venus is visible it always appears reasonably bright, but at times it is dramatically brighter than at other times (at its brightest, Venus appears as bright as the landing lights of an approaching plane). The difference in brightness of the other planets is not as dramatic as that of Venus, but nonetheless, the five planets visible with the naked eye all differ noticeably in brightness from time to time.

In general, these are the only clear observational differences between stars and planets. The thousands of points of light we call stars have remained in fixed positions relative to one another at least since the beginning of recorded history, and generally the brightness of each star appears to be pretty much the same over time, whereas five points of light – those we call planets (after the Greek word for wanderer) – drift relative to the stars and also appear more and less bright at different times.

Any adequate astronomical theory must be able to account for these observations. For example, such a theory must be able to take into account the differing brightness and drifting of Jupiter, and be able to predict where Jupiter will appear in the night sky a year from now.

The drifting nature of the planets makes predicting their positions much more difficult than, say, predicting the positions of the stars. But the situation is more complicated yet. For example, although Jupiter will generally drift slightly eastward each night, relative to the stars, about once a year it will stop drifting for a few days, and then start drifting the "wrong" way, that is, westward. Then after drifting westward for a few weeks, it will again stop drifting for a few days, and then resume its usual eastward drifting for about another year.

This curious, "backwards" drifting of the planets is termed *retrograde motion*. All of the planets exhibit retrograde motion, though not at the same intervals. Jupiter and Saturn exhibit retrograde motion about once a year, Mars about once every two years, Venus about every year and a half, and Mercury about three times a year.

The motion of the planets, especially this curious retrograde motion, would make planets the most troublesome item in terms of developing an astronomical theory that was good at explanations and predictions. As we will see shortly, however, theories were developed that did an outstanding job at explanation and prediction.

Before closing, a few final empirical facts about the planets are worth mentioning. These facts seem fairly minor, and in a sense they are, but later they will play substantial roles in deciding between competing astronomical theories. First, Mercury and Venus never appear far from the sun. That is, wherever the sun is in the sky, Mercury and Venus will be close. If you hold a foot-long ruler at arm's length, that will be about the maximum distance Venus will ever be from the sun, and Mercury will be even closer.

A corollary of this fact is that you can see Mercury and Venus only either just before sunrise or just after sunset. For example, sometimes Venus trails behind the sun, so that when the sun sets, Venus will not be far behind in the western sky. Again, Venus will never be much more than about a ruler's length above the western horizon, and Venus then sets in the western horizon within a few hours after sunset. Or at some times of the year Venus is leading the sun, in which case you can see Venus rising in the early morning before sunrise; it will be visible for at most a few hours, until the sun also rises.

Another seemingly minor fact, but one that would play a role later in arguments for and against competing astronomical theories, concerns a correlation between the apparent brightness of Mars, Jupiter, and Saturn, and the times at which these planets are exhibiting retrograde motion. As mentioned, all the planets vary in how bright they appear. Mars, for example, appears noticeably brighter about every two years. Recall from the discussion above that Mars exhibits retrograde motion about every two years. As it turns out, Mars' retrograde motion, and the times at which Mars appears brightest, are closely correlated. That is, Mars always appears brightest around the same time as it is in retrograde motion. Likewise for Jupiter and Saturn – they also appear brightest during the same period as when they exhibit retrograde motion.

Different astronomical systems will explain these seemingly minor facts in different ways. As we will see in later chapters, some systems will explain these facts in a more natural way, and this will factor into the debate over which astronomical system was best.

Concluding Remarks

The empirical facts that astronomical theories must respect are not simple, by any means, but they are relatively straightforward. These facts have been known for a long, long time – some of the earliest major civilizations, dating back thousands of years, were well acquainted with them. As it turns out, these facts are not at all easy to account for in an astronomical theory. That is, it proved to be quite difficult to develop a theory that accurately predicted and explained them. Before moving on to consider such theories, we need to explore some other facts these theories needed to respect. These are certain philosophical/conceptual facts, especially concerning the motion of the moon, sun, stars, and planets, that play important roles in the debate over competing astronomical theories. These philosophical/conceptual facts will be the main topic of the next chapter.

Chapter Twelve

Astronomical Data: The Philosophical/Conceptual Facts

In this chapter, we look at some of the key philosophical/conceptual facts relevant to the astronomical theories with which we are concerned. The two such facts playing the most important roles are what we earlier called the perfect circle fact and the uniform motion fact. We discussed these in Chapter 3, and here we explore them more carefully.

The perfect circle and uniform motion facts are reasonably easy to state. The perfect circle fact is that heavenly bodies such as the moon, sun, stars, and planets move in perfect circles (rather than having some other sort of motion, such as elliptical). The uniform motion fact is that the motion of those heavenly bodies is uniform, that is, they neither speed up nor slow down, but rather always move at the same rate.

Although the facts are relatively easy to state, one cannot really understand them, or the depth of commitment of our predecessors to those facts, without understanding the context in which those facts lived. So in this chapter our main goal is not just to understand the perfect circle and uniform motion facts, but also to see how those facts fit into the broader context of beliefs. Exploring this topic will also provide a better idea as to how many of the pieces of the Aristotelian jigsaw puzzle fit together.

We will begin by considering a substantial scientific problem faced by our predecessors. We will then look at how the perfect circle and uniform motion facts fit into the solution to this problem.

A Scientific Problem with the Motion of the Heavenly Bodies

At some point in our education (or at multiple points, as is often the case), most of us were required to memorize the principle of inertia, also known as Newton's

Worldviews: An Introduction to the History and Philosophy of Science.   Richard DeWitt
© 2010 Richard DeWitt

first law of motion. This memorization process is so effective that most people I know can still, years later, flawlessly recite the principle. It is usually stated as follows:

Principle of inertia: An object in motion remains in motion in a straight line, and an object at rest remains at rest, unless acted upon by an outside force.

This principle was not recognized until the 1600s, and it took quite a lot of effort, over quite a number of years, before it was clearly and accurately stated. Galileo came close to getting it right, and then Descartes was the first to clearly state the principle. Newton later followed Descartes' formulation, incorporating it into his science as his first law of motion.

Why did this principle, which currently is probably the most widely known (or at least the most widely memorized) principle of science, take so long to work out? In large part, the answer is that it runs counter to all our experience. Think about your everyday experience with objects in motion. Can you think of a single case in which an object in motion remained in motion? The fact is, in everyday experience, objects in motion never remain in motion. They always come to a halt. Thrown baseballs, thrown frisbees, dropped objects, bicycles, cars, airplanes, acorns and apples that fall off trees, and in general, every object with which we are familiar, comes to a halt unless something keeps it in motion.

In short, our everyday experience would lead us to a very different principle of motion, and in fact, this principle of motion was the view of motion that seemed obviously correct from the time of Aristotle to the 1600s. For convenience, I will simply refer to this as the pre-1600s principle of motion.

Pre-1600s principle of motion: An object in motion will come to a halt, unless something keeps it moving.

This principle of motion seems correct, based on everyday experience, and it fits in well with the views, within the Aristotelian worldview, that objects will naturally tend to move toward their natural place in the universe. For example, consider a falling rock. The rock is composed primarily of the earth element, which again has a natural tendency to move toward the center of the universe. So while a dropped rock is falling, what keeps it moving is its internal tendency to move toward its natural place. It will, though, come to a halt, usually because it is stopped by the surface of the Earth. And even if there were nothing such as the surface of the Earth to stop the rock, it would still eventually come to a halt when it reached its natural place at the center of the universe. In short, the view that an object in motion will come to a halt unless something keeps it moving, is not only well supported by everyday experience, but also fits in well with other beliefs within the Aristotelian jigsaw puzzle of beliefs.

So far, so good. But there is one type of movement that is problematic, and that is the movement of the heavenly bodies such as the moon, sun, stars, and planets. In our everyday experience, these are the only objects that continue to move, never coming to a halt. And these objects have been in motion, in a regular and repeating pattern (as described in the previous chapter), since the beginning of recorded history.

But if the heavenly bodies are in continuous motion, and if objects in motion come to a halt unless something keeps them moving, it follows immediately that there must be something keeping the heavenly bodies moving. What could this source of motion be? What could keep the moon, sun, stars, and planets continually moving?

We can draw one immediate conclusion about this source of motion. Whatever it is, it cannot supply us with a full understanding of the movement of the heavenly bodies if the source of motion is itself moving. To understand this point, consider a typical cause of motion, as when I cause my pen to move across my desk by pushing it with my finger. In this case, the source of motion – my finger – is itself moving. But in this case, the movement of my finger must itself have a cause, so to fully understand the motion of the pen, we must not only understand that the pen is being moved by my finger, but also understand the source of motion for my finger.

In general, if we have a source of motion that is itself moving, then that source of motion must itself have a source of motion. And so a full understanding would require understanding how the source of motion is moving.

Given this, we cannot explain the motion of the heavenly bodies by appealing to any sort of cause that is itself moving. What must be causing the heavenly bodies to move, then, must be a source of motion that is *not* itself moving. And there is only one type of source of motion that does not itself move. Such a source of motion is not one that jumps immediately to mind, and it is probably best approached by way of example.

Suppose I am in a park, and I see my wife on the far side of the park. Exclaiming "Ah, my beloved!" I begin to move toward my wife. Notice that she herself may not be moving, and she may not even be aware that I am present. But even though she is not moving, and is not even aware of my presence, she nonetheless is the cause of my movement. She causes my movement by being an object of desire, and in so doing she is a source of motion that is not itself in motion.

Another example: suppose you see a $20 bill on the floor across the room, and wanting it, you move toward it. Your desire for that bill is the source of your motion, and in this way the money acts as a source of motion that does not itself move.

Of course, the stars and planets do not move out of a desire for my wife. Or a desire for money. But this sort of account seems to be the only way we can have a source of motion that is not itself moving, and so this – an account involving an object of desire – must be the same sort of source of motion that applies to the heavenly bodies.

What kind of object of desire could be the source of motion for the heavenly bodies? Aristotle had inherited a tradition, one that goes back so far that we cannot be sure of its origins, of viewing the heavens as a place of perfection. The perfection is rooted in the almost unchanging nature of the heavens, with the only change being in the positions of the moon, sun, stars, and planets. As Aristotle notes in *On the Heavens*, "throughout all past time, according to the records handed down from generation to generation, we find no trace of change either in the whole of the outermost heaven or in any one of its proper parts."

Whereas the heavens are a place of almost unchanging perfection, the only sort of *absolute* perfection would be the perfection of the gods. So in something like the way I move out of a desire to be near my wife, the heavenly bodies must move out of a desire to emulate the perfection of the gods. The best way for the heavenly bodies to emulate the perfection of the gods would be through perfect motion, and perfectly circular motion, at uniform, unchanging speed, is the most perfect sort of motion.

In summary, the gods provide a source of motion for the moon, sun, stars, and planets. And the gods do so not by themselves moving, but rather by being an object of desire. In particular, the heavenly bodies move in perfect circles at uniform speed out of a desire to emulate the perfection of the gods. Given the context of the time, and especially the need for a source of movement that itself does not move, this seems the best explanation for the unchanging, perpetual motion of the heavenly bodies.

Three cautionary notes

Three quick notes are in order before closing this section. First, the notion of desire, as used above, is a tricky one for the modern ear. Just as we have notions and concepts the Greeks did not have, the Greeks likewise had notions and concepts that we do not have. One of these is a certain concept of an unconscious desire, or what I prefer to think of as a natural, internal, goal-directed tendency. This concept is unlike anything we currently have. The main sense of unconscious desire we have is a Freudian sort of unconscious desire. But a Freudian unconscious desire is used only in the context of a conscious agent, and so is not at all the same notion as the Greek concept. In general, do not think of the "desire" of the planets to move in perfect, uniform motion as in any way involving the planets thinking to themselves "Gosh, I'd sure like to be like the gods, so I think I'll move around in perfect circles." Rather, the "desire" is this unconscious sort of desire, or, better, a natural, internal, goal-directed tendency.

Second – and this ties in to the point just made – earlier we discussed that there were four basic elements (earth, water, air, and fire) in the sublunar region (that is, the region below the moon, including the Earth), with the fifth element, ether, found only in the superlunar region (that is, in the heavens, in the region from the moon outward). It is really the element ether that has the unconscious desire, or natural tendency, to emulate the perfection of the gods. That is, the essential nature of ether is to move in perfect circles with uniform speed. In this way, the general story involving ether is not much different from the general story involving the four basic sublunar elements, earth, water, air, and fire. For example, recall that the earth element has a natural, goal-directed tendency (or an unconscious desire) to reach the center of the universe, and this is the essential nature of that element. And that is why, for example, rocks fall downward. So likewise ether has an essential nature, and its essential nature is to move in perfect circles at uniform speed. And that is why objects in the heavens move the way they do.

Finally, Aristotle's own notion of the gods should not be interpreted in any sort of religious sense. Aristotle's gods are real "things," as they must be in order to be a source of movement for the heavenly bodies. The gods are "unmoved movers," in the sense that they are the source of motion, yet without themselves moving. Aristotle's own discussion of the gods is complicated, and there is controversy over how his writings on the gods should be interpreted. But it is clear that Aristotle thought of the gods as being a sort of intellectual perfection, and even more clear that his gods are not in any way religious gods. For example, these gods had nothing to do with the origin of the universe, they are unaware of anything happening on the Earth, they are unaware of our existence, and as such it would be pointless to pray to them. In later centuries, after Aristotle, when Judaic, Islamic, and Christian philosophers and theologians more or less mixed religion with Aristotelian views, Aristotle's nonreligious gods were transformed into the religious God of Judaic, Islamic, and Christian tradition. This religious God continued to provide the needed explanation for the continual motion of the heavenly bodies.

Could This Account be Used for a Moving Earth?

In Chapter 10, we explored the reasons behind viewing the Earth as spherical, stationary, and at the center of the universe. Recall that some of the reasons for a stationary, central Earth involved the claim that there was nothing that could keep the Earth in motion.

I am often asked whether an account, similar to that above, could have been used to explain a moving Earth. If the explanation of the constantly moving heavenly bodies is rooted in the view that their internal essence is to continually move in perfect circles, why not tell a similar story for the Earth? Why couldn't our predecessors have said the essence of the earth element was to move in circles, say in circles about the sun, and use that as an explanation for a moving Earth?

The question is a good one, and seeing the answer to it helps illustrate the interconnections between beliefs in the Aristotelian jigsaw puzzle of beliefs. The answer is not that viewing the earth element as having an internal essence to move in circles (again, say in an orbit around the sun) is inherently contradictory. After all, if it is all right to view ether as having an internal essence to move in circles, it cannot very well be inherently contradictory to say something similar about the earth element.

So the answer to the question is not a matter of a circular-moving earth element being inherently contradictory. Rather, the answer involves considering whether a circular-moving earth element would fit into the overall jigsaw puzzle of beliefs. And as it turns out, it will not fit, and that is why viewing the earth element as having an internal essence to move in circles is not a viable option.

To see this, suppose we try to accept that the earth element has an internal tendency toward circular movement. Notice that right away we lose our explanation for a wide range of the most obvious daily phenomena. For example, we can no longer explain why rocks fall. Rocks are presumably composed mainly of the earth element, and when released, they move in a straight line toward the Earth. But they should not move like this if the tendency of the earth element is to move in a circular fashion.

Likewise, we lose our explanation for what keeps us on the surface of the Earth. Again, the Aristotelian explanation is that we are composed mainly of the heavy elements earth and water, and the natural tendency of those elements is to move downwards, and that is what keeps us on the Earth. But again, if we view the earth element as having a natural circular motion, then we lose this explanation.

Moreover, the idea that the earth element has a natural tendency toward continual circular motion clashes with the simple observation that the earth element is the heaviest element. The Earth, being a massive body composed of the heaviest element, would be by far the heaviest object in the universe. In contrast, ether was viewed as a special, very light (perhaps weightless) element. As Ptolemy notes in the preface to the *Almagest*, it does not make much sense to have the heaviest, most difficult to move object in the universe be in continual motion. Instead, it makes more sense to view the heaviest object in the universe as being stationary, and have the lightest objects – those composed of ether – be in constant motion.

In short, viewing the Earth as moving due to the earth element having a natural tendency toward circular movement is not a viable option, in that this view is inconsistent with the overall jigsaw puzzle of beliefs. More generally, adopting the notion that the Earth is in motion at all, for whatever reason, would require constructing an entirely new jigsaw puzzle, a new worldview. Eventually such a new jigsaw puzzle would be constructed, but not until new discoveries were made, mostly in the 1600s. And as mentioned several times above, constructing this new jigsaw puzzle would take a great deal of work, time, and talent.

Concluding Remarks

As noted above, and also as noted by Aristotle, the idea that the heavens are a place of perfection goes back to the beginnings of recorded history. The idea is understandable, in that the special, beautiful objects in the sky move in unchanging, repeating patterns, and these patterns have not changed for thousands and thousands of years. Aristotle himself inherited the tradition of the heavens as a region of perfection, and when he developed his views, much more fully than those before him, this idea of perfection remained.

As we saw above, the idea of the heavens as a region of perfection provided a way to understand how the heavenly bodies could be in continual motion. But the explanation carried with it the notion that the heavenly bodies must move in perfect circles and at uniform speeds. These facts – perfect circles and uniform motion – became extremely well entrenched. It is difficult to overemphasize how obvious these facts would have seemed to our predecessors. It was a matter of common sense – everyone simply knew – that planets move in perfect circles and at uniform speeds. These were obvious facts, and facts that did not appear to be appreciably different from the empirical facts we explored in the previous chapter.

In hindsight, we can recognize that the perfect circle fact and the uniform motion fact were not facts at all. They were philosophical/conceptual "facts," beliefs that seemed like empirical facts but instead turned out to be rooted more in the overall system of beliefs. In subsequent chapters, we will explore how these facts – both the empirical facts discussed in the previous chapter, as well as the perfect circle and uniform motion facts – were incorporated into the Ptolemaic and Copernican systems.

Finally, as a bit of foreshadowing, in Part III of this book we will see how certain facts that we hold – facts that seem to us to be obvious, empirical facts – have turned out, in light of recent discoveries, to be mistaken philosophical/conceptual "facts." In a sense, then, we are in a similar situation to that of our predecessors in the 1600s. Just as new discoveries forced them to rethink beliefs they had long taken as obvious facts, so too recent discoveries force us to rethink some of our basic beliefs about the sort of universe we inhabit.

Chapter Thirteen
The Ptolemaic System

The transition from the Aristotelian worldview to the Newtonian worldview involved, in important ways, competing theories as to the structure of the universe. In the next few chapters, we look at the central astronomical theories involved in this transition, some Earth-centered and some sun-centered. We will begin with a look at the Ptolemaic system.

The main goal of this chapter is to provide an overview of this system, as it was presented in Ptolemy's *Almagest*, published around AD 150. As noted earlier, the *Almagest* is a substantial and technical work, composed of 13 books spanning about 700 pages. We will begin with some background material on Ptolemy's system, and then look at some of the details of the system.

Background Information

As with any theory, Ptolemy's system needed to respect the relevant facts. In this case, the relevant facts consist largely of the empirical facts discussed in Chapter 11, as well as the philosophical/conceptual facts of circular, uniform motion discussed in the previous chapter.

Generally speaking, Ptolemy's system succeeds at respecting these facts. His system clearly respects the perfect circle fact, in that his entire approach is based on using only perfect circles for the motions of the heavenly bodies. As we will see below, he has some difficulties with the uniform motion fact, but he manages at least in some sense to respect this fact.

With respect to the empirical facts, his system does an especially nice job. That is, when it comes to explaining and predicting the facts discussed in Chapter 11, although his system is not perfect (few theories are), the margin of error is low.

For example, if we use Ptolemy's system to predict, say, where Mars will appear in the night sky on this date a year from now, or if we use the system to predict when and for how long Mars will next exhibit retrograde motion, the prediction will be very close to what we observe. It is very much worth emphasizing that no other theory of the universe before Ptolemy, or for 1,400 years after Ptolemy, was anywhere close to his with respect to prediction and explanation. It is fitting that the name by which we know this work, *Almagest*, was given it by Arabic translators and derives from a phrase meaning simply "the greatest." Although the theory may appear somewhat archaic to our eyes, the Ptolemaic system was a spectacularly impressive accomplishment.

We should take a moment to clarify what Ptolemy did and did not do. Ptolemy's approach is a mathematically based one, and he makes intricate use of various mathematical devices. Most of the mathematical devices he employs, however, were not original to him, but had been discovered in earlier centuries.

Nor, of course, was Ptolemy the first to develop an Earth-centered view of the universe. As we saw earlier, the view that the Earth was spherical, stationary, and at the center of the universe goes back to before Aristotle, 500 years before Ptolemy.

So Ptolemy did not originate the general Earth-centered approach that he takes, nor did he originally develop the mathematical devices that he employs. But what he did was take these rough notions and develop them into a precise theory, and a theory capable, for the first time in history, of providing accurate predictions concerning astronomical events. Or to put the point another way, before Ptolemy there were at best rough sketches, rather than anything that could properly be used in making predictions about astronomical events. With Ptolemy came a precisely crafted theory, capable of impressively accurate predictions and explanations.

As a final note in this section, you will sometimes hear that Ptolemy's system is not really a system, in the proper sense of the word. In a sense, this is correct, in that Ptolemy treats each of the heavenly bodies in an isolated, rather than unified, way. For example, one book of the *Almagest* deals exclusively with Mars, another exclusively with Venus, another with the sun, and so on, without ever providing a unified system of the entire universe. In that sense, one might say that Ptolemy's approach is not, strictly speaking, a system of the universe, but rather a collection of independent treatments of the various components of the universe. I will, however, continue to use the word "system" to describe Ptolemy's theory, since all these independent treatments add up to an approach that can be used to make predictions for all of the components of the universe.

With these background observations in mind, we will move on to consider an overview of the Ptolemaic system. For ease of discussion, we will focus on just one planet, in this case, Ptolemy's treatment of Mars. Let's begin with a description of the components involved in Ptolemy's treatment of Mars, and then discuss the rationale behind those components.

A Brief Description of the Components
of Ptolemy's Treatment of Mars

Figure 13.1 illustrates the key components of Ptolemy's treatment of Mars. Here, Mars moves around a point, labeled point A in the diagram. The circle this movement traces out, that is, the small circle that has A as its center, is called an *epicycle*.

The center of the epicycle, that is, point A, itself moves around on a larger circle that has point B as its center. A larger circle such as this is called either a *deferent* or an *eccentric*, depending on whether B is located at the center of the system (in this case, the center of the Earth), or is instead displaced from the center of the system. In this particular case, this is an eccentric, since as you can see, the center of the movement of the epicycle, point B, is not centered on the center of the Earth.

To clarify the distinction between deferents and eccentrics, note the Earth is at the center of Ptolemy's system. That is, the outermost boundary of the system is the sphere of the fixed stars (this is the periphery of the universe), and since the Earth is at the center of that sphere, the Earth is at the center of the system. If point B coincided with the center of the Earth (that is, the center of the system), then the large circle centered on B would be termed a deferent. On the other hand, if, as in the picture above, point B is not at the center of the system, then the larger circle is termed an eccentric.

In short, deferents and eccentrics are basically the same, in that both are the larger circles on which epicycles revolve. Think of an eccentric as an off-centered deferent.

The equant point is a point involved in the speed at which Mars' epicycle moves. The equant point is the most difficult component to explain, and so I will hold off on the details until we consider the rationale for these components.

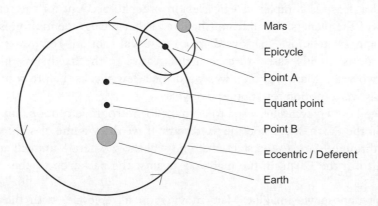

Mars

Epicycle

Point A

Equant point

Point B

Eccentric / Deferent

Earth

Figure 13.1   Treatment of Mars on the Ptolemaic system

Finally, this sort of a structure – that is, an epicycle moving on a larger circle – is termed an *epicycle–deferent system*. For convenience, we will call such an arrangement an epicycle–deferent system even if, strictly speaking, the system employs eccentrics rather than deferents.

The Rationale behind These Components

The Ptolemaic treatment of Mars is clearly somewhat complicated, with circles moving on circles, some circles off-centered, and with the still mysterious equant point involved. What were the reasons for these components?

First, some comments on epicycle–deferent systems in general. Epicycle–deferent systems are enormously flexible, in the sense that an extremely broad range of motions can be produced simply by varying the size, speed, and direction of motion of the components. That is, in any epicycle–deferent system, one has a broad range of options in how large or small one makes the epicycle and the deferent, one likewise has a broad range of options in how fast one has the planet moving on the epicycle, also how fast one has the epicycle moving on the deferent (or on the eccentric, as the case may be), and one also has the option of having the movement on the epicycle and deferent be either clockwise or counterclockwise.

This flexibility allows one to produce a wide range of motions, simply by adjusting these options. For example, all of the motions shown in Figure 13.2 were produced by an epicycle moving on a deferent. The dotted lines represent the path traced out by Mars as it moves on its epicycle, as the epicycle itself is moving around the Earth. All of these motions (and a wide variety of other motions as well) can be produced simply by varying factors such as the size of the epicycle, the size of the deferent (or eccentric), the speed at which Mars moves on its epicycle, the speed at which the epicycle moves, and so on.

So epicycle–deferent systems are useful for the amount of flexibility they provide. But in addition, any Earth-centered approach needs epicycles (or some method at least as complex as epicycles) in order to account for the retrograde motion of the planets. Recall from Chapter 11 that retrograde motion is when a planet appears to move "backward" from its usual motion. For example, Mars usually drifts slightly eastward each night relative to the fixed stars, but about every two years Mars will drift westward for a few weeks, before resuming its usual eastward motion for another two years.

To see how epicycles are used to account for retrograde motion, suppose we focus on the Earth, Mars, and the fixed stars. If we draw a line of sight from the Earth, through Mars, to the stars, that line will show where Mars will appear, as viewed from the Earth, in the night sky against the backdrop of the stars (see Figure 13.3).

Now suppose we imagine Mars moving on its epicycle, with the epicycle moving around the Earth. If we draw continual lines of sight from the Earth

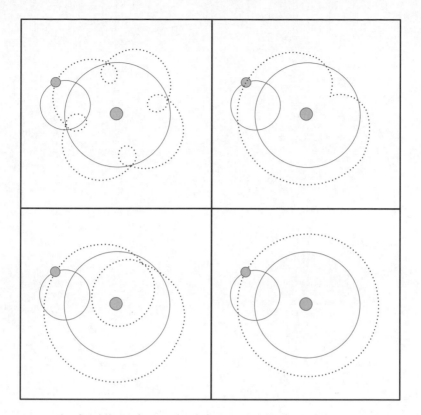

Figure 13.2 The flexibility of epicycle–deferent systems

through Mars, this will indicate where Mars will appear in the night sky, over the course of time, against the backdrop of the stars (see Figure 13.4). The numbers in Figure 13.4 represent the consecutive positions of Mars. As you see, Mars usually appears to move in one direction against the backdrop of the stars. That is, numbers 1 to 7 represent a steady motion eastward relative to the fixed stars. Then at 8, Mars has just begun to drift westward. Mars continues drifting westward at 9 and 10, and then from 11 to 15 resumes its usual eastward drifting. And in general, this is how an epicycle–deferent system accounts for retrograde motion. In fact, if one is committed to an Earth-centered system, with uniform circular motion, then epicycles turn out to be the best way to account for retrograde motion.

Incidentally, it should be noted that the sizes and speeds of the epicycle and deferent in these diagrams are not the correct sizes and speeds for Mars. These sizes and speeds were chosen to make for a simpler illustration. But by adjusting the sizes and speeds (and using eccentrics, as described below), one can get the "backward" appearance of Mars' motion to work out so that the model accurately predicts and explains when Mars is actually exhibiting retrograde motion.

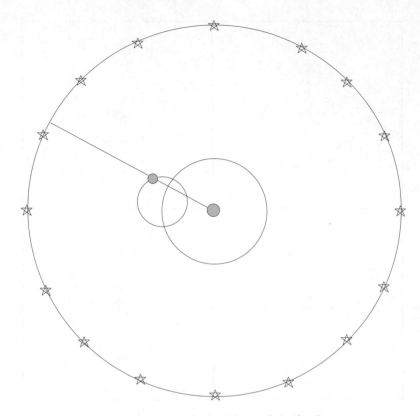

Figure 13.3   Position of Mars against the backdrop of the fixed stars

Let's now turn to the question of why Ptolemy used an off-center deferent, that is, an eccentric. The reason is simply that, if one uses a simple epicycle and deferent (again, the deferent would be a circle centered on the Earth), then one cannot get the model to make accurate predictions and explanations. That is, the model simply will not do what you need it to do, namely, make accurate predictions and explanations. But either of two modifications to the simple combination of epicycle and deferent will produce a model that quite accurately predicts and explains the motion of Mars.

The first option is to introduce an additional, small epicycle on the epicycle shown in Figure 13.1 above. The result would be as pictured in Figure 13.5. This additional epicycle adds even more flexibility to the model. With this additional flexibility, one can now adjust the model such that the predictions and explanations concerning Mars will be extremely accurate.

Such additional epicycles are sometimes called *minor epicycles*, to distinguish them from the *major epicycles*, such as the single epicycle pictured in Figure 13.1 and the larger epicycle pictured in Figure 13.5. The difference between major and minor epicycles is that major epicycles are the ones needed to handle retrograde

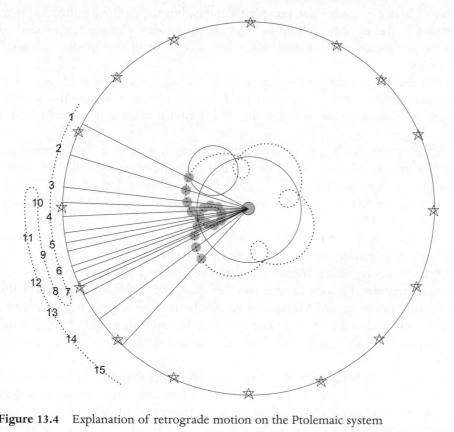

Figure 13.4 Explanation of retrograde motion on the Ptolemaic system

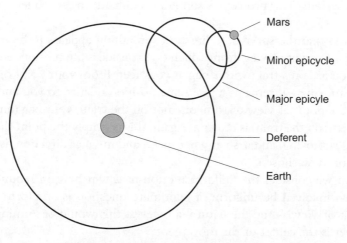

Figure 13.5 Minor and major epicycles

motion. Major epicycles also provide flexibility, but their primary function is to account for retrograde motion. In contrast, minor epicycles are not needed to handle retrograde motion, but rather are used to give the model additional flexibility.

As mentioned, adding minor epicycles is one way to get the predictions and explanations for Mars to come out right. The other option is to move the deferent off-center, that is, to use an eccentric. This option is what is pictured in Figure 13.1.

Either option – adding a minor epicycle or using an eccentric – will work to get the predictions and explanations back in line with the observed data. In fact, the two options are mathematically equivalent, so either will work equally well. Ptolemy chose to use eccentrics, and thus the construction for Mars looks as it does in Figure 13.1.

The final component to explain is the equant point. This also involves a problem with getting the model to correctly predict and explain the observed data. In particular, this problem is tied to the philosophical/conceptual fact of uniform motion. Recall that the two key philosophical/conceptual facts the Ptolemaic system needed to respect were the perfect circle fact (that all motions of heavenly bodies are perfectly circular) and the uniform motion fact (that the motion of heavenly bodies is uniform, that is, neither speeding up nor slowing down).

If you look at the figures in this chapter, you can note that the Ptolemaic system clearly respects the perfect circle fact. That is, all motions are in terms of perfect circles. We have actually looked only at the treatment of Mars, but in all of Ptolemy's constructions, all minor and major epicycles, deferents, and eccentrics are perfect circles. So clearly, there is no problem respecting the perfect circle fact.

The uniform motion fact is a different matter, and provides a problem for Ptolemy's system. This problem is sometimes difficult to see, so let's approach it slowly.

First, note that the speed and direction something appears to be moving will depend on the point of view from which one considers the motion. For example, suppose you are on a train with a bag at your feet. From your point of view, that bag is not moving – it remains in the same position relative to you and your feet. But from the point of view of someone not on the train, your bag (and you, and everyone else on the train) is moving. Again, this is simply the point that whether something is moving, and if so at what speed and in what direction, is relative to the point of view chosen.

So when we consider the uniform motion fact, which again requires that the movements involved be uniform, a legitimate question is "uniform relative to what point of view?" And the natural answer to this would be "uniform relative to whatever is the center of the movement."

If we look at just the movement of Mars on its epicycle, there is no problem. That movement is indeed uniform, that is, as Mars moves around the center of the epicycle, it moves with uniform speed relative to that center.

But now consider the motion of the center of the epicycle. If we ask "What should this motion be uniform relative to?" there are two natural answers. The first would be that the center of the epicycle moves with uniform speed relative to the center of the entire system, that is, relative to the center of the Earth. The second would be that the center of the epicycle moves with uniform speed relative to the center of the eccentric on which it moves.

The problem is, if you take either of these options – that is, if you make the motion of Mars' epicycle uniform relative either to the center of the Earth or to the center of Mars' eccentric – the system won't work. When I say it won't work, I mean simply that your predictions and explanations will not be accurate. In other words, if Ptolemy tries to respect the uniform motion fact in the most straight-forward way, his system will not handle the data in an acceptable way. That is, the predictions and explanations are no longer accurate.

One option for addressing this problem would be to abandon the uniform motion fact. But again, this was a well-established fact, and had been for centuries before Ptolemy, and even before the time of Aristotle. Moreover, as discussed in the previous chapter, the uniform motion fact is closely tied to the understanding of how the heavenly bodies move, so abandoning this fact would likewise mean abandoning the long-held understanding of what accounts for the motion of the heavenly bodies. In short, giving up the uniform motion fact was not really a viable option.

The other option for Ptolemy was to make the motion of Mars' epicycle uniform relative to some point other than the center of the Earth or the center of the eccentric, and this was the option Ptolemy adopted. As it turns out, one can calculate a point, within the eccentric on which Mars' epicycle moves, such that if Mars' epicycle moves with uniform speed relative to this point, then the model will come back in line with the data. And this point is what is called the *equant point*.

In summary, the equant point for Mars is the point with respect to which Mars' epicycle moves with uniform speed. But that point is a somewhat contrived point, calculated so as to make the predictions come out accurate, rather than being either of the places from which you would expect the movement to the uniform.

This, then, completes the overview of the main components needed to handle the motion of Mars. Clearly, this is a complex apparatus. But, to a remarkable degree of accuracy – it works.

Concluding Remarks

Above, we described only the parts of the Ptolemaic system that concern Mars. This treatment of Mars should be sufficient to provide a flavor for the Ptolemaic system. As noted, Ptolemy treats each of the five planets, the moon, the sun, and stars separately. The treatment of the other planets, and to some extent the moon and sun, bear similarities to the treatment of Mars. That is, generally speaking, the constructions needed to account for the motions of the other planets are similar (though not identical) to those for Mars, with the planets requiring their own epicycles, eccentrics, and equant points. The apparatus needed to account for the motions of Mercury and the moon are somewhat more complex than those described above for Mars, while handling the motions of the sun is somewhat less complex. In general, it should be clear that the Ptolemaic system is a quite complex collection of constructions for handling the sun, moon, stars, and planets.

But – and this is a crucial point – in spite of its complexity, the Ptolemaic system did a marvelous job of handling the data, providing for the first time in history the ability to accurately predict and explain an extraordinarily wide range of astronomical data.

Chapter Fourteen

The Copernican System

In the previous chapter we looked at the Ptolemaic system. As we saw, Ptolemy's system was quite successful in terms of predicting and explaining the relevant data. Although the theory was modified in the centuries following Ptolemy's death, the modifications were relatively minor, and the dominant astronomical theory for the next 1,400 years was essentially that of Ptolemy.

In the 1500s, Nicolas Copernicus (1473–1543) developed an alternative theory of the universe. Copernicus developed his system in the early 1500s, and published it the year he died. One of our main goals in this chapter will be to see how the Copernican system works. In addition, we will look at a brief comparison of the Copernican and Ptolemaic systems, including a discussion of which system provides the more plausible model of the universe. Finally, we will explore the question of what motivated Copernicus, with particular emphasis on ways in which certain philosophical/conceptual beliefs influenced his work.

Background Information

The Copernican system is a sun-centered system. Today we view the sun as the center of our solar system, but, notably, Copernicus' system did not merely have the sun at the center of the revolution of the planets; rather, he placed the sun at the center of the entire universe.

In many ways the Copernican system is like the Ptolemaic system, but with the position of the Earth and sun swapped. For example, like Ptolemy, Copernicus viewed the stars as all being equidistant from the center of the universe, embedded in the so-called sphere of the fixed stars. As it did for Ptolemy, this sphere defined the outermost boundary of the universe. Copernicus' universe was larger than

Worldviews: An Introduction to the History and Philosophy of Science. Richard DeWitt
© 2010 Richard DeWitt

Ptolemy's, that is, the sphere of the fixed stars was larger and further than generally believed by advocates of the Ptolemaic system, but the Copernican universe,
like Ptolemy's, was relatively small compared to our conception of the size of
the universe. And also as with the Ptolemaic system, the Copernican system used
epicycles, deferents, and eccentrics, though notably it did not require equant
points. Again, generally speaking, the Copernican system had a great many similarities to the Ptolemaic system, with the most obvious difference being the position of the sun and Earth.

It is also worth noting that Copernicus was dealing with essentially the same
empirical facts as Ptolemy (again, the main such facts are covered in Chapter 11).
The data was not exactly the same – in the 1,400 years separating Ptolemy and
Copernicus, some new astronomical observations had been made, some existing
observational mistakes had been corrected, and quite a few new observational
mistakes had been introduced (either by mistaken observations or by mistakes in
copying records). But, generally speaking, the empirical data available during
Copernicus' time was still based on naked-eye observation, and this data was
similar to the data with which Ptolemy worked.

In addition, Copernicus was firmly committed to the same key philosophical/
conceptual facts as Ptolemy. That is, Copernicus firmly believed (as did almost all
his contemporaries) that an acceptable model of the universe must respect the
perfect circle and uniform motion facts.

It is often claimed that the Copernican system is vastly simpler than the
Ptolemaic system, and that the Copernican system is superior at prediction and
explanation. But as we will see shortly, this is simply a mistake. The Copernican
system is easily as complicated as the Ptolemaic system, and no better (or worse)
at prediction and explanation than the Ptolemaic system. When authors claim the
Copernican system is simpler than Ptolemy's, and superior at prediction and
explanation, they most likely are thinking of Kepler's system, which was not
developed until 70 years after Copernicus' death, and which is the subject of a
later chapter.

With this background material in mind, let's look at an overview of the
Copernican system.

Overview of the Copernican System

As we did with the Ptolemaic system, we will simplify matters by focusing on the
motion of a single planet. We will again use Mars as an example, and again begin
with a picture. It should be noted that, in Figure 14.1, the circles are not drawn
to scale, but rather drawn so as to be more easily distinguishable. On the
Copernican system, Mars moves in a circle around point A (again, a small circle
such as this is called an *epicycle*). Point A moves in a circle around point B (again,
such circles are known as *deferents* or, if off-center, *eccentrics*). Point B also moves,

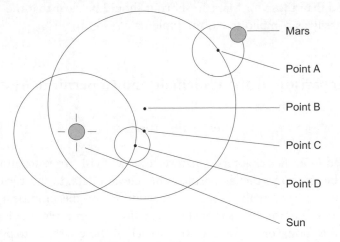

Figure 14.1 The treatment of Mars on the Copernican system

but as it moves it remains in a fixed position relative to point C. Point C is the center of the eccentric on which the Earth moves (to simplify the picture, the Earth is not shown in this diagram, but if the Earth were in the picture, point C would be the center of its eccentric). Point C moves in a circle around point D, and finally, point D moves in a circle around the sun. Not for nothing did I say the Copernican system was as complicated as the Ptolemaic.

Again, much like the Ptolemaic system, the Copernican system employs epicycles, deferents, and eccentrics, in a complicated system of circles on circles. Notice, however, that there is no equant point in the diagram, and in fact the Copernican system does not use equant points. Also, although the Copernican system does require epicycles, the epicycles are used for the flexibility they provide, but they are not needed to account for retrograde motion, as is the case with the Ptolemaic system.

If we ask the question "Why did Copernicus need this complex apparatus?" the answer, in a nutshell, is that without it the predictions and explanations do not work out. In other words, as with the Ptolemaic system, by using these complicated devices, Copernicus was able to work out a system that does a quite good job at explanation and prediction (as good as, though not better than, the Ptolemaic system). And without such devices, Copernicus was unable to get the model to match the known data. In short, just like the Ptolemaic system, the Copernican system is complicated, but when all is said and done, it works – that is, it explains and predicts the relevant data to a remarkable degree of accuracy.

So far, we have discussed only the motion of Mars. On the Copernican system, the apparatus needed to account for the other outer planets, that is, Jupiter and Saturn, is similar to that shown in the diagram above. The apparatus needed for the Earth is somewhat less complicated, as is that for the moon. Finally, the devices used for the movement of the inner planets, Mercury and Venus, are more complicated than those for Mars. In short, it should be clear that the Copernican system is easily as complicated as the Ptolemaic system.

Comparison of the Ptolemaic and Copernican Systems

Respecting the facts

As discussed in earlier chapters, whatever else we wish from scientific theories, they must be able to predict and explain the relevant data. In this regard – that is, in terms of accuracy with respect to accounting for the empirical data – the Ptolemaic and Copernican systems are essentially the same. Neither is perfect, but both are quite good. For example, if we use each of these systems to predict where Mars will appear in the night sky exactly a year from now, or to predict exactly when the summer solstice will occur for the next 10 years, or to predict any of a vast range of astronomical events, both systems will provide predictions that closely match the facts.

With respect to the philosophical/conceptual facts of perfectly circular and uniform motion, the Copernican system is slightly better. Both systems respected the perfect circle fact, that is, both systems modeled the motion of the planets and stars using only perfect circles. But, as discussed in the previous chapter, the Ptolemaic system respects the uniform motion fact only by using the rather strained device of equant points. In contrast, Copernicus was able to eliminate this hedge, and was able straightforwardly to respect the uniform motion fact. Again, even though these "facts" sound quite alien to our ears, most of Ptolemy's and Copernicus' contemporaries were committed to them, and so respecting them is a matter of some importance. In this respect, it is worth noting that Copernicus himself considered the elimination of equant points to be one of the most important reasons for preferring his theory.

In short, there is little difference between the Ptolemaic and Copernican systems in terms of predicting and explaining the empirical facts. With respect to the relevant philosophical/conceptual facts, the Copernican system respects the uniform motion fact in a somewhat more straightforward way.

Complexity

There is little difference between the two systems in terms of complexity. For example, if we look at the types of devices required (such as epicycles, deferents, eccentrics, and the like), as well as the number of such devices employed, the Copernican and Ptolemaic system are about equally complicated. Even though the complexity of systems such as this cannot be precisely quantified, and so it is not possible to compare exactly the complexity of the two systems, I think we can all agree on this point: both systems are very complex, and with respect to complexity, there is little to distinguish them.

Retrograde motion and other more "natural" explanations

Recall the Ptolemaic explanation of retrograde motion, that is, the occasional "backward" motion of the planets. In the Ptolemaic system, each planet required a major epicycle, the primary purpose of which was to account for the retrograde motion of the planet.

In contrast, retrograde motion receives a quite different explanation on the Copernican system. Again we will use Mars as an example, but similar accounts apply for the retrograde motion of the other planets as well.

On the Copernican system, the Earth is the third planet from the sun, and Mars is the fourth planet. Moreover, the Earth completes about two revolutions about the sun for every one revolution Mars completes. As a result, about every two years the Earth catches up to and then passes Mars. During the period in which the Earth is passing Mars, Mars appears, from the Earth, to move backward against the backdrop of the stars. Figure 14.2 may help illustrate this point. The lines are again the lines of sight drawn from the Earth, through Mars, out to the stars, and will show where Mars will appear against the backdrop of the stars. Note that the lines usually move in one direction, representing the usual eastward drift of Mars relative to the fixed stars. For example, in 1 to 3, Mars is shown in this usual eastward motion, then in 4 to 6, Mars is drifting westward, and then in 7 and 8 Mars has resumed its usual eastward drifting.

On the topic of retrograde motion, recall the seemingly minor empirical fact discussed at the end of Chapter 11, that Mars, Jupiter, and Saturn all appear brightest around the same time as they exhibit retrograde motion. Looking again at Figure 14.2, we can see why this would be expected. On the Copernican system, Mars will undergo retrograde motion only when the Earth catches up and passes Mars. Note that this will be the time at which Earth and Mars are the closest together, and so one would expect Mars to appear brighter at these times. The same story goes for Jupiter and Saturn as well – that is, they too will undergo retrograde motion only around those times when they are the closest to the Earth. So the correlation between the retrograde motions of Mars, Jupiter, and Saturn, and the times at which those planets appear brightest, has a quite natural explanation on the Copernican system.

Speaking of more natural explanations, recall also the other seemingly minor piece of empirical evidence discussed at the end of Chapter 11, that Venus and Mercury never appear far from the sun. On the Copernican system, Venus and Mercury are inner planets (that is, they are between the Earth and the sun). So no matter where Venus and Mercury are in their motions around the sun, when viewed from the Earth they must appear to be in the same region of the sky as the sun.

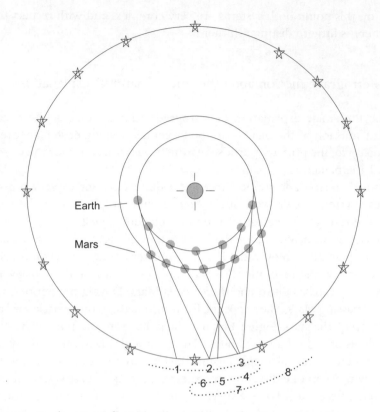

Figure 14.2   Explanation of retrograde motion on the Copernican system

In short, the Copernican system has a more natural explanation for retrograde motion, for the correlation between retrograde motion and the apparent brightness of Mars, Jupiter, and Saturn, and for the fact that Venus and Mercury always appear to be close to the sun. And these are all advantages of the Copernican system.

From a realist standpoint, which system is the more plausible model of the universe?

Recall our earlier discussion on instrumentalism on the one hand, and realism on the other. Again, instrumentalism is an attitude toward a theory in which one is primarily concerned with how well the theory predicts and explains the relevant data. Realism, on the other hand, is an attitude in which a theory is expected not only to predict and explain, but also to model or picture the way things really are.

Almost everyone took the various devices of these systems, such as epicycles, with an instrumentalist attitude. That is, these were generally viewed not as physically real, but rather as mathematical devices necessary to make accurate predictions and explanations. So the issue of realism generally does not arise for devices such as epicycles.

But the realist issue is very much relevant for the Earth-centered versus sun-centered parts of these two theories. So a legitimate question is, from a realist perspective, which model of the universe – Ptolemy's Earth-centered approach, or Copernicus' sun-centered approach – is the more plausible model of the universe?

With respect to this question, the data available at the time strongly supports the Ptolemaic system. Recall the arguments from Chapter 10, supporting the conclusion that the Earth was stationary and at the center of the universe. These are all strong arguments (though they eventually turned out to be mistaken, albeit for subtle reasons), and so with respect to the question of which system was better matched to the best science of the day, the answer is clear: the Ptolemaic system is far better than the Copernican.

In summary, the Ptolemaic and Copernican systems are comparable in terms of prediction, explanation, and complexity. In eliminating equant points, the Copernican system more straightforwardly respects the uniform motion fact, and it more straightforwardly accounts for retrograde motion, for the correlation between the differences in brightness of the planets and their times of retrograde motion, and for the fact that Venus and Mercury always appear near the sun. These seem to be relatively small advantages, however, compared to the evidence available at the time that pointed to a stationary Earth, which was more consistent with the Ptolemaic system.

What Motivated Copernicus?

As noted in the discussion above, the Copernican system was much like the Ptolemaic system. For example, both systems make extensive use of epicycles, deferents, and eccentrics. In most respects (except for the elimination of equant points and the explanation of retrograde motion) the Copernican system was no better than the Ptolemaic system, and in some important respects (for example, the issue of whether it is more reasonable to believe the Earth is stationary or in motion), the Copernican system is much worse off than the Ptolemaic system.

So if the Copernican system had only a handful of minor advantages, and had the substantial disadvantage of being incompatible with the current best physics, then what in the world would have motivated Copernicus to develop his system? Life is short, yet Copernicus devoted much of his life to working out his system. If there were good reasons to think that the Earth could not be in motion, then why would Copernicus spend so much of his life developing a system in which the sun was the center of the universe, with the Earth in motion around it?

This question is a good one to ponder, and one worth re-emphasizing: Copernicus spent an enormous amount of time, over the course of decades, working out his system. Yet his system is clearly at odds with all the evidence pointing toward a stationary Earth. Nor was there any new empirical evidence available to Copernicus that would have supported his view of a moving Earth. So what in the world would have motivated Copernicus to devote his life to develop a theory that seems like it could not possibly be correct?

I do not intend, in this section, to attempt a full answer to this question. I do wish, though, to suggest how philosophical and conceptual issues might motivate the work of a scientist. For some time now various scholars have argued that Copernicus' leanings toward Neoplatonism, and his commitment to the philosophical/conceptual beliefs about perfectly circular, uniform motion, were key motivating factors in the development of his sun-centered system. What follows is an outline of these views.

Neoplatonism

In a nutshell, Neoplatonism is sort of a "Christianized" version of Plato's philosophy. Plato lived about 400 BC, and, roughly speaking, he believed there is a wide variety of objectively existing, nonphysical, eternal "forms." These forms are the objects of knowledge, that is, when we acquire a piece of knowledge, as opposed to having a mere belief or opinion, our knowledge is knowledge of one or more of these objectively existing, nonphysical, eternal forms. For example, when we come to know the Pythagorean theorem, or other truths of mathematics, we have acquired knowledge not of objects here on earth (for example, a drawing of a right triangle), but rather we have acquired knowledge of an objectively existing, nonphysical, eternal form.

According to Plato, the forms involve not only truths of mathematics, but "higher" forms as well, such as forms of truth and beauty (such forms are "higher" not just in the sense that they are more difficult to grasp, but more important as well). The highest form of all is the form of the Good. Plato says little directly about the form of the Good. But he does make clear that this form is the highest, most important form.

Instead of trying to describe directly the form of the Good, Plato speaks metaphorically about this form. In particular, Plato always uses the sun as his metaphor for the Good. For example, Plato says that, just as the sun is the source of all life, so too the form of the Good is the source of all truth and knowledge. Likewise, in his allegory of the cave, Plato describes a prisoner who has escaped the cave and is finally able to gaze upon the sun. In this allegory, the prisoner represents the lover of wisdom who has completed his or her intellectual journey, escaping ignorance (represented by the cave), and eventually coming to understand the highest truth of all, the form of the Good (represented by the sun). In short, in the allegory of the cave, as always, the sun is Plato's metaphor for the Good.

Several hundred years after the death of Plato, the movement called Neoplatonism incorporated Plato's philosophy into Christianity. I will ignore most of the details of Neoplatonism, and just emphasize that, for a Neoplatonist, Plato's form of the Good becomes identified with the Christian God. And the sun – Plato's metaphor for the Good – now comes to represent God.

As a philosophy, Neoplatonism has come and gone at various times in western history. During the time of Copernicus, it was a not uncommon philosophy; however, the evidence linking Copernicus to Neoplatonism is not as clear as one would like. It is very likely that Copernicus would have been exposed to Neoplatonic ideas during his student years, and some of what Copernicus writes sounds as if it is coming from someone with Neoplatonic leanings. Some scholars have been fairly convinced that Copernicus was heavily influenced by Neoplatonism; others are less convinced. The usual account linking Neoplatonism to the development of Copernicus' sun-centered view is straightforward: if Copernicus was a Neoplatonist, and viewed the sun as the physical representation of God in the universe, then the appropriate place for the representation of God would be the center of the universe. On this account, a main reason why Copernicus pursued a sun-centered view of the universe stemmed from philosophical beliefs that were substantially influenced by Neoplatonism.

Copernicus' commitment to uniform, circular movement

I have discussed, at numerous points, how deeply committed most astronomers were to the belief that the motion of the stars and planets had to be perfectly circular, and uniform in the sense of never speeding up or slowing down. In hindsight, this commitment was primarily a philosophical/conceptual commitment. Although there is a small amount of empirical evidence supporting the belief (for example, the stars do appear to move in a circular fashion), the degree of commitment to this belief far outstripped the empirical evidence for it.

As described in the previous chapter, Ptolemy was able to respect the uniform motion fact only by using the rather strained device of the equant point. By way of quick review, the epicycle of a planet such as Mars moves with uniform speed relative to an imaginary point, called the equant point. A line drawn from the equant point to the center of Mars' epicycle will sweep out equal angles in equal time, and in this sense, Mars' epicycle moves with uniform speed relative to the equant point. But Mars' epicycle most decidedly does not move with uniform speed relative to the Earth, or relative to the center of the circle around which that epicycle moves.

Given the fact that the Ptolemaic system was able to account quite well for the empirical data, and as such was a very useful and valuable model, almost all astronomers were willing to accept the fudge factor of the equant point. Copernicus, however, was not. He was simply too committed to the uniform motion view to accept a device such as the equant, and this commitment also helped motivate him to develop a system that did not require equant points.

This is a good illustration of the way in which it was not empirical data, but rather, philosophical/conceptual "data" that helped motivate Copernicus to develop his theory. As it turns out, this is not a particularly unusual event. In the history of science, it is often (though not always) philosophical/conceptual commitments that in part motivate scientists to develop new theories. So in this respect, Copernicus was not an unusual scientist at all.

As a final point in this section, it is worth noting that we all have such philosophical/conceptual beliefs, many of which are so embedded in our way of thinking that they appear to be straightforward empirical facts. When we look back in history, it is relatively easy to identify beliefs, such as the perfect circle and uniform motion facts, that were primarily philosophical/conceptual in nature. It is also relatively easy to see how such facts motivated scientists such as Copernicus. In contrast, it is very difficult to put our fingers on the philosophical/conceptual commitments of ours that are masquerading as empirical beliefs. Later in this book, when we turn to some examples from more recent science, we will attempt to flesh out some of our own philosophical/conceptual commitments.

The Reception of the Copernican Theory

Recall that all the evidence of the time pointed to a stationary Earth, and so it seemed that Copernicus' theory could not possibly be correct. Given this, one might think that his theory would have been immediately dismissed, and would certainly not have been widely read or discussed.

But in fact, in the years following Copernicus' death (the same year his system was published), and continuing through the remainder of the 1500s, his theory was widely read, discussed, taught, and put to practical use. Part of the reason for this was that Copernicus' system was the first thorough, sophisticated astronomical system published in the 1,400 years since Ptolemy. People of his time were justifiably impressed, and Copernicus was widely referred to as a "second Ptolemy."

Another reason involved the production of astronomical tables. Such tables were the primary way that an astronomical system, such as Ptolemy's, was put to practical use. An analogy might help clarify this. Suppose I need to find out about some astronomical event – for example, suppose I am planning a late-afternoon outdoor social, and I need to know what time the sun will set. It would be *possible* for me to compute the time of sunset from our current best astronomical theories, but it would be extremely burdensome to do so. What I would do instead is take the much simpler route – I would probably go to the Internet and search for information on what time the sun sets.

The data on sunset times I would find on the Internet (or in other sources, such as a current almanac) is derived from our current astronomical theories, but the people who put together this data have done all the hard work. Astronomical tables were somewhat similar. They were derived from the current best theory – for most of our history, this would have been the Ptolemaic theory – and then those who needed astronomical data would use the tables as a source.

In the 1500s, a new set of astronomical tables was badly needed (the previous set had been produced in the 1200s, and were out of date). As it turns out, the astronomer who produced these new tables based them on Copernicus' theory. Again, since the Copernican and Ptolemaic systems were essentially equivalent with respect to prediction and explanation, this astronomer could have used either system and arrived at about equally good tables. But he used the Copernican system, and this both publicized it and gave it added prestige.

So in the second half of the 1500s, the Copernican system was widely known, widely read, and widely taught in European universities. Importantly, though, it was taken with an instrumentalist attitude by almost everyone. That is, with few exceptions (there were some Neoplatonists who took it realistically, as well as a few others) the Copernican system was used as a practical device, but not one that people thought reflected the way the universe really was. In short, in the late 1500s the Ptolemaic and Copernican system coexisted peacefully. (At least, this was true among astronomers – there were some attacks by religious leaders who vigorously opposed the Copernican system, but for religious, not empirical, reasons.) Generally speaking, among astronomers the Ptolemaic system was taken with a realist attitude (or at least, the Earth-centered part of the theory was taken realistically), and the Copernican system was taken with an instrumentalist attitude. That is, the Copernican system was taken as a system that was useful, though not one that reflected the way the universe really was.

Concluding Remarks

In this chapter, we have looked at an overview of the Copernican system, compared this system with the Ptolemaic system, discussed Copernicus' motivation for developing his system, and noted that the Copernican system was well received, albeit with an instrumentalist attitude, by astronomers in the late 1500s. This presentation was rather brief, covering a lot of ground in a fairly short space, but it should convey at least a good flavor of the Copernican system and some of the key issues surrounding it.

This relatively peaceful situation would change dramatically in the early 1600s. At this time the telescope was invented, and this produced, for the first time since before recorded history, new astronomical data. In the next two chapters we will look briefly at two more key astronomical systems, and then turn to the new data generated by the telescope.

Chapter Fifteen

The Tychonic System

In this brief chapter, we look at an overview of the Tychonic astronomical system. This system is, in a sense, a blending of parts of the Ptolemaic and Copernican systems. Since it is mainly a rearranging of familiar components rather than an employment of new components, we can get a good overview of it fairly quickly. We will begin with some brief introductory material.

Tycho Brahe (1546–1601) was a respected astronomer of the second half of the 1500s. His principal contributions were the development of an alternative astronomical system, now called the *Tychonic system*, and his decades of extraordinarily accurate empirical observations. His astronomical observations were crucial in the development of Kepler's system, but this is an issue we will discuss in the next chapter. For now, let's take a brief look at Tycho's astronomical system.

As with most astronomers of the time, Tycho was well acquainted with the Copernican system. And he respected the fact that in some ways the Copernican system had advantages over the Ptolemaic system. As noted earlier, aspects of the Copernican system were more elegant than the Ptolemaic system. For example, as discussed in the previous chapter, the Copernican system's explanation of retrograde motion was much better than the Ptolemaic explanation.

But as was also the case with most astronomers of the time, Tycho recognized that the preponderance of the evidence pointed toward a stationary Earth, and so, speaking from a realist standpoint, the Copernican system could not be the correct model of the universe. Tycho was, however, able to develop a system that included most of the perceived advantages of the Copernican system, but at the same time keeping the Earth as the center of the universe.

On the Tychonic system, the Earth is the center of the universe, and the sphere of the fixed stars again defines the periphery of the universe. The moon and sun revolve around the Earth. The planets, however, revolve about the sun. That is,

Worldviews: An Introduction to the History and Philosophy of Science. Richard DeWitt
© 2010 Richard DeWitt

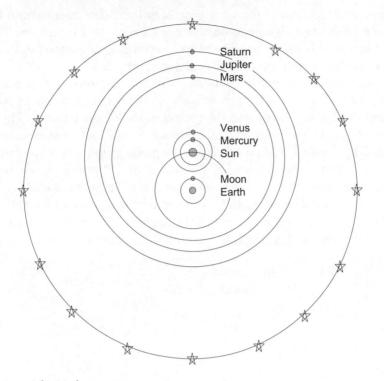

Figure 15.1   The Tychonic system

although the Earth is stationary and at the center of the universe, and the moon and sun revolve about the Earth, the center of the movement of the planets is the sun. A simplified picture of the Tychonic system, leaving out components such as the various epicycles, is given in Figure 15.1. It is not obvious from the description or the picture, but in fact, the Tychonic system is mathematically equivalent to the Copernican system. That is, it is possible to mathematically transform the Copernican system into the Tychonic system (or vice versa). Given this, the Tychonic system is equivalent to the Copernican system in terms of predicting and explaining the empirical data discussed in Chapter 11. In particular, exactly as with the Copernican system, the Tychonic system is good, but not perfect, when it comes to handling this empirical data.

As mentioned, one of the perceived advantages of the Tychonic system was that it allowed one to keep the advantages of the Copernican system, while at the same time keeping the Earth as the center of the universe. So although the Tychonic system looks somewhat odd to our eyes, it incorporates the most appealing features of the Ptolemaic and Copernican systems.

Moreover, shortly after Tycho's death, the newly invented telescope would provide new evidence suggesting that at least some of the planets revolve around the sun. The Tychonic system then became an alternative that would allow one to accept that the planets moved around the sun, in keeping with the newly discovered evidence, but at the same time keep the Earth stationary, in keeping with the evidence and arguments discussed in Chapter 10. In short, the Tychonic system became a viable compromise way of keeping certain attractive features of both the Earth-centered and the sun-centered approaches.

As a brief aside, it is worth mentioning that the Tychonic system is still alive today. In the past 10 years, there have been no fewer than four books published defending the Tychonic system as the correct model of the universe. There is no shortage of people (most, including the authors of the books just mentioned, are motivated by a literal reading of certain religious scriptures) who continue to believe the Earth is the center of the universe. And if one is committed to an Earth-centered view, then the Tychonic system is the best alternative.

This, then, completes our brief overview of the Tychonic system. We will discuss one more important system, that of Kepler, and then move on to consider the new evidence that became available with the invention of the telescope.

Chapter Sixteen

Kepler's System

In this chapter, we will explore the system developed by Johannes Kepler (1571–1630), as well as explore some of the factors that motivated Kepler. As we will see, Kepler essentially "got it right." That is, Kepler was the one who finally developed a system that not only was completely accurate with respect to prediction and explanation, but was also vastly simpler than any alternative system. Moreover, speaking from a realist point of view, his system seems to describe the way the moon and planets really do move. In fact, Kepler's view of the movements of the moon and planets is basically the same model we have today.

Kepler was an interesting individual. We will explore not only the system he developed, but also some of the reasons that motivated him to work on developing a new system. As we saw with Copernicus, some of these reasons involved factors that seem to us more philosophical/conceptual than scientific. But before exploring this, we will begin with some background material.

Background Information

Kepler was born in 1571, several decades after the publication of the Copernican system, and several decades before the invention of the telescope would provide new empirical data supporting a sun-centered view. When he was in his late twenties, Kepler worked for the astronomer and observer Tycho Brahe, whom we discussed briefly in the previous chapter. His collaboration with Tycho was brief, because Tycho died less than two years after Kepler joined him. But Tycho's influence on Kepler was crucial in the eventual development of Kepler's system. Given this, a few words on Tycho's influence on Kepler are in order.

Worldviews: An Introduction to the History and Philosophy of Science. Richard DeWitt
© 2010 Richard DeWitt

Tycho Brahe's empirical observations

As noted in the previous chapter, Tycho's principal contributions were the development of the Tychonic system, and his extraordinarily accurate empirical observations. In the previous chapter, we explored briefly his astronomical system. But Tycho's astronomical observations proved much more important in Kepler's eventual development of his system.

In a nutshell, Tycho was by far the most careful, accurate, and diligent observer of his time, and for that matter, probably the best naked-eye observer of all time. Over the course of 20 years, Tycho collected extremely accurate data (the accuracy was essentially to the limit of what is possible with naked-eye observation) on the movement of the sun, moon, and planets. He collected an especially large amount of data on the observed position of Mars, and it was largely the data on Mars that would prove to be a crucial factor in Kepler's development of his system.

Tycho and Kepler

Following Tycho's death, Kepler took possession of key parts of the data Tycho had accumulated, and it was largely this access to Tycho's data that made it possible for Kepler to work out his system.

By no means should this be taken as suggesting that Kepler's work was any sort of easy or straightforward extrapolation from Tycho's data. Kepler worked extraordinarily hard developing his system, and it took him many years to hit upon the right approach.

Tycho's data convinced Kepler that none of the alternatives – neither the Ptolemaic, Copernican, nor Tychonic systems – was able to make completely accurate predictions and explanations. The data involving Mars was especially problematic, in that none of these systems handled the data on Mars particularly well. Given this, it was clear to Kepler that none of these systems was quite right.

Kepler began working on an alternative approach, focusing on the movement of Mars. Notably, his approach was a sun-centered one. Part of his preference for a sun-centered view stems from his student days, during which he was taught by an ardent supporter of the Copernican system. Partly as a result of this, from an early age Kepler adopted a sun-centered view of the universe, and so when he began work on a new approach, that approach was a sun-centered one.

Incidentally, at this point it is something of a misnomer to speak of Kepler's "system," since his early work focused only on providing a treatment of Mars. However, eventually his success with Mars was extended to the remaining planets, the sun, and the moon, and so could then properly be called a system. For convenience, I will continue to speak of Kepler's system, with the understanding that the complete account would be some years off.

As with almost all of his contemporaries, Kepler was originally committed to the perfect circle and uniform motion facts. Thus, he spent a fair amount of time trying to modify the Copernican system, with the sun at the center, and with all movement given in terms of perfect circles and uniform motion. And Kepler did, in fact, make some important improvements to the Copernican system.

But by the very early 1600s, Kepler realized that no system based on uniform motion could account for the observed motion of Mars. He then began exploring alternatives, allowing Mars to move with varying speeds during different parts of its orbit. Not much later, Kepler likewise concluded that no system based only on perfect circles would be able to account for the observed motion of Mars, and so he began exploring orbits of varying shapes.

Importantly, note that Kepler had, at this point, abandoned the two key philosophical/conceptual facts of perfectly circular and uniform motion. Below, we will explore some of the factors that made it easier for Kepler, in contrast to most astronomers who preceded him, to consider movements other than uniform, circular movements. For now, though, we will continue with the outline of Kepler's progress.

Kepler eventually discovered that elliptical orbits, with the planets moving at varying speeds during the course of their orbits around the sun, could account perfectly for the data on Mars. In 1609 he published his model for the motion of Mars, with an elliptical orbit and with Mars moving at varying speeds, and soon afterward extended this approach to cover the remaining planets. We will now move on to look more carefully at Kepler's approach.

Kepler's System

Let's begin by looking at Kepler's key innovations – elliptical orbits and varying speeds – in more detail.

You may know that an ellipse is sort of an elongated circle. An ellipse can be given an exact mathematical characterization, but the easiest way to visualize an ellipse is to imagine two pins holding the ends of a string to a piece of paper. Now imagine pulling the string taut with a pencil, and moving the pencil around the pins, keeping the point of the pencil on the paper. The resulting figure will be an ellipse. Figure 16.1 may help. The points occupied by the pins are called the *foci*

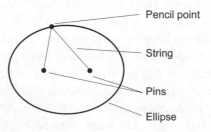

Figure 16.1   An ellipse

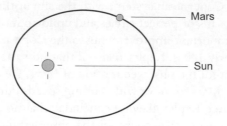

Figure 16.2   Orbit of Mars on Kepler's system

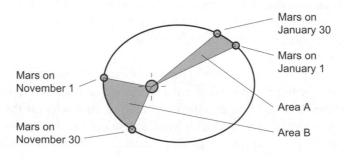

Figure 16.3   Illustration of Kepler's second law

of the ellipse. As mentioned, Kepler's first innovation was to have the planets move about the sun in elliptical orbits, with the sun occupying one of the foci of the ellipse. For example, the orbit of Mars would be as shown in Figure 16.2. In the diagram, the ellipse is greatly exaggerated for the purposes of illustration.

This description of planetary orbits – that planets orbit the sun in elliptical orbits, with the sun occupying one of the foci of the ellipse – is known as *Kepler's first law of planetary motion.*

Kepler's other major innovation was to have the planets move at varying speeds during the course of their orbits around the sun. More specifically, on Kepler's system, a line drawn from the planet to the sun will sweep out equal areas in equal time. This description of the speed of planets is known as *Kepler's second law of planetary motion*, and it is most easily illustrated by the diagram in Figure 16.3. To understand Kepler's second law, suppose there is a line connecting Mars and the sun. In the 30 days between January 1 and January 30, that line will sweep out a certain area (area A in the diagram). According to Kepler's second law, during any other 30-day period, that line will sweep out an area of equal size. For example, in the 30 days between November 1 and November 30, the line will sweep out a certain area (area B in the diagram). According to Kepler's second law, area A and area B will be equal. And in general, a line connecting a planet to the sun will sweep out equal areas in equal times.

Kepler's second law has an important implication for the motion of planets. Since a planet, for example Mars, is closer to the sun at certain points in its orbit (in this illustration, Mars is closest on the left side on the diagram), Mars must move faster during this part of its orbit, and move more slowly when it is in the part of its orbit furthest from the sun. In other words, given Kepler's second law, the motion of planets is not uniform. Rather, the speed of a planet varies at different places in its orbit.

Kepler's system, with its use of elliptical orbits and nonuniform planetary motion, was able to predict and explain the empirical facts perfectly. Moreover, it is far simpler than either the Ptolemaic or the Copernican system. With respect to simplicity, notice that Kepler's system uses no epicycles, deferents, eccentrics, equant points, or the like. Instead, there is simply one elliptical orbit for each planet, and that is all.

But note also that Kepler's system abandons both the perfect circle fact and the uniform motion fact. Recall that these two facts were central beliefs for over 2,000 years. So although Kepler's system handled the empirical data wonderfully, it would require substantial conceptual changes in the Aristotelian worldview.

What Motivated Kepler?

From what we have discussed so far, one might get the impression that Kepler was a reasonably straightforward investigator, motivated mainly by a desire to develop theories to handle the empirical facts. Kepler was actually a much more complex figure than this. As with our earlier discussion concerning Copernicus, I will not attempt a full exploration of the various factors involved in Kepler's development of his system, but instead provide enough information to get a flavor of the sort of philosophical/conceptual issues involved in Kepler's discoveries. In particular, we will focus on an aspect of Kepler that motivated him throughout his life, namely, his desire to read the mind of God.

Kepler's desire to read the mind of God

Throughout his life, Kepler was convinced God had a definite plan, a blueprint, so to speak, in constructing the universe. And Kepler was passionately driven to discover this blueprint, to read the mind of God and know the plan God put into place when the universe was created. Kepler's desire would manifest itself in a variety of ways, a few of which will suffice to illustrate.

Recall that when he was in his late twenties, Kepler went to work for Tycho Brahe. Why did Kepler want to work with Tycho? Much of the answer to this lies in a "discovery" Kepler had made a few years earlier, and exploring Kepler's discovery will help illustrate the sort of blueprint Kepler envisioned God as having, and what reading the mind of God would entail.

Kepler's first substantial publication came about four years before he went to work for Tycho, and in it Kepler announced a discovery he would consider important throughout his life. Kepler was concerned with questions such as why God created a universe with exactly six planets (Mercury, Venus, Earth, Mars, Jupiter, and Saturn), rather than, say, a universe with five or seven or some other number of planets. And why did God space the planets the way he did, rather than using some other spacing? Kepler was convinced there were answers to questions such as these.

For some years, Kepler had tried a variety of answers to such questions. For example, he tried answers that appealed to various mathematical ratios and functions, none of which provided a satisfactory answer. But in the mid-1590s, Kepler hit upon the idea of answering such questions by appeal to what are called the "perfect solids." A brief explanation of these solids is in order. Incidentally, bear with me through this discussion, as this will take a bit of time to explain. When we are finished, though, I think you will have a much better idea of the rather unusual person Kepler was.

Consider a cube, which is probably the clearest example of a perfect solid. A cube is a three-dimensional figure, where each face is identical; in particular, each face is a square. Note that a square is itself a two-dimensional figure, where again each component (that is, each side) is identical; in particular, each side is a line identical in length to all the other sides. And in general, a perfect solid has these same characteristics of the cube: each face of a perfect solid is an identical two-dimensional figure, and the two-dimensional figure is itself composed of identical components, that is, lines of equal length.

Since the ancient Greeks, it has been known that there are only five perfect solids. These five are the following: (a) a cube, with 6 faces, each face being a square; (b) a tetrahedron, with 4 faces, each face being an equilateral triangle; (c) an octahedron, with 8 faces, each face being an equilateral triangle; (d) a dodecahedron, with 12 faces, each face being a pentagon; and (e) an icosahedron, with 20 faces, each face being an equilateral triangle.

Now suppose we take a sphere of arbitrary size, and inside it we just fit a cube. That is, we put a cube that is just the right size so that the corners of the cube just touch the sides of the sphere. Then suppose that, inside this cube, we just fit another sphere, that is, we put in a sphere that is just the right size so that the sides of the sphere just touch the sides of the cube. Although we are dealing with three-dimensional figures, a two-dimensional drawing of this would look like Figure 16.4.

Now suppose we continue to nest solids and spheres, in the following order. We next place a tetrahedron of just the right size inside the innermost sphere in Figure 16.4, then another sphere inside this, then a dodecahedron, then another sphere, then an icosahedron, then another sphere, then an octahedron, and finally one last sphere. The resulting construction will look like Figure 16.5 (again, this

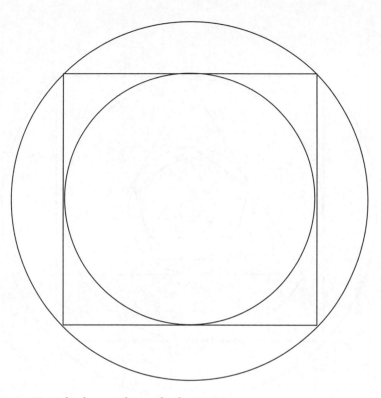

Figure 16.4   Nested sphere, cube, and sphere

is a two-dimensional representation of what is really a three-dimensional construction, but it will give a good idea of the result).

Again, bear with me as we make some final observations about this construction. Let's make some notes about the spheres. Note that the size of the sphere we begin with will determine the size of the cube we place in it, which in turn determines the size of the sphere we put in the cube, and so on. That is, the size of the original sphere will determine the size of all the other spheres, and will likewise determine the actual distance between each sphere.

But although the actual distance between each sphere depends on the choice of the size of the first sphere, the *relative* distances between the spheres do not. That is, regardless of what size sphere we begin with, the relative distances between the spheres remain the same. This is perhaps easier to see if we look at a drawing of just the spheres. That is, Figure 16.6 is just like Figure 16.5, except that the five solids have been removed to make the spacing of the spheres more obvious. Again, no matter what size sphere we begin with when we build this construction of spheres and perfect solids, the relative distances between the spheres will be as illustrated in Figure 16.6.

Now, at this point a good question is: What does any of this have to do with astronomy? The answer is this: On the Copernican (or any other sun-centered

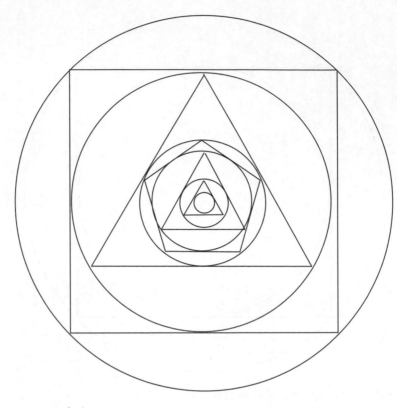

Figure 16.5   Kepler's construction

system), it is possible to figure out the relative distances between the planets. And as it turns out, the relative distances of the planets are rather close to the relative distances between the spheres in Kepler's construction.

This is an interesting fact, and there is little doubt in my mind that this fact is merely a curious coincidence about our solar system. But not so in Kepler's mind. To Kepler, this was his first big breakthrough in reading the mind of God. The construction in Figure 16.5 is what God had in mind in constructing the universe – God wanted to model this construction, reflecting relationships between spheres and the perfect geometric solids. This is why God created the universe with six planets – one for each sphere in the above construction – rather than with five or seven or some other number of planets. This is why God created the universe spacing the planets the way he did – the spacing of the planets reflects the spacing of the spheres in Kepler's construction. In addition, every corner of the outermost and hence most prominent of the solids, the cube, has three lines going off at right angles to one another, and in this God was reflecting the three dimensions of space in the universe. And so on.

As mentioned, this was Kepler's first major "discovery," and it was, in fact, one of the primary reasons he wished to work for Tycho Brahe. That is, he wanted

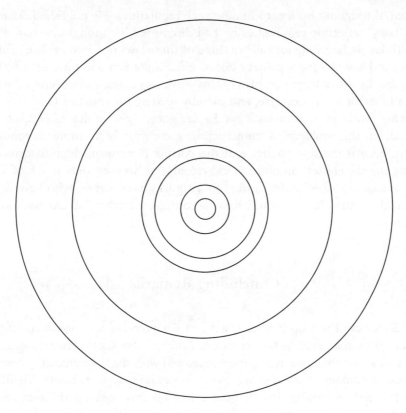

Figure 16.6 Kepler's construction with the solids removed

to work with the best observer of the time, partly to help confirm his discovery. This discovery was announced in his first major publication. But his desire to read the mind of God, and his conviction that this system of nested spheres was a key component in God's blueprint, was a lifelong commitment for Kepler. For example, much later, near the end of his life, Kepler would expand on the approach described above, incorporating into it musical harmonies such that God, in constructing the universe, was reflecting not merely geometrical structures, but musical ones as well. In short, Kepler's construction with the solids, and his desire to read the mind of God, did not merely reflect the youthful Kepler. As Thomas Kuhn put it, in a turn of phrase I rather like, Kepler's construction with the perfect solids "was not simply a youthful extravagance, or if it was, he never grew up" (Kuhn 1957, p. 218).

Kepler's construction with the perfect solids did not directly lead him to the discoveries for which he is best remembered, namely the elliptical orbits with varying speeds. But his passion for discovering the regularities God included in constructing the universe – the kind of passion reflected in his lifelong commitment to the construction involving the solids – did lead to the key discoveries that we remember as Kepler's first and second laws of planetary motion. Kepler was devoted, throughout his life, to finding such regularities. He published dozens of such "laws" reflecting regularities he had discovered, or thought he had discovered. Today we largely ignore all but three of these laws (the two explained above, plus a third law relating a planet's distance from the sun and time of orbit), but to Kepler, this was a large part of what his work was about – discovering the regularities inherent in the universe, and thereby reading the mind of God.

Science works in mysterious ways. Kepler got it right, in that after 2,000 years of work on the problem of constructing a completely accurate astronomical system, he was the first to arrive at the correct picture of elliptical orbits and varying speeds. He was an unusual individual. But his approach, much of which (such as that described above) strikes us as quirky, was part of who he was. And it is unlikely that Kepler could have accomplished what he did without the quirkiness.

Concluding Remarks

After discussing the Copernican system in Chapter 14, we took a moment to consider how that system was received. And we saw that, generally speaking, almost all astronomers soon became acquainted with the Copernican system, and that quite a number used the system, with most taking an instrumentalist attitude toward it. Before closing this chapter, a few observations on the reception of Kepler's system are in order.

The reception of Kepler's system is somewhat less straightforward to discuss than the reception of Copernicus' system. This is due partly to the fact that a number of astronomers attempted to replicate Kepler's success, but using perfect circles and uniform motion. That is, these astronomers recognized Kepler's accomplishment in producing a system that would account for the empirical data far better than any existing system, but wrongly believed they could use Kepler's work to gain new insight into how to modify existing systems of perfect circles and uniform motion, and thereby achieve the same level of accuracy as Kepler's system. In a sense, then, such astronomers were acknowledging Kepler's success, while not fully accepting his methods.

A second complicating factor concerns the timing of Kepler's work. Kepler published his system (at least, his work with the motion of Mars) in 1609. Astronomical issues were generally issues for the specialists – generally, mathematically trained astronomers – and not so much issues for public debate. The very next year, however, Galileo published his discoveries with the telescope. Galileo's discoveries are discussed in more detail in the next chapter, but at this point, suffice it to say that Galileo's discoveries were easily accessible to a much wider audience. Moreover, they were very exciting discoveries, and somewhat overshadowed Kepler's work, and certainly Galileo's work gained a wider audience than did Kepler's.

One last complicating issue is worth mentioning. Shortly after Galileo published his discoveries with the telescope, the Catholic church took a formal stand against the sun-centered view, and restricted the sorts of discussions and writings allowed on the subject. Kepler's 1609 publication mentioned above, concerning the motion of Mars, as well as some of his later writings, were placed on the *Index* of forbidden books (essentially a list of publications that Catholics were forbidden to read). Given such developments, some who had planned to write on the Earth-centered versus sun-centered debate decided to withhold publishing anything on the subject. As a result, Kepler was publishing his most important discoveries at a time when there was less open debate and discussion concerning the alternative systems.

So although it is more difficult to get a clear sense of the reception of Kepler's views, it is clear that eventually, the advantages of Kepler's system – its simplicity together with the fact that it handled the empirical data far better – coupled with Galileo's new evidence from the telescope supporting a sun-centered over an Earth-centered approach, became widely recognized. In addition, toward the end of his life, in the late 1620s, Kepler produced a set of astronomical tables, based on his system, that were far superior to any produced using any competing system. As a result, by the mid-1600s it was clear to all who followed these issues that the planets, including the Earth, did indeed move in elliptical orbits, at varying speeds, around the sun. The philosophical/conceptual "facts" of uniform, perfectly circular motion were, by the mid-1600s, finally recognized not to be facts at all.

Chapter Seventeen

Galileo and the Evidence
from the Telescope

Another central player in the transition from an Earth-centered view to a sun-centered view was Galileo (1564–1642). Galileo contributed in important ways to astronomy, physics, and mathematics, but it is his work affecting astronomy that most concerns us here. The main goals of this chapter will be to understand the new data available from the telescope, to see how that data impacts the debate between advocates of the various astronomical systems, and to explore the reception of Galileo's discoveries.

As we will see, Galileo's work with the telescope provided, for the first time, new empirical data relevant to the Earth-centered versus sun-centered debate. However, we will also see that the new evidence did not, by itself, settle the issue. Galileo thought the new evidence supported the sun-centered view, but others, equally familiar with the evidence, did not. As usual, we will begin with some background material.

Background Information

Galileo and the Catholic church

The telescope was invented just before 1600, and Galileo began using it for astronomical observations in 1609. Galileo was one of the first to use the telescope for astronomical observations, and in doing so he discovered interesting new data that very much influenced the debate between advocates of the Earth-centered and sun-centered systems. He published his first set of findings in 1610, and published additional findings over the next several years.

Worldviews: An Introduction to the History and Philosophy of Science.   Richard DeWitt
© 2010 Richard DeWitt

The new data would eventually involve Galileo in a well-known dispute with the Catholic church. Given this, a few words on the religious situation during Galileo's time are in order.

Although we have not discussed this at any length, it will come as no surprise that the church preferred the Earth-centered view. One of the reasons for this (but not the only reason) concerns various passages in the Christian scriptures that suggest the Earth is stationary and that the sun moves around the Earth. Thus, the dispute between Galileo and the church would necessarily involve scriptural interpretation.

Also worth noting is that the Catholic church, by and large, had a history of being tolerant of new scientific views. For example, for the most part the church was not opposed to the Copernican system. Of course, up until the new evidence from the telescope, the Copernican system was generally taken with an instrumentalist attitude, and as such was not contrary to scripture. But the point is that the church did not generally oppose new scientific views, and was generally willing to reinterpret scripture when required by new discoveries.

This was, however, a rather touchy time for the Catholic church. The Protestant Reformation had begun the previous century, and the church was actively engaged in trying to suppress the spread of what it considered heretical views. So Galileo's work with the telescope came at a time in which the church was not as tolerant as it might have otherwise been.

It is worth mentioning that Galileo himself was a devout Catholic. He certainly had no desire to undermine the church, nor would he take lightly the eventual concerns that some of his views might be heretical. As we will see below, Galileo had genuine differences of opinion involving the interpretation of scripture, and these differences would play a role in his dealings with the church.

A note on the nature of the evidence from the telescope

With respect to the debate between advocates of the Earth-centered and sun-centered systems, it is important to keep in mind that there is no data available from naked-eye observations that can settle the debate. In fact, as emphasized all along, naked-eye observations support an Earth-centered view.

Notably, even with the telescope there is no way to tell directly whether an Earth-centered or sun-centered view is correct. I want to take a moment to discuss this, because I think it is a point that is often misunderstood, and because appreciating this point will help one get a better sense of the nature of the evidence provided by the telescope.

Suppose we leave Galileo for a moment, and jump ahead 400 years to the present time. Even today, with all the technological advances of the past 400 years, we have no technology that directly shows whether the Earth moves around the sun, or whether the sun moves around the Earth. Our most direct evidence that the Earth moves about the sun is the stellar parallax that was finally documented

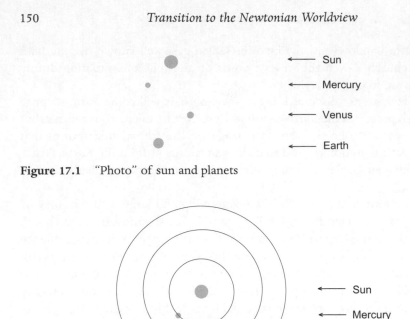

Figure 17.1   "Photo" of sun and planets

Figure 17.2   Sun-centered interpretation of "photo"

for the first time in the 1800s. But the evidence from stellar parallax is not direct observational evidence, in the sense we discussed in Chapter 3.

Not even photographs from space directly settle the issue of whether the Earth moves around the sun. To see this, suppose we had a photograph, such as Figure 17.1 showing the sun, Mercury, Venus, and the Earth. Incidentally, we have no such photograph. Such a photograph would have to be taken from a vantage point above or below the axis of the Earth, but as a matter of fact, we do not send spacecraft in that direction. The interesting features of our solar system – the planets, asteroids, and so on – tend to lie more or less on a plane that cuts roughly through the equator of the Earth. So our spacecraft are generally sent out along that plane, and not in the direction of the axis of the Earth. But the point is that, even if we did have such a photograph, it would not show whether the Earth or sun is the center of our solar system.

To see this, notice that the "photograph" in Figure 17.1 is equally compatible with either an Earth-centered or a sun-centered system. That is, the photograph is compatible with the sun-centered view illustrated in Figure 17.2. But the photograph is equally compatible with the Earth-centered view illustrated by Figure 17.3.

In short, even if we could take such photographs of our solar system, they would not show whether the Earth or sun was the center. Even if we took, say, long-running videotapes, so as to plot the position of the sun and planets over the course of time, such videotapes would show only the relative motions of the

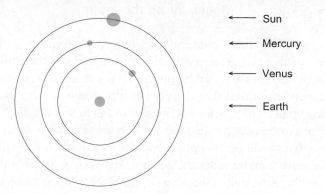

Figure 17.3 Earth-centered interpretation of "photo"

sun and planets. That is, such videotapes would show only the way in which the sun and planets move relative to one another. But the relative motions of the sun and planets on an Earth-centered Tychonic system are the same as the relative motions of the sun and planets on a sun-centered system. In other words, even if we had such videotapes, they would turn out to be compatible with an Earth-centered Tychonic system. (I should note that such a videotape would not be compatible with the original Tychonic system, with its epicycles, perfect circles, and uniform motion, but rather with a modified and "modernized" Tychonic system. Such a modified Tychonic system is similar to what is shown in Figure 15.1, but incorporates elliptical orbits with planets moving at varying speeds. It is this modified Tychonic system that is preferred by the modern-day advocates of an Earth-centered system mentioned at the end of Chapter 15.)

I am going on at some length on this issue, but it illustrates a general, and important, point: the evidence we gain from our technology is rarely as direct as it is often assumed to be. This is an important point to keep in mind as we consider Galileo's evidence from the telescope.

So Galileo's evidence from the telescope, although fascinating and important, does not directly settle the debate between advocates of the sun-centered and Earth-centered systems. But the evidence does provide a range of indirect evidence that certainly impacts the debate. Let's now move on to consider Galileo's evidence.

Galileo's Evidence from the Telescope

Galileo reported a variety of new observations gathered from his use of the telescope. Some of this data raised only relatively minor issues for Earth-centered views, whereas other data raised more substantial issues. In each case, we will discuss not only what Galileo reported seeing, but also how the new data influences the issue of a sun-centered versus an Earth-centered view.

Mountains on the moon

Galileo was one of the first to use the telescope to observe some of the features of the moon's surface, including mountains, plains, and what we now know to be craters. To some extent these can be seen with the naked eye, and others before Galileo had even speculated that there were mountains on the moon, but only with the telescope can one make out these features in any detail.

The fact that the moon has features such as mountains does not in any way directly show that the Earth moves around the sun. Rather, this fact figures into the debate because it undermines the general Aristotelian picture of the universe. Recall that on the Aristotelian worldview, the objects in the heavens were composed only of ether, and this fact figured into the general Aristotelian explanation of the movements of the heavenly bodies. So the fact that the moon appears to be a large rocky body, similar in appearance in many ways to the Earth, shows pretty clearly that this Aristotelian belief cannot be correct.

It is worth noting that this evidence, by itself, would never have been sufficient to seriously undermine the Aristotelian worldview. While it is true that the Aristotelian worldview included the belief that heavenly bodies were composed of ether, that particular belief – that part of the jigsaw puzzle – could have been modified without seriously changing the overall set of beliefs. For example, the moon is on the border of the sublunar and superlunar regions, so it would not be unreasonable to envision the moon containing both sublunar and superlunar elements. To put it another way, the Aristotelian beliefs about the moon are not core beliefs. But without question, the existence of features such as mountains on the moon shows that the Aristotelian worldview could not remain unchanged in the face of the new evidence from the telescope.

So the existence of mountains on the moon largely factored into the debate by showing there were some cracks in the Aristotelian worldview. But there was another way this data helped make the sun-centered view more plausible. Recall the argument for a stationary Earth (discussed in Chapter 10) that was based on the idea that nothing could keep the Earth in motion. Again, the Earth is a large, rocky body, and, much like the large boulder in my front yard, it will remain stationary unless something continually moves it. The argument seems compelling enough. But with the telescope we can now see that the moon appears to be a large rocky body, and it clearly is in continual motion. So if the large rocky moon can be in continual motion about the Earth, perhaps the large rocky Earth could be in continual motion about the sun.

Sunspots

Galileo was also one of the first to use the telescope to observe sunspots, that is, dark regions that can be seen when looking at the sun. One cannot observe the sun directly through a telescope, as this would make short work of one's retina, but the telescopic image of the sun can be observed by projecting it on a piece of paper.

Galileo used this method to observe sunspots. Using his observations, Galileo was able to argue convincingly that such spots must be regions on the sun itself, rather than, for example, being the images of small planets moving in front of the sun.

As with the mountains on the moon, this data also does not provide direct evidence for a sun-centered view. But the sun is clearly in the superlunar region (not merely bordering the region, as with the moon). So if sunspots are on the sun, as Galileo convincingly argued, then the superlunar region must not be the region of unchanging perfection it was believed to be on the Aristotelian worldview. So as with the mountains on the moon, this data proves to be another crack in the Aristotelian worldview.

The rings, or "ears," of Saturn

Galileo's evidence concerning Saturn had a similar consequence to that of the evidence concerning the moon and sun. Galileo was the first to observe that, at times, Saturn has bulges on the side, appearing to be something like handles or ears. We now know that Galileo was observing the rings of Saturn, although the resolving power of his telescope was not sufficient to see the rings as anything but bulges on the sides of Saturn. (It would be about another half-century before the bulges were correctly hypothesized to be a ring structure around Saturn.)

Once again, this data provides another small crack in the Aristotelian worldview. Recall that the heavenly bodies were composed of ether, which naturally has a perfectly spherical shape. The planets, then, since they are composed of ether, must be perfectly spherical. Galileo's observations showed that Saturn, along with the moon and sun, did not fit the expectations of the Aristotelian worldview.

The moons of Jupiter

Of all the phenomena capable of being viewed with Galileo's telescope, the moons of Jupiter were probably the most enjoyable to observe. Through a telescope, Galileo was able to observe four small points of light, which vary their position around Jupiter over time, and which Galileo correctly reasoned were moons orbiting Jupiter. Even today, the moons of Jupiter are perhaps the most pleasant sight available through a small telescope (as enjoyable to observe as the rings of Saturn).

In a wise career move, Galileo named the moons of Jupiter the "Medicean Stars," in honor of the Medici family (one of the most powerful families of Italy). Galileo hoped to become a member of the Medicean court, and he would soon be successful, shortly afterwards being named the chief mathematician and philosopher ("philosopher" more in the sense of what we would call a scientist) of the Medicean court.

Galileo spent a great deal of time carefully observing and plotting the position of the moons, and was able to establish that they were indeed bodies moving about Jupiter. This too was evidence that did not fit easily into the Aristotelian worldview. Recall that, on the Aristotelian worldview, and in particular on the Ptolemaic system, the Earth is the single center of all rotation in the universe. All heavenly bodies – the moon, sun, stars, and planets – move in circles about the center of the universe, that being the center of the Earth. But Galileo's discovery of bodies orbiting Jupiter shows conclusively that, contrary to the Aristotelian belief, there is not only one single center of rotation in the universe.

As a corollary, it is worth mentioning that supporters of the Earth-centered view had argued against the sun-centered view on the grounds that the movement of the Earth's moon was rather awkward. That is, it is somewhat inelegant to have a body moving around the Earth, and then have the Earth moving around the sun. But Galileo's discovery of the moons of Jupiter puts this argument to rest, since even the advocates of the Earth-centered view would have to accept that at least one body – Jupiter – has bodies moving about it while it itself moves.

The phases of Venus

The phases of Venus provide some of the most dramatic evidence relevant to the debate. With the naked eye, one cannot observe the fact that Venus, like our moon, goes through a full range of phases. But with a telescope, the phases of Venus are easily observable, and Galileo was the first to discover them. Moreover, not only does Venus go through a full range of phases, but it differs in size depending on what phase it is in. Figure 17.4 illustrates Venus as it appears in its full, three-quarter, half, quarter, and crescent phases. To understand the importance of this data, we need to understand why Venus appears in different phases at different times. Since the explanation will be essentially the same as the explanation for why the moon goes through phases, let us first discuss the moon, and then turn to Venus.

The phases of the moon are a consequence of the relative positions of the sun, moon, and Earth. At any given time, one half of the moon will be illuminated by the sun, and the other half will be dark. When the moon and Earth are positioned such that we see the full illuminated side, then we have a full moon. When we see only half the illuminated surface, we have a half-moon, and when

Figure 17.4   Phases of Venus

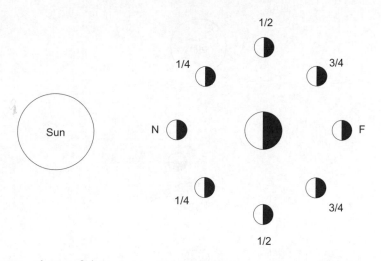

Figure 17.5 Phases of the moon

we see only a small portion of the illuminated half, then we have a crescent moon. Figure 17.5 may help. The various moons labeled ¼, ¾, and so on, represent the various positions, relative to the Earth and sun, that the moon occupies during its approximately 27-day orbit around the Earth. Incidentally, the figure is not at all drawn to scale. For example, from the figure one might get the impression that the moon is often in the Earth's shadow, but that is an artifact of having to condense the figure so it will fit on a page, and of having to put the figure on a two-dimensional sheet of paper. In fact, because the sun is large enough, and the moon and Earth far enough apart, and the orbit of the moon is "tilted," the moon is only occasionally in the shadow of the Earth (and this is why lunar eclipses occur relatively rarely, rather than occurring on every revolution of the moon about the Earth).

When the moon is at the position labeled F, with the illuminated half facing the Earth, we will see a full moon. When the moon is at the point labeled ¾, we see a three-quarter moon, and so on for the half-moon, quarter (or crescent) moon, and for the new moon, where we see no moon in the night sky (the position labeled N).

If Venus goes through a range of phases, as Galileo discovered it does, then like the moon, the phases must be the result of the relative positions of the sun, Earth, and Venus. Importantly, a sun-centered system (either that of Copernicus or of Kepler) leads to a very different prediction concerning the phases of Venus than does the Ptolemaic system. In particular, on a sun-centered view we would expect Venus to go through a full range of phases. In contrast, if the Ptolemaic system is correct, then Venus should at most be seen as a crescent, but never in a half, three-quarter, or full phase.

These differing predictions are best illustrated through diagrams. Let's first consider the Ptolemaic picture of the Earth, sun, and Venus, shown in Figure 17.6.

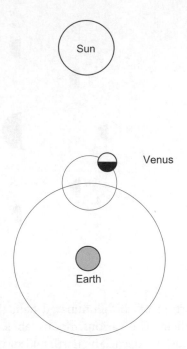

Figure 17.6   Sun, Venus, and Earth on the Ptolemaic system

Here is where a key empirical fact, discussed at the end of Chapter 11, comes into play. Recall that, as a matter of fact, Venus never appears far from the sun. That is, wherever the sun is in the sky, Venus will never be far from it. Again, that is why Venus can be seen only just after sunset (at certain times of the year), or just before sunrise (at other times of the year). The remainder of the day and night Venus cannot be seen, either because (at night) it is below the horizon along with the sun, or else it is in the daytime sky near the sun, and the light from the sun prevents us from seeing it.

On the Ptolemaic system, there is one and only one way to explain this fact, and that is to have the sun and Venus complete a revolution about the Earth in the same amount of time (or more precisely, the sun and Venus' epicycle complete a revolution in the same amount of time). In other words, the Earth, sun, and Venus' epicycle must always be arranged in a line, as they are shown in Figure 17.6.

But notice this entails that the illuminated half of Venus is always facing away from the Earth. And so, like the moon when its illuminated half is facing away from the Earth, Venus will appear (at most) as a crescent. In other words, on the Ptolemaic system, we could at most see only a small portion of the illuminated half of Venus. We could never see a full Venus, or a three-quarter Venus, or a half-Venus. All these would require configurations that are impossible on the Ptolemaic system.

Galileo's discovery of the phases of Venus, then, provides straightforward disconfirming evidence for the Ptolemaic system. In contrast, as explained below, on

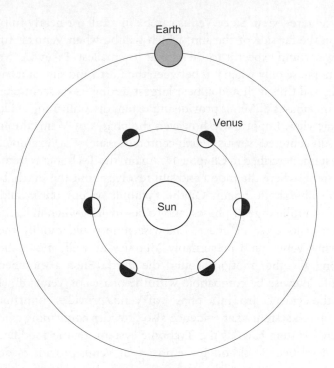

Figure 17.7 Sun, Earth, and Venus on a sun-centered system

a sun-centered view one would expect Venus to go through a full range of phases, and so the phases of Venus provide confirming evidence for the sun-centered view.

Before explaining how a sun-centered view accounts for the phases of Venus, note first that, on a sun-centered view, the empirical fact that Venus always appears in the sky close to the sun is explained by the fact that Venus is an inner planet. That is, Venus is closer to the sun than the Earth is. In Figure 17.7, note that from the Earth, no matter where Venus is in its orbit, it can never appear to be very far away from the sun. Moreover, since on a sun-centered view Venus completes a revolution around the sun in a shorter time than the Earth does (225 days compared to the Earth's 365), Venus will at some times be on the far side of the sun from us, and thus appear as a full Venus, sometimes to the side of the sun and so appear as a half-Venus, sometimes between us and the sun and so appear not at all or as a crescent, and so on. In short, on a sun-centered view we would expect Venus to go through a full range of phases, and so Galileo's discovery provides confirming evidence for the sun-centered view.

Not only does the sun-centered view correctly predict that Venus should go through a full range of phases, but in addition, it quite naturally accounts for the correlation between the phases of Venus and the size Venus appears to be.

Notice in Figure 17.4 that Venus appears smallest when it is in its full phase, and largest when it is in its crescent phase. This is exactly what we would expect from a sun-centered view. Since Venus can be in a full (or nearly full) phase only when it is on the far side of the sun, which will be when Venus is furthest from the Earth, we would expect it to then be at its smallest. Likewise, Venus can be in its crescent phase only when it is between the Earth and sun, at its nearest point to the Earth, and thus it should appear largest during its crescent phase.

In short, the phases of Venus provide important disconfirming evidence against the Ptolemaic view. Importantly, however, the phases of Venus do not suffice to settle the sun-centered versus Earth-centered issue. For example, recall the Tychonic system, described in Chapter 15. Again, the Tychonic system is an Earth-centered system, where the moon and sun revolve about the Earth, but in which the planets revolve about the sun. On the Tychonic system, one would also expect Venus to show a full range of phases, and to be smallest when in its full phase and largest when in its crescent phase. Likewise, one could modify the Ptolemaic system, having Venus (and presumably Mercury as well) move about the sun while keeping all other motion around the Earth. Such a modified Ptolemaic system would likewise be compatible with the phases of Venus. In short, to the extent that the evidence from the phases of Venus provides confirming evidence for a sun-centered system, that evidence also provides confirming evidence for an Earth-centered system such as the Tychonic system, or the modified Ptolemaic system just described. So although the phases of Venus provide evidence against the original Ptolemaic system, they do not settle the issue between a sun-centered approach and an Earth-centered approach.

As a brief point, note how this example nicely illustrates the underdetermination of theories discussed in Chapter 5. That is, even new evidence as dramatic as the discovery of the phases of Venus turns out to be perfectly compatible with both a sun-centered approach (including both Copernicus' and Kepler's theories) as well as an Earth-centered approach such as the Tychonic system, or even a modified sort of Ptolemaic system as described above. This is typical in science – new evidence, even dramatic new evidence, is typically compatible with two or more competing theories. In other words, the available evidence usually does not uniquely determine a particular theory as being correct.

Finally, it is worth noting that, although the phases of Venus do not settle the issue, they nonetheless require that substantial changes be made. That is, the Ptolemaic system had been the default system for the past 1,500 years, and now that system had to be replaced. So whether one switches allegiance to a sun-centered view, or to the Tychonic system, or to a modified Ptolemaic system, one is required to make substantial changes in one's beliefs about the structure of the universe.

The stars

One final discovery is worth discussing, even if only briefly. With the telescope, Galileo discovered there were countless more stars than those visible with the naked eye. This at least suggests the possibility that the universe is much larger than previously suspected, and even perhaps that the universe is infinite, with an infinite number of stars. Galileo himself did not advocate this view, but, in the decades to follow, the idea of a huge, perhaps infinite universe took hold, and Galileo's discovery of the vastly greater number of stars would fit into this new view of the universe.

The Reception of Galileo's Discoveries

Galileo's discoveries with the telescope were understandably viewed as exciting new discoveries, and they made Galileo one of the best-known scientists of the time. He published most of his discoveries with the telescope between 1610 and 1613, and in these publications we can see that Galileo is now thinking that the sun-centered view must be the correct model of the universe. Recall that, by this time, the sun-centered Copernican system had been used and taught for about 70 years, but that during this time the Copernican system was generally taken with an instrumentalist attitude. Galileo, now, is suggesting a realist attitude toward the sun-centered view.

The church (that is, the Catholic church, which had strong influence in the Catholic countries of Europe) had no problems with the sun-centered Copernican system, so long as that system was taken with an instrumentalist attitude. But it began to have substantial problems with the idea that the sun-centered view might really be the way the universe is constructed.

Galileo traveled to Rome in late 1615, to try to prevent the sun-centered view from being condemned by the church. By this time Galileo realized the evidence from the telescope alone would probably not be sufficient to convince those on the other side of the debate, and he developed another argument, based on the tides, to support the sun-centered view. This new argument was presented in a work he distributed in 1615. Incidentally, Galileo's account of the tides turned out to be wrong – he was essentially arguing that the oceans sloshed around due to the Earth's movement, much as water on the deck of a boat will slosh around due to the boat's movement.

In spite of Galileo's efforts, in early 1616 the view that the sun was stationary at the center of the universe was judged to be heretical, and the teaching or written defense of such a view was prohibited. Notably, however, the teaching of the sun-centered view was not prohibited outright. Rather, what was prohibited was teaching the reality of the sun-centered view. The sun-centered view could still be written about and taught as a "hypothesis" – that is, it could be treated with what we would term an instrumentalist attitude. Copernicus' own book on the subject, published back in 1543, could still be taught, but only after corrections were made so as to edit out the passages where Copernicus suggests the reality of the sun-centered view.

Given that Galileo was by this point taking a realist attitude toward the sun-centered view, and openly advocating that realist attitude, the church's judgment against the reality of the sun-centered view could have been much worse for Galileo. Neither Galileo himself, nor any of his writings, were mentioned in the church's formal judgment against the sun-centered view. However, Galileo was summoned to a meeting with Cardinal Bellarmine (Bellarmine was one of the leading church figures involved in the decision to judge as heretical the view that the sun was stationary and at the center of the universe). At that meeting, Galileo was instructed clearly and in writing that he was not to hold or to teach the reality of the sun-centered view. But at bottom, although the judgment in 1616 was bad news for Galileo, it could have been much worse.

What can we say about the church's judgment against the reality of the sun-centered system and, given the relevance of Galileo's evidence from the telescope, what can we say about the church's attitude toward that evidence? Was this a case of refusing even to consider such evidence? Was the church unwilling to accept any evidence against their view, no matter how compelling that evidence was, so that they were treating the Earth-centered view as unfalsifiable?

As is often the case, these questions are more complex than they at first appear. As noted, Bellarmine was one of the leading church figures involved in this decision, so let's focus on his views, and contrast them with Galileo's.

First, there was no issue about accepting Galileo's evidence from the telescope. Bellarmine was a competent astronomer, and he and other church astronomers, including the notable mathematician and astronomer Christopher Clavius, replicated Galileo's observations and certified them as accurate. The church astronomers not only verified Galileo's discoveries, but had high words of praise for them.

The issue was what Galileo's discoveries say about the Earth-centered versus sun-centered issue. As discussed above, Galileo clearly thinks the evidence from the telescope shows the sun-centered system is correct, in spite of what scripture says. Bellarmine, on the other hand, is convinced that what scripture says is correct, and he seems to think that no evidence could possibly run counter to this. So, is Bellarmine treating the Earth-centered view as unfalsifiable?

This is not an easy question. To explore this further, let's look a bit further into Galileo's and Bellarmine's views. As this issue was unfolding, Bellarmine and Galileo had both circulated letters in which they articulated their respective views. Bellarmine's views are best summarized in a relatively short piece of writing from 1615, generally referred to as the *Letter to Foscarini*. Galileo's views are most clearly articulated in portions of a rather longer piece, also circulated in 1615, generally referred to as the *Letter to the Grand Duchess Christina*. (Christina, by the way, was a prominent member of the Medici family, and it was important that he remain in good standing with the family by trying to reassure her and others that his views were not contrary to scripture or Catholic teaching.)

In the *Letter to the Grand Duchess*, Galileo makes it clear that he believes every word of the Bible is correct. But, he says, the Bible was written for everyone, including people who lived in far earlier, less advanced times, and people who have had little or no education. As a result, the Bible was written in such a way that its true meaning is often difficult to determine. So, Galileo argues, if we are dealing with empirical/scientific matters for which we can gather empirical/scientific evidence and proofs, we should never use the Bible to make a final judgment about such matters. First, Galileo says, such matters (for example, whether the sun moves around the earth or vice versa) are not matters relevant to salvation (that is, Galileo thinks that whatever you believe about this matter, your belief will not affect your salvation or lack of salvation). Second, it would be bad for the church to make a final pronouncement on an empirical matter, based on what the Bible says, only to have that pronouncement later be shown by empirical evidence to be definitely wrong. So as a general policy, Galileo is advocating not making any judgments about empirical matters based on scriptural passages.

Bellarmine, in the *Letter to Foscarina*, makes clear his disagreement with Galileo on these points. Bellarmine begins by noting that, with respect to the key question here (whether the sun moves around the earth or vice versa), the relevant biblical passages seem quite clear. Moreover, Bellarmine points out that there is no disagreement about how to interpret the relevant portions of the Bible; for example, Bellarmine notes that all the writers about these biblical passages agree that the Bible says, clearly, that the sun moves about the earth. So contrary to what Galileo suggests, this does not seem to be a situation involving difficult scriptural interpretation.

Bellarmine also clearly rejects Galileo's claim that these are not matters relevant to salvation. Bellarmine grants that, in general, matters involving scientific questions may not be relevant to salvation. But in this case it is relevant to salvation, because the Bible says the sun moves around the earth, and so one cannot reject this belief without rejecting the authority of the Bible, and doing so would be to reject something that God has spoken. So this, on Bellarmine's view, makes the sun-centered versus Earth-centered view relevant to salvation.

Finally, it is worth noting that in the letter, Bellarmine is clear that if it could be demonstrated that the Earth moves around the sun, then we would have to accept that demonstration. But at the same time (presumably for the reasons outlined above), Bellarmine indicates that he thinks such a demonstration would not and could not be forthcoming. Nonetheless, Bellarmine does at least consider the possibility of such a demonstration, and notes that, should such a demonstration be made, church leaders would need to carefully consider how it was they had so badly misunderstood the scriptures on this point.

Note that Galileo and Bellarmine agree on a number of issues. Both accept the data from the telescope, and both accept the authority of scripture. Both agree that scripture suggests the sun moves about the Earth, and both agree that the data from the telescope suggests otherwise.

However, Galileo and Bellarmine differ on how to weigh various pieces of evidence. Galileo's view is that scripture is undeniably correct when it comes to matters involving salvation. But on other matters, those not involving salvation, the scriptures need not be taken as undeniably correct. And since, on Galileo's view, whether the Earth or sun is the center of the universe is not a matter relevant to salvation, this is a situation where the evidence from the telescope can trump the evidence from scripture.

In contrast, Bellarmine's view is that all aspects of scripture are undeniably correct. We can, Bellarmine believes, misunderstand scripture, but in such cases it is our misunderstanding that is at fault, not the scriptures. Since it is unlikely we are misunderstanding the scriptural passages that speak of a stationary Earth and moving sun, in this dispute the scriptural evidence trumps the evidence from the telescope. In short, I think it is fair to say that, if asked, Bellarmine would have agreed that he was willing to give up the Earth-centered view if provided with sufficient evidence. But it is clear that Bellarmine has a different notion than Galileo on which evidence counts the most, and given the evidence Bellarmine favors, he thinks it extremely unlikely, perhaps impossible, that such evidence will ever be forthcoming.

This situation is exactly what we discussed earlier, when we first explored the issue of falsifiability back in Chapter 7. As we saw then, issues involving falsifiability often come down to the issue of what evidence is weighed most heavily. And the issue of what evidence is weighed most heavily is usually a matter of one's overall view. Bellarmine was quite respectful of scientific discoveries, but he was first and foremost a church leader, and for him scriptural evidence trumps scientific evidence. Galileo, on the other hand, was very respectful of religious matters, but he was first and foremost a scientist, and for him evidence gathered from new scientific discoveries trumps religious evidence.

So where does this leave us with respect to the question of whether Bellarmine was treating the Earth-centered view as unfalsifiable? I think that if Bellarmine were advocating these views today, following 400 years of extraordinarily successful and productive science based on respect for empirical evidence, then his attitude would be as unreasonable and unfalsifiable as Steve's views (from Chapter 7) are. But in the context of the early 1600s, there was no good reason to think that the sort of empirically based approach Galileo was advocating would be as successful as it has been. So I think the only fair answer is that there simply is not a clear yes or no response to the question of whether Bellarmine was treating the Earth-centered view as unfalsifiable. As we investigate these sorts of cases, we find they are far more complex than they at first appear. And this complexity is, I think, much of what makes the history and philosophy of science so interesting.

Concluding Remarks

As noted above, Galileo came through the church's 1616 judgment against the reality of the sun-centered view relatively unscathed. Some years later he was less fortunate. In early 1632 Galileo published *Dialogues Concerning the Two Chief World Systems*, which was a substantial book discussing the arguments for and against the Earth- and sun-centered systems. Recall that merely discussing the sun-centered system was not prohibited; only advocating the reality of it was.

This book was not well received by the church, and the view of the church was that it crossed the line between discussion of the sun-centered view, and advocacy of that view. There were a number of issues that complicated this situation. The book is written as a dialogue, and so technically Galileo could claim that it was his characters in the dialogue, rather than he himself, that were advocating various positions. But this really convinced no one. The book is pretty clearly advocating the reality of the sun-centered view. And given the meeting mentioned earlier (including written documentation) in 1615 with Bellarmine, in which he was told he could not teach or hold the reality of the sun-centered view, there is not much question that Galileo did cross the line. However, as another factor to complicate the situation, the book had gone through the church's standard review process, and had been approved for publication. Also relevant is the fact that, over the years, Galileo had managed to offend a number of influential people. In his writing, he could at times be extremely sarcastic and unflattering toward various individuals, and he managed to accumulate a number of enemies, some powerful, who disliked him. Moreover, Galileo seemed at times to be rather politically tone-deaf. Then as now, there are always various political realities that need to be recognized. Today, for example, it would not be politically wise to put forth a grant request to, say, the National Science Foundation, and begin the request by insulting the panel that will judge the request. For Galileo, it would not have been politically wise, say, to publish a book with sections that offended the Pope. Yet his 1632 book did seem to offend the Pope. At the very least, there were those who tried to convince the Pope that he should be offended, and this surely did not help the situation.

The details of Galileo's trial are complex and in some respects controversial, but the eventual outcome was that his book was prohibited by the church, Galileo was judged as suspected of heresy and sentenced to imprisonment, and he was made to formally declare that the sun-centered view was false. He spent the remainder of his life under house arrest, dying in 1642. He did, however, manage to continue working while under house arrest, returning to some of his earlier work on the mechanics of bodies in motion, and producing some important writing on this subject.

Given the issues involving the church, many of Galileo's contemporaries were understandably hesitant to openly advocate the sun-centered view. But eventually the cumulative effect of discoveries – such as Kepler's discovery that a system using elliptical orbits and varying speeds could account for the data much better than any alternative; Kepler's publication in 1627 of astronomical tables far superior to any others, based on his system; and Galileo's evidence from the telescope – would convince most of those who paid attention to such matters that the Earth and planets do indeed revolve about the sun, and do so in elliptical orbits with varying speeds. This would in turn raise a host of problems for the existing world-view. We turn next to these issues.

Chapter Eighteen

A Summary of Problems Facing the Aristotelian Worldview

Suppose you and I are living in the first half of the 1600s, and that we are keeping up on the new discoveries. We are familiar with Galileo's evidence from the telescope, and we recognize that it provides substantial problems for the traditional Ptolemaic Earth-centered view. We are also familiar with Kepler's vastly simpler model of the universe, employing elliptical orbits with planets moving at varying speeds at different points in their orbits, and we appreciate the fact that Kepler's system is much better at prediction and explanation. And we recognize the superiority of the astronomical tables Kepler published in the late 1620s, which were based on his new system. In general, suppose we have become convinced that Kepler's sun-centered view is correct (and it seems likely that most of those who followed these developments similarly became convinced, at least by the mid-1600s, of the correctness of Kepler's view). In short, we are now convinced that the Earth and other planets move about the sun in elliptical orbits at varying speeds, and that the Earth revolves on its axis, completing a revolution once a day. Such views would raise a variety of problems for the Aristotelian worldview, and the main goal of this chapter is to summarize such problems.

I should note that not everyone became convinced of the correctness of the sun-centered approach. As we saw in earlier chapters, no matter how much evidence accumulates against a view (in this case, the Earth-centered view), it is always an option to hold onto one's view by rejecting various auxiliary hypotheses or, in some cases, by modifying parts of the view while still retaining its key elements. For example, as noted in the previous chapter, Galileo's evidence from the telescope, especially the discovery of the phases of Venus, is quite strong evidence against the Earth-centered Ptolemaic system. However, the Earth-centered Tychonic system is compatible with the phases of Venus, and for that matter, the Tychonic system is compatible with all of the evidence from the telescope. For those committed to retaining belief in an Earth-centered view, the Tychonic

Worldviews: An Introduction to the History and Philosophy of Science.   Richard DeWitt
© 2010 Richard DeWitt

system became the preferred theory. And as noted in Chapter 15, the Tychonic system (or more precisely, a modernized version of the system) is still the preferred system for those few remaining advocates of an Earth-centered view of the universe.

Nonetheless, it seems clear that by about the mid-1600s, the majority of those keeping up with the new discoveries became convinced of the correctness of the sun-centered view. However, accepting this view of the Earth and universe would raise a host of problems for the Aristotelian worldview, and the main goal of the remainder of this chapter is to summarize such problems. In addition, we will look briefly at a closely related issue, namely, the fact that this new view of the universe requires a new science, and in particular, a science that is compatible with a moving Earth. We will begin with a summary of problems for the Aristotelian worldview.

Problems for the Aristotelian Worldview

If the Earth is in motion, both revolving on its axis and revolving about the sun, what keeps us on the Earth, and why do heavy objects fall downwards? Recall that, on the Aristotelian worldview, heavy objects had a natural tendency to move toward the center of the universe, and that was what kept us on Earth and why heavy objects fell. But this piece of the Aristotelian jigsaw puzzle cannot remain if the Earth is not the center of the universe.

Moreover, what keeps the Earth in motion? Again, on the Aristotelian worldview, objects in motion tend to come to a halt, unless something keeps them in motion. This belief coincides with our everyday experience, but this piece of the jigsaw puzzle must also be wrong if the Earth is in motion.

In a similar vein, why, when we throw an object straight up, does it land back in our hand? The standard belief on the Aristotelian worldview was that, if the Earth is in motion, then we should move out from under a thrown object while that object is in the air. So the behavior of thrown objects is another piece of the Aristotelian jigsaw puzzle that no longer works.

And if the Earth rotates on its axis once a day, we should be moving at about 1,000 miles per hour due to this motion. And if the Earth is moving about the sun, then it must be moving at incredible speeds in its orbit about the sun (we now know this speed to be close to 70,000 miles per hour). But on the common-sense view of the Aristotelian worldview, we would expect to notice the effects of such dramatic motions. So why doesn't it feel like we are in motion? Why isn't there a strong wind in our face, and why can't we feel the usual vibrations and other effects that we would expect to accompany such motion?

And what about the elliptical orbits and varying speeds of Kepler's system? On the Aristotelian worldview, the explanation for the continual movement of, for example, the planets depended on the view of the heavens as a place of perfection.

Continual perfect, uniform circular motion would be appropriate for the heavens. But the elliptical orbits and varying speeds are not the sort of motion one would expect if the heavens are a region of perfection. So the new beliefs substantially challenge the beliefs, on the Aristotelian worldview, that the heavens are a place of perfection.

Likewise, the part of the Aristotelian jigsaw puzzle that represents the belief in a fairly small, cozy universe no longer works. If the Earth is in motion about the sun, then our position in space varies by a huge distance (we know it to be almost 200,000,000 miles) from one part of our orbit to the point at the farthest opposite side of our orbit. Given this extreme distance we travel as the Earth moves about the sun, the only explanation for the lack of observed stellar parallax is that the stars are unbelievably far away, and so the universe must be incredibly vast, and perhaps even infinite.

But an infinite universe will not fit into the Aristotelian picture, especially into the Aristotelian view of natural motion. An infinite universe has neither a center nor a periphery. So the account of the natural motion of the basic elements, either motion toward the center of the universe or away from the center of the universe, cannot be made to work in an infinite universe.

Moreover, by this time (especially in Europe) Christian theology and the Aristotelian view of the universe had become so tightly merged that a challenge to one was a challenge to the other. And in this combined Christian / Aristotelian picture, even the idea of a vast, although finite, universe could not be made to fit comfortably. For example, recall that the superlunar region had long been viewed as a region of perfection, and this perfection had now come to be envisioned as the perfection associated with the Christian God. Heaven itself was viewed as existing in this region of perfection, in fact, heaven was often viewed as existing beyond the sphere of the fixed stars. That sphere (as with the spheres that carried the planets) revolved out of a desire for perfection, albeit now the perfection associated with the Christian God. But note that on the sun-centered view, the sphere of the fixed stars no longer moved. Worse yet, there is no longer any need for such a sphere at all. After all, if the stars do not move, then such a sphere plays no role. And so on the sun-centered view, the stars came to be envisioned as not embedded in a single sphere, but rather scattered throughout a vast, perhaps infinite space. And this undermines not only the view of heaven as located beyond the sphere of the fixed stars, but also the view of the Christian God as playing an immanent role in the workings of the universe. In short, the huge universe required by the sun-centered view was problematic for these religious views as well.

Moving to more general aspects of the Aristotelian worldview, consider the Aristotelian view of the universe as teleological and essentialistic. For example, recall the Aristotelian explanations for why objects fell, for what kept us on the Earth, and for the motion of the planets and stars. These explanations were based on the view of the universe as essentialistic and teleological. But these explanations, as conceived of on the Aristotelian worldview, no longer work if the Earth is moving about the sun. This calls into question the general view of the Aristotelian worldview of an essentialistic and teleological universe.

And what is the role of humans in a universe where we are basically on a speck of dust moving through a universe consisting of mostly empty space? Recall that Aristotle's and Ptolemy's Earth-centered models were not developed for religious reasons, but, nonetheless, the Earth-centered model did fit in well with western religious views. Such religious views tended to view humans as the center of creation, and this fit in well with the belief, on the Aristotelian worldview, that the Earth was the center of the universe. So the new discoveries call into question the religious aspects of the Aristotelian worldview, especially the view of humans as special.

Note that these problems for the Aristotelian worldview are not small, isolated problems involving peripheral pieces of the jigsaw puzzle. Rather, the problems are with the core beliefs of the Aristotelian worldview. As such, mere tinkering with the Aristotelian worldview, replacing old pieces of the jigsaw puzzle with new pieces while still keeping the overall puzzle, is not a viable option. Instead, what is needed is a new worldview. And that new worldview will, importantly, need some new scientific views. In other words, the new worldview will require, as part of it, a new science. We will turn now to a brief look at what sort of new science was needed.

The Need for a New Science

As emphasized above, the implications of the new discoveries of the early 1600s went far beyond the issue of whether an Earth-centered or sun-centered view is correct. For example, almost the whole of Aristotelian science, which had been the dominant science for 2,000 years, depends on an Earth-centered universe. Thus, the demise of the Earth-centered view meant also the demise of Aristotelian science. Worse yet, there was no science to replace it.

For example, the Aristotelian explanation of falling bodies – that heavy bodies naturally tend toward the center of the universe, which was the center of the Earth – is no longer tenable with an Earth in motion around the sun. So in the early 1600s, there was no explanation of something as simple as why rocks fall. Likewise there was no explanation for how the Earth could be moving at immense speeds in its orbit around the sun, and yet it feels as if we are stationary. In a similar vein, there was no explanation for why, when we throw a rock straight up in the air, it lands back at the spot where it was thrown. Nor is there any explanation for Kepler's elliptical orbits, or for that matter, what keeps planets moving in the first place. In short, the science that had been in place for two millennia was no longer acceptable, and there was no science to replace it.

The new view of the universe, with the Earth in motion around the sun, required a new science to go with it. And most fundamentally, the new science needed to be a science that was compatible with a moving Earth. Key pieces of this new science would eventually be supplied by Newton, and will be discussed in the next two chapters.

Concluding Remarks

I do not want to give the impression that the Aristotelian jigsaw puzzle remained intact from the time of Aristotle to the 1600s, and that then, almost overnight, that jigsaw puzzle had to be abandoned. In fact, the Aristotelian jigsaw puzzle underwent various modifications during the 2,000 years from Aristotle to the 1600s. For example, the major religions of the western world were not part of the original Aristotelian jigsaw puzzle, but those religious views were added to the Aristotelian worldview during the middle ages. Likewise, there were modifications to the original Aristotelian views of motion, with some of these changes helping to pave the way for the discovery of the principle of inertia in the 1600s.

Nonetheless, throughout the changes the worldview remained an Aristotelian worldview, with a corruptible Earth at the center of the universe, perfect heavens in the region beyond the moon, and the universe conceived of as thoroughly essentialistic and teleological. It was a relatively small, cozy universe, and a universe that fit well with the dominant religious views of the time.

The discoveries of the early 1600s brought a change not just in particular beliefs about the universe, but rather, in the general view of the *sort* of universe we live in. The teleological, essentialistic, Earth and human-centered universe had to go, and with it went the general views of the type of universe we inhabit. The next two chapters focus on the view of the universe that replaced the Aristotelian view.

A word of caution

This is perhaps a good place to briefly repeat a word of caution first mentioned in the Introduction. We are looking at a broad-brushstroke picture of people and events spanning a huge period of time. In such a broad picture, there is always the danger of coming away with an overly simplified and misleading view of certain influences and connections between people and events.

An example will help illustrate. We have just seen that the Aristotelian worldview was no longer viable, given certain discoveries in the 1600s. In addition, we saw that a new science, one compatible with a moving Earth, was very much needed. And in the next two chapters, we will see how Newton's contributions provided major pieces of this much needed new science.

It is true that Aristotle's science was no longer viable. And it is true that a new science was needed. And it is true that Newton provided key pieces of that new science. But what is not true – or at best, it would be misleading to come away with this idea – is that Newton himself was deliberately and directly attempting to fill the void left by the demise of Aristotelian science. In other words, it would be easy to come away from a broad-brushstroke picture such as this with the idea that, for example, Newton was directly responding to, and directly motivated by, the problems we have just been discussing. But doing so would be a mistake.

As always, the real story is much more complicated. Newton, like all of us, was a complex individual motivated by a complex array of factors. Rivalries with contemporaries advocating competing theories, the reception of his early work, personality clashes, disputes over who discovered what first, and even, arguably, his childhood relationship with his mother influenced and motivated Newton. In general, a large array of complex factors played important roles in the development of Newton's work.

The general moral is that there are a lot of complexities underlying broad pictures of the sort we are looking at here. As mentioned in the Introduction, I think there is substantial value in general approaches such as the one taken in this book, but I encourage you to keep in mind that there are a lot of subtleties underlying the broad picture. And also, as mentioned in the Preface, when you are finished with this book I hope you find yourself intrigued enough to want to further explore some of these complexities.

With that brief reminder behind us, let's move on to consider the development of the new science and the new worldview.

Chapter Nineteen

Philosophical and Conceptual Connections in the Development of the New Science

The 1600s were a fascinating time, with an enormous number of changes in the air – scientific, conceptual, religious, and political changes, to name just a few. Contrary to what I think is often assumed, there is a surprising amount of interplay and cross-fertilization between such areas. Philosophical/conceptual changes in the 1600s influenced scientific discoveries and vice versa; likewise, religious, political, and scientific changes all influenced one another, and so on.

The main goal of this chapter is to illustrate how such areas can influence one another. Given space limitations, we will not be able to explore this in any detail, but it will be possible to get at least a sense of the sorts of influences that such seemingly different areas can have on one another. We will look in particular at how some of the religious and philosophical views of Nicholas de Cusa and Giordano Bruno influenced developments in the 1600s, and also at how the largely metaphysical view of atomism played a role. These are but two examples, but they should be sufficient to convey a flavor of the interplay and interconnections between seemingly diverse areas. We will begin with some issues concerning the size of the universe.

The Size of the Universe

Recall that, on the Aristotelian worldview, the universe was thought to be relatively small. The stars were envisioned as being embedded in a sphere, with the center of the Earth being the center of that sphere. The sphere of the fixed stars, as it is generally referred to, was thought to be the outermost periphery of the universe. And although our predecessors thought of the universe as being large, they had no idea how incredibly huge the universe would turn out to be. In fact, compared to our conception of the size of the universe, their view was of a relatively small universe.

This conception of a relatively small universe would change in the 1600s. With the telescope, Galileo had observed countless more stars than previously known, which itself suggests the universe might be larger than previously conceived. More directly, the growing acceptance that the Earth travels about the sun requires a much larger universe. Recall one of Ptolemy's strongest arguments for why the Earth must be stationary: we do not observe stellar parallax. That is, if the Earth moves about the sun, then our movement should result in an apparent shift in the relative positions of the stars. But since we do not observe such a shift – that is, we do not observe stellar parallax – the Earth must not be in motion.

As discussed in Chapter 10, this is a very powerful argument for a stationary Earth, and the only way the Earth could be in motion without there being observable stellar parallax would be if the stars are unimaginably further away than they were thought to be on the Aristotelian worldview. So now that it has become clear that the Earth does indeed move about the sun, we are forced to accept the view that the universe is vastly larger than we have ever imagined it to be.

In the late 1500s and early 1600s, the idea of a large, perhaps infinite universe was a difficult one to accept. Even today, the size of the universe is pretty mind-boggling. It is worth taking a moment to get a sense of just how large a universe we are speaking of. Let's start just by thinking about our own sun and solar system. To get a sense of the size, suppose we build a model of our solar system. To establish a scale, let us imagine that the Earth is the size of a typical globe (about a foot in diameter). If the Earth is the size of a typical globe, then the sun would be the size of a 10-story building, and located about two miles away. Pause for a moment to visualize the scale: the Earth as a globe, and the sun located two miles away. In just considering the distance from the sun to the Earth, we are already speaking of quite a substantial distance. The outermost planet in our solar system would be about the size of a tennis ball, and would be located 80 miles away. Again, take a moment to visualize this: a 10-story building representing the sun, a foot-wide globe two miles away representing the Earth, and 80 miles away a tennis ball, representing the outermost planet in our solar system. Again, our own solar system is huge. In fact, our conception of the size of our solar system is much larger than our predecessors' conception of the size of the entire universe.

Now consider the fact that our sun is just one star among the *hundreds of billions* of stars in our galaxy. On the scale described above, that is, if the Earth is the size of a typical globe, the next nearest star would be 500,000 miles away. And this star is, on a cosmic scale, our immediate next-door neighbor. That is, if the Earth were the size of a globe, our next-door neighbor in our galaxy would be 500,000 miles away. In short, the stars are separated by an enormous amount of space.

Moreover, at this point we are still just talking about our own galaxy, that is, the Milky Way galaxy. The Milky Way galaxy consists of hundreds of billions of stars in our region of the universe. Incidentally, even on the darkest of nights you can see only a small fraction – about 3,000 – of these stars, and all of the stars you see are in our Milky Way galaxy.

With its hundreds of billions of stars, it is difficult to comprehend the sheer enormity of our galaxy alone. But our galaxy is itself but one of hundreds of billions of galaxies in the visible universe, each containing as many stars as our own. Even for us, when we stop to think about it, the universe is an unimaginably huge place.

Now try to put yourself in the shoes of someone living in Europe in the early 1600s. You would likely have been raised with the belief that God made the universe primarily for humans, and you would likely believe, quite reasonably, that the universe is a relatively small, cozy place, with the Earth at the center. This picture is a comfortable picture in which the universe seems to make sense. But now, virtually overnight, there is reason to believe that the Earth is not the center of the universe, that the universe is not a small, cozy place, but is actually huge beyond all imagination. Our Earth is like a speck of dust lost in an infinite sea. It is easy to understand how this idea would have been a very difficult one to accept.

In earlier centuries, a few philosophers and theologians had argued, primarily on philosophical grounds, that an infinite universe, with infinitely many stars, would be the only universe compatible with an infinitely great God. Most notable among these people were Nicholas de Cusa (1401–64) and Giordano Bruno (1548–1600). It should be stressed that neither de Cusa nor Bruno were scientists – their views were based almost exclusively on philosophical and religious grounds.

During their lives, de Cusa's and Bruno's views on an infinite universe were not widely shared. (Bruno's views, for example, led to his condemnation by the Inquisition, and he was burned alive for heresy in 1600). However, in the early 1600s, when it began to become clear that the universe was huge, and perhaps infinite, the ideas of de Cusa and Bruno helped make the idea of an infinite universe more palatable. Their ideas of an infinitely large universe reflecting the infinite greatness of God helped with the acceptance of a difficult new idea.

In a sense, we are talking here of a sort of conceptual Band-Aid. In the 1600s we are forced to accept that the universe is much, much larger than we ever imagined. Conceptually, this huge universe makes no sense, and we need some way of making this new belief fit into our way of conceptualizing the universe. De Cusa's and Bruno's idea helps – the notion that an infinite universe would reflect the infinite greatness of God helps make conceptual sense of the new view of the size of the universe.

Not only did de Cusa's and Bruno's views help make conceptual sense of a huge, perhaps infinite universe, but, notably, their views were tied in with an old philosophy called atomism. Atomism dates back to the ancient Greek philosophers Leucippus and Democritus (fifth century BC), and their successors Epicurus (341–270 BC) and Lucretius (99–55 BC). As it turns out, atomism became a popular view in Europe in the late 1500s and the 1600s (during the 1600s, this atomistic view was more often referred to as a "corpuscular" view). There are a number of reasons behind the revival of atomism during this period, but part of this revival was tied to the increasing popularity of the philosophies of de Cusa and Bruno.

According to the theory of atomism, reality ultimately consists of atoms and the void. Atoms were envisioned as tiny, indivisible particles – the smallest possible particles, in fact. The void, on the other hand, was thought of in much the same way as we think of a vacuum, that is, as completely empty space. Some atoms are stuck together, and these form the objects we see around us. Other atoms are simply flying through empty space (that is, through the void). The atoms that are moving through the void behave something like pool balls – they travel in a straight line unless they collide with another atom or atoms, in which case they bounce off each other much as pool balls bounce off each other.

Atomism is much more of a metaphysical, philosophical/conceptual belief than an empirical belief. One cannot observe atoms moving through the void, nor was there any good empirical evidence supporting the view that reality ultimately consists of atoms and the void. But although atomism is more of a philosophical/ conceptual view, it is a view that fit in well with the emerging ideas of this time, and one that was quite fruitful for developing new scientific ideas.

For example, consider the principle of inertia. According to the principle of inertia, an object in motion will remain in motion in a straight line forever, unless acted upon by an outside force. As we noted earlier, Descartes was the first to get a clear statement of what we now consider the principle of inertia. And it is not a coincidence that Descartes was influenced by the atomistic (or corpuscular) view. Recall that inertia, as discussed earlier, is an extremely counter-intuitive principle, and one of the more difficult principles for those in the 1600s to arrive at. But think about atomism, and think about an infinite universe. Focus on an individual atom moving through space, and let us suppose that this atom never collides with another atom. According to atomism, how will this atom behave? Answer: It will keep moving in a straight line forever. But this is essentially the principle of inertia. In other words, if you have the concept of an infinite universe, and you conceptualize the universe along the lines of atomism, then the principle of inertia is much easier to grasp. We find, then, the idea of an infinite universe, and the philosophy of atomism, helping to smooth the road for one of the central scientific principles discovered in the 1600s, namely, the principle of inertia.

Importantly, I do not want to leave the impression that discovering the principle of inertia was simply a matter of accepting an infinite universe together with the philosophy of atomism. Discovering the principle of inertia involved a combination of experimental work and new ways of conceptualizing the universe, together with a lot of effort by a lot of people over a long period of time. But as with the situation involving the size of the universe, there was a surprising amount of interplay between areas that are often thought to be isolated and separated from one another.

This is just a brief part of a complex story concerning developments and changes in the 1600s. The acceptance of the belief in a huge, perhaps infinite universe, and the recognition of the principle of inertia, are mainly scientific beliefs. But as we have seen, the recognition and acceptance of these new scientific beliefs involved a surprising number of metaphysical, philosophical/conceptual, and religious beliefs.

Concluding Remarks

As emphasized at the outset of this chapter, the 1600s were a time of enormous change, including philosophical/conceptual change, religious change, political change, and of course scientific change. Our goal in this brief chapter was to get a sense of some of the ways that philosophical/conceptual ideas, on the one hand, and more straightforward scientific ideas, on the other, can influence and feed off one another. The interactions between such areas are complex and fascinating, and in this chapter we have at least gotten a flavor of this.

Chapter Twenty

Overview of the New Science and the Newtonian Worldview

The development of the new science in the 1600s was the cumulative effort of many researchers. The work that capped these efforts, however, was Newton's 1687 *Mathematical Principles of Natural Philosophy*. The work is generally referred to as the *Principia* (after its Latin title *Principia Mathematica Philosophiae Naturalis*). The *Principia* presented a new physics, compatible with a moving Earth, and provided the core of what we now think of as Newtonian science. This work also provides a convenient means by which to explore the Newtonian worldview, that is, the new jigsaw puzzle of beliefs that would be the successor to the Aristotelian jigsaw puzzle.

Our main goal in this chapter will be to take a look at both Newton's science and the new (Newtonian) worldview. We will begin with an overview of Newton's science.

The New Science

As discussed in Chapter 18, core pieces of the Aristotelian worldview cannot be made to fit with the idea of a moving Earth, and thus the acceptance that the Earth moves about the sun requires an entirely new science. The new science that emerged was the product of a great deal of work over a number of decades. As noted, this new science culminated in the work of Newton. As such, we will look primarily at Newton's science, although we should keep in mind that his work owes a debt to a number of other scientists. (Also worth mentioning is that Newton developed calculus, independently of but at the same time as Gottfried

Worldviews: An Introduction to the History and Philosophy of Science. Richard DeWitt
© 2010 Richard DeWitt

Leibniz (1646–1716). The calculus was an important mathematical tool in the development of Newton's science, and today remains one of the most widely used mathematical tools.)

The *Principia* is a substantial work consisting, in the most recent English translation, of about 600 pages. Newton's science, however, is often characterized as consisting, at heart, of three laws of motion together with the principle of universal gravitation. Certainly, in 600 pages Newton does more than merely present a handful of laws of motion and the notion of universal gravity. Nonetheless, there is a sense in which gravity and the laws of motion are the heart of Newton's science. In what follows, then, we will consider these, as well as discuss some general issues about Newton's science.

The three laws of motion

Newton begins the *Principia* with a section of definitions, in which he explains how he will be using various terms found throughout the book. His next section is a brief (about 10-page) section in which he presents the three laws of motion.

The first law is what we now commonly term the principle of inertia. We first discussed the principle of inertia in Chapter 12, in which it was presented in the way it typically is today: An object in motion remains in motion in a straight line, and an object at rest remains at rest, unless acted upon by an outside force. Newton's phrasing of the principle is slightly different, but his phrasing and the usual modern phrasing are equivalent in meaning.

As discussed earlier, the principle of inertia runs contrary to everyday experience, and it was one of the more difficult principles to work out in the 1600s. Various precursors to it had been widely discussed in the previous century, and in the early 1600s Galileo made a variety of investigations of bodies in motion, in which he almost, but not quite, correctly characterized the key idea of inertia. By the mid 1600s, Descartes had an accurate characterization of inertia, and Newton's first law of motion borrows substantially from Descartes' characterization.

To understand the second law of motion, consider the behavior of a hit baseball. The harder you hit the ball, the faster and further it goes. That is, the change in the motion of the ball is proportional to the force applied (how hard you hit it). More fully, the second law of motion states that a change in the motion of an object is proportional to the force applied to the object, and takes place along the straight line in which the force is applied. The law is often summarized as $F = ma$, that is, force equals mass times acceleration. As with the baseball example, this entails that the acceleration of an object will be equal to the force applied divided by the mass of the object.

The third law states that for any action there is always an opposite and equal reaction. The standard illustration of this law is the recoil of a gun, in which the action of the bullet being propelled in one direction results in an equal and opposite reaction, namely, the recoil of the gun in the opposite direction.

Universal gravitation

The three laws of motion, which again are central ingredients in Newton's science, take only about two pages to present in the *Principia*. The other key ingredient, the notion of universal gravitation, is somewhat more complicated to explain. In this section, I want to explain the slow way in which Newton develops the idea of universal gravitation in the *Principia*, and then in the final section of this chapter (just before the concluding comments) give an idea of why he takes this slow, careful approach. Let's begin with the way universal gravitation is typically presented currently.

Universal gravitation is generally presented as a mutually attractive force between any two objects. For example, the gravitational attraction of the sun attracts the Earth toward the sun, and at the same time, the gravitational attraction of the Earth attracts the sun toward the Earth. Likewise, when I drop a book, the Earth's gravity attracts the book toward the Earth, but at the same time, the gravitational attraction of the book attracts the Earth toward the book. The gravitational attraction of the book has virtually no effect on the Earth because the Earth is so enormously more massive than the book; likewise, in the sun–Earth example, the fact that the sun is so much more massive than the Earth explains why the Earth's gravity has relatively little effect on the sun as compared to the sun's effect on the Earth.

More particularly, the gravitational attraction between two objects is proportional to the product of their masses. That is, the more massive the objects, the greater the gravitational attraction. Also, the attraction is inversely proportional to the square of the distance between the objects, so that as the distance between objects increases, the force of the gravitational attraction between them diminishes rapidly.

Such is the way universal gravitation is typically presented these days. And in fact, this characterization of universal gravitation is presented in the *Principia*. But unlike the laws of motion, which are stated fully and concisely at the beginning of the book, this characterization of gravity emerges slowly.

Not counting the preface, Newton first discusses gravity in the first few pages of the *Principia*, in the section on definitions. At this point, though, Newton uses "gravity" only to refer to whatever it is that attracts objects to the Earth, and he clearly is not using the term in the sense of universal gravitation. Much later in the book (400 pages later, in fact), Newton shows that the Earth's gravity must extend at least to the moon, and is responsible for the orbit of the moon. He also shows that whatever force keeps the moons of other planets (for example, the moons of Jupiter) in their orbits, that force must have the same characteristics as the Earth's gravity (that is, the attractive force is proportional to the masses of the bodies, and is inversely proportional to the square of the distance between the bodies). He also shows that whatever force keeps the planets in orbit about the sun must likewise have the same characteristics as the Earth's gravity. At this point, at Proposition 7 of Book 3, he is ready to generalize the notion of gravity: Gravity exists in all bodies universally.

Here, then, we finally have the radical notion of universal gravitation. And by the end of the *Principia*, Newton has treated us to an impressive display of the explanatory power of universal gravitation, coupled with the laws of motion. The *Principia* is a revolutionary work, and the range of phenomena that can be handled by this small number of ingredients (the three laws of motion plus universal gravitation) is truly impressive.

Overview of the Newtonian Worldview

Again, the Aristotelian worldview is an Earth-centered worldview. And the belief that the Earth is at the center of the universe is not merely a peripheral belief, but a core belief, one that cannot be replaced without replacing most of the pieces of the jigsaw puzzle. Newton's science provided many of the scientific pieces of a new puzzle – in particular, Newton provided a science with extraordinary explanatory power and, importantly, one compatible with a moving Earth. Notably, most of the pieces of the Aristotelian jigsaw puzzle – and not just the scientific pieces, but the philosophical/conceptual pieces as well – are not compatible with the new science. In other words, we needed a range of new philosophical/conceptual pieces to accompany the scientific pieces supplied by Newton.

For example, on the Aristotelian worldview the universe was viewed as teleological and essentialistic. Objects behaved as they did because of internal, essential natures. But with Newton's science, objects no longer behave because of internal essences; rather, objects behave as they do largely because of the influence of external forces. The entire Aristotelian view of the universe as a universe full of goals and purposes does not fit with the new science, and indeed, the universe begins to be viewed as more like a machine. In the same way that parts of a machine push and pull against one another, and the behavior of the various parts is due to the forces applied to them by other parts, so likewise the behavior of objects in the universe comes to be viewed as due to the push and pull of other objects and to forces acting on them.

This machine metaphor becomes the dominant metaphor for the new worldview. And this sort of universe, in which the push and pull of external forces is central to understanding the behavior of objects in the universe, is almost completely the opposite of the overall Aristotelian view. In short, the teleological and essentialistic view of the universe, which went hand in hand with the science of the Aristotelian worldview, is replaced by a mechanistic, machine-like view of the universe, which goes hand in hand with the new science.

Accompanying the machine metaphor, the view of God changed as well. Again, for Aristotle himself, the gods were not religious gods at all, but rather were needed to explain what kept the stars and planets in motion. And as previously mentioned, in later centuries Aristotle's conception of the gods was replaced with the Christian/Judaic/Islamic conception of God. So although the details of the conception of God changed during the Aristotelian worldview, one central Aristotelian conception remained: the idea that God was a necessary component in the minute-to-minute workings of the universe. In other words, in the Aristotelian worldview, God, or something like God, was needed for scientific reasons, that is, as a constant source of the motion of the heavenly bodies.

But with the new science, nothing like this is needed to run the universe. The motion of the planets, for example, is explained as a consequence of inertia (a body in motion remains in motion, so the planets, being in motion, will remain in motion) together with gravity (which explains why the planets move about the sun rather than going off in a straight line). In short, with the new science, God is no longer needed to run the universe.

Religious beliefs tend to be deeply entrenched, so not surprisingly, people did not abandon their religious beliefs. But the concept of God changed considerably. In particular, God came to be viewed as a sort of watchmaker-God, that is, one who designed and constructed the universe, and set the universe in motion. But thereafter the universe runs along without the constant intervention needed in the previous worldview.

Likewise, the general conception of an individual's role in society changed. The Aristotelian worldview included what might be considered a hierarchical outlook. That is, much as objects had natural places in the universe, so likewise people had natural places in the overall order of things. As an example, consider the divine right of kings. The idea was that the individual who was king was destined for this position – that was his proper place in the overall order of things. It is interesting to note that one of the last monarchs to maintain the doctrine of the divine right of kings was the English monarch Charles I. He argued for this doctrine – unconvincingly, it might be noted – right up to his overthrow, trial, and execution in the 1640s. It is probably not a coincidence that the major recent political revolutions in the western world – the English revolution in the 1640s, followed by the American and French revolutions – with their emphasis on individual rights, came only after the rejection of the Aristotelian worldview.

In general, the Aristotelian worldview included a conception of a small, cozy universe, with the Earth at the center. The universe was full of natural goals and purposes, and the outlook was teleological and essentialistic. This view extended to people as well, who had their natural places in the overall order of things, much as objects have their natural places in the universe. And God, or something like God, was needed on a day-to-day, minute-to-minute basis, to keep the universe in motion.

All of these views change with the emerging new worldview. The universe is now viewed as huge, and perhaps infinite, with our sun merely as the center of the revolution of the planets in our solar system. The universe was now viewed as machine-like, with no purposes or goals to explain the behavior of objects. Rather, objects behaved as they did as the result of external, purposeless forces. Nor was God, or anything like God, needed to run the universe. Rather, the universe ticks on, day after day, much like a watch ticks on.

Philosophical Reflections: Instrumentalist and Realist Attitudes toward Newton's Concept of Gravity

Before closing this chapter, it is worth taking a minute to discuss a rather interesting aspect of the usual Newtonian view of gravity, and one that ties in with some of the key philosophical issues we have discussed. Moreover, this will go some way toward explaining why, as discussed above, Newton took such a slow and careful approach to presenting the notion of gravity in the *Principia*.

In particular, I want to spend a moment discussing how, when viewed a certain way, the notion of gravity is quite an odd notion. Let me begin with an example I will return to later in the book. Suppose I put a pen on the table, and I ask you to move the pen, but without having any sort of contact whatsoever with the pen. You are not allowed to touch the pen, blow on the pen, throw objects at it, shake the table it is on, or have any sort of contact at all with the pen. Yet I am asking you to move the pen in spite of not being allowed to have any contact with it. You will almost certainly think I am asking you to do something impossible. And our sense that I'm asking you to do something impossible comes from a common conviction, and one that goes back at least to the ancient Greeks, that one thing (for example, you) cannot influence another thing (for example, the pen) unless there is some sort of contact or communication between the two. This conviction is often summarized by saying that there can be no "action at a distance," to use a common phrase.

Now let's return to the notion of gravity. Gravity is usually conceived of as an attractive force between bodies. To use a typical example, the gravitational force of the Earth attracts my pen, so that when I release the pen, it falls toward the floor. And if we ask "Why did the pen fall?" the usual answer would be that it fell because it was under the influence of this gravitational force.

Likewise, if we ask whether gravity is a real force, that is, whether gravity really exists, the usual answer is "Of course it does." That is, people typically take a realist attitude toward gravity, viewing it as a force that really exists, and that largely explains much of the everyday phenomena we observe around us.

I suspect that most of us take gravity with a realist attitude largely because we have been raised with the notion of gravity from an early age, and so we tend not to notice that there are some rather odd features of gravity, at least gravity taken with a realist attitude. To see these odd features, contrast gravity with other cases where there is attraction between objects. For example, suppose I put a rubber band around two pens, and pull the pens apart, thereby stretching the rubber band connecting them. In this case, the pens are, in a sense, attracted toward each other. And if I release the pens, they will quickly move toward each other. But in this case, the nature of the attraction is easily understood – the pens are connected by a stretched rubber band, and it is exactly this stretched rubber band that is the cause of the attraction between the two pens.

We can easily understand the nature of the attraction in the case of the pens attached by a stretched rubber band. But now return to the dropped pen example, and note that there seems to be no connection between the pen and the Earth. There is no rubber band connecting the Earth and pen, no strings, no nothing. Yet in spite of the fact that there seems to be no connection whatsoever between the pen and Earth, the pen moves toward the Earth when released. Viewed this way, gravity doesn't sound like science; it sounds like magic.

In short, if taken with a realist attitude – that is, if gravity is thought of as a really existing force – then the effect of gravity sounds a great deal like some sort of mysterious action at a distance. And in fact, when Newton first published the *Principia*, there were quite a number of critics who attacked him for introducing a force that required mysterious action at a distance. Some of these critics were quite influential, including (to name just one among many), Gottfried Leibniz (noted earlier as the co-developer of the calculus). Leibniz criticized Newton for introducing "occult" forces into science, and the basis of this criticism was exactly the problem that gravity seemed to involve mysterious action at a distance.

One solution to this problem was to take gravity with an instrumentalist attitude, and indeed Newton himself typically professed to take an instrumentalist attitude toward gravity. To get a better sense of what this would amount to, consider again the dropped pen. Newton's equations, including those concerning gravity, can be used to make excellent predictions about how the pen will fall (for example, the rate of acceleration of the pen). Taking an instrumentalist attitude would essentially involve viewing such equations as providing an excellent account of the way *that* objects behave, while remaining agnostic on the issue of *why* they behave that way. In other words, one can use the equations, especially those involving gravity, to provide excellent predictions, while remaining silent on the issue of whether gravity is a "real" force.

Newton did hold out the hope that a realistic account of gravity could be given, consistent with the mathematical treatment he provided in the *Principia*, and in a way that involved only mechanical interactions with no action at a distance. But although the next two centuries would see somewhat different treatments of gravity (for example, the notion that objects are responding to a gravitational field acting locally, that is, without requiring action at a distance, would be one alternative approach), no completely unproblematic account would be forthcoming. (At least, the accounts are problematic when viewed from a realist perspective. None of the accounts, including Newton's, is problematic so long as one adopts a purely instrumentalist stance.) Eventually, as we will discuss in later chapters, Einstein's general theory of relativity will provide an account of gravity that does not involve action at a distance. But as we will see, Einstein's account of gravity is very different from the Newtonian view of gravity that most of us were raised with.

Concluding Remarks

The old Aristotelian worldview was incompatible with the new discoveries of the 1600s. Its replacement certainly did not develop overnight, but eventually the new outlook described above emerged, and it is this outlook that we will refer to as the Newtonian worldview. As with the Aristotelian worldview, the Newtonian worldview developed over time, but it retained the key view of a mechanistic, machine-like universe.

One of the features of science as it developed in the 1600s was an increasing appeal to laws, for example, Kepler's laws of planetary motion or Newton's laws of motion. The increasing prominence of scientific laws raises some interesting philosophical questions, for example, what is a scientific law? In the next chapter, we take a brief look at some of the puzzling issues that arise with the notion of such laws. Then in the following chapter, we will sketch some of the ways the Newtonian worldview developed over the next two centuries.

Chapter Twenty-One

Philosophical Interlude: What is a Scientific Law?

Since the 1600s, the notion of scientific laws has played an increasingly prominent role in science. In earlier chapters we discussed, for example, Kepler's laws of planetary motion, Newton's laws of motion, and Newton's characterization of the principle of universal gravitation. And in the next chapter, we will briefly see other examples that emerged in the centuries following the general acceptance of the Newtonian picture, for example, principles involving electrical attractions expressing relationships between electrical and magnetic phenomena, and many others. All of these are often taken as scientific laws. These sorts of laws seem to capture something fundamental about physical phenomena, and the search for, and characterization of, these sorts of laws has been an important part of science since the scientific changes of the 1600s.

But what is a scientific law? The main goal of this brief chapter is to illustrate, as happens so often, that once we start to look into this question, we quickly run into some deeply puzzling issues. Attempts to address these and other issues surrounding scientific laws, especially in the past 40 or so years, have resulted in a quite complex set of proposals, arguments, counter-arguments, counter-proposals, and the like. One thing that is clear is that, after decades of debate by knowledgeable and articulate defenders of various accounts, there is no consensus on the question of what scientific laws are or how best to characterize them.

The details of this debate are beyond the scope of this chapter, though as usual the Chapter Notes will point to additional material for those who would like to delve into this in greater detail. But it is not particularly difficult to get a sense of some of the puzzling issues that quickly arise when exploring the notion of a scientific law. So the more modest goal of this chapter will be to convey a sense of some of these puzzling issues.

Worldviews: An Introduction to the History and Philosophy of Science. Richard DeWitt
© 2010 Richard DeWitt

Scientific Laws

Philosophers, especially for the past 50 years, have tended to distinguish between scientific laws on the one hand, and laws of nature on the other. Much has been written on the distinction between the two, but here is one way to briefly characterize that distinction. What we tend to think of as scientific laws – for example, Kepler's laws of planetary motion, Newton's laws of motion, the principle of universal gravitation, and the like – tend to only approximate the way objects behave (more on this below). Kepler's second law, for example, at best characterizes how a planet would orbit in a two-body system, that is, if the only bodies in existence were the planet and the sun. In our actual solar system, in which every planet is influenced by all sorts of factors, including the gravitational influence of other planets, Kepler's second law only approximately characterizes the orbit of a planet.

But because scientific laws such as Kepler's tend to closely approximate the behavior of objects, it is common to think that such laws are somehow reflecting, even if only approximately, some deeper feature of the world. And the deeper feature reflected by a scientific law would presumably be a law of nature. So roughly, it is common to characterize a law of nature as being a fundamental feature of the universe that is responsible for the way the universe works, while scientific laws are often thought of as principles that approximately reflect those laws of nature.

In what follows I will focus primarily on scientific laws, though issues involving the fundamental features of the world that such laws presumably reflect will often arise. Let's begin with two features commonly associated with scientific laws.

Common features associated with scientific laws

It is common to think of a scientific law as reflecting some fundamental, exceptionless aspect of the universe, that is, as reflecting some way in which things must behave, as opposed to merely reflecting how some things happen to behave. Consider, for example, what we often refer to as Kepler's second law of planetary motion. This is the law we first discussed in Chapter 16, and it is often referred to as the "equal areas" law. By way of a brief reminder, recall that this law basically states that a line drawn from a planet to the sun will sweep out equal areas in equal time. (It might help to refer back to Figure 16.3.)

As suggested, we tend to view this law as reflecting, or at least partly reflecting, some basic exceptionless regularity in the universe. Notably, I say "at least partly reflecting" because, strictly speaking, such a law would at best be completely accurate only under ideal circumstances, for example, if the planet were not influenced by any other forces such as the gravitational attraction of other bodies in the solar system. More will be said below on issues involving ideal circumstances.

The key point I want to draw attention to here is that the law is usually viewed as reflecting (or at least approximately reflecting) an exceptionless regularity, that is, planets always behave in this way, and presumably always have and, so long as there are planets, always will. In this way, laws are typically viewed as different from most other regularities we observe. For example, it is a regularity that my local diner usually has hot coffee available during hours they are open. But this is not an exceptionless regularity – they occasionally, though not often, run out. Likewise, it is a regularity that the average temperature (in the northern hemisphere, at least) in June is higher than the average temperature in May. But this too is not an exceptionless regularity. Not often, but occasionally, May is hotter than June.

But a statement like Kepler's second law is usually viewed as saying something about the way planets always behave, not merely about how they usually behave. And this is generally viewed as a feature of scientific laws – they reflect exceptionless regularities. For now, let's log this idea, that is, that reflecting an exceptionless regularity seems to be a key feature of a scientific law.

Another key feature we usually associate with scientific laws is that we view them as reflecting objective features of the world. Although we have occasionally touched upon the idea of objectivity in this book, we have not discussed it in any detail. So a few words on this are in order.

The key idea in the way I am using "objective" here involves whether something is or is not dependent on humans. More specifically, we tend to view something as objective if we think it would have existed even if humans had never existed, and otherwise, we tend not to view it as objective. I should note that this is not the only sense of the word "objective," but it is the sense in which I am using the term here.

Consider, for example, some popular dessert dish, say chocolate mousse. According to food historians, chocolate mousse appears to have first been invented in France in the 1700s, and thereafter became increasingly popular around the world. Chocolate mousse is clearly a human invention, and had humans never existed – for that matter, probably had the French never existed – chocolate mousse likewise would never have existed. In this sense, then, chocolate mousse is not an objective feature of the world (again, in the sense of the word "objective" we are using here).

In contrast, we usually think of the planet Jupiter as being objective. That is, most of us are of the view that even if humans had never existed, Jupiter would still have existed. For example, there is good reason to think that the evolution of large mammals, including humans, may not have happened had certain events not occurred in the past, for example, a massive meteor impact about 65 million years ago that contributed to the extinction of the dinosaurs and thereby opened the way for large mammals to appear. But imagine a situation in which that meteor missed the earth, dinosaurs continued to dominate the landscape as they had for the previous 100 million years, and large mammals, including humans, never appeared. In such a scenario, one without humans, we still tend to think

Jupiter would have existed. That is, unlike chocolate mousse, we tend to think of Jupiter as existing independently of humans. In other words, we tend to view Jupiter as being an objective feature of the world.

Incidentally, the name "Jupiter" is of course not objective. This word is clearly a human invention. But we typically think that the object that word names, the planet we call Jupiter, would have existed even if humans had never existed.

Moreover – and here is where we tie this back in with Kepler's second law – in a scenario such as that just described, where there are no humans, we tend to think that Jupiter would still have orbited the sun just as it does now. In particular, we tend to think that, if humans had never existed, Jupiter would still have orbited in accordance with Kepler's second law. In other words, we tend to view Kepler's second law as capturing an objective feature of the world.

As with the case of the word "Jupiter," the phrase "Kepler's second law" would of course not have existed had humans not existed. But just as we tend to think that the object our word "Jupiter" names would have existed had humans not existed, so too we tend to think that the regularity captured by our phrase "Kepler's second law" would still have been a feature of the universe even if humans had not existed. And again, this is just to say that we tend to view Kepler's second law, and other scientific laws, as capturing objective features of the world.

There is much more that can be said by way of characterizing common views about scientific laws. But for the sake of our discussion, let's just focus on the two features of laws we have identified above. First, that we tend to view such laws as reflecting exceptionless regularities, and second, we tend to view such laws as reflecting objective features of the universe. In what follows, let's explore these two features. We will see that we quickly run into difficult and puzzling issues.

Exceptionless regularities

Let's begin by exploring the first of the presumed features of scientific laws discussed above, that is, the notion that laws reflect exceptionless regularities. This seemingly simple feature quickly becomes surprisingly problematic.

First, let's note that exceptionless regularities are everywhere, but most such regularities are ones we do not want to consider candidates for scientific laws. Consider two examples. All English sentences ever written have consisted of fewer than a million words, and so this is an exceptionless regularity about English sentences. But we would never consider "All English sentences consist of fewer than a million words" as a candidate for a scientific law. As a second example, as far as I can remember (and for no good reason), I have always put my pants on left leg first. Assuming my memory is correct, this too is an exceptionless regularity, but certainly not something about which we want to consider proposing a scientific law. We could sit here and generate thousands of examples of similar sorts of exceptionless regularities, the vast majority of which we would not want to consider candidates for scientific laws.

It seems, then, that although capturing an exceptionless regularity is an important part of being a scientific law, it turns out that we do not want to consider the overwhelming majority of exceptionless regularities to be candidates for scientific laws. This observation raises a simple but difficult question: What is the difference between exceptionless regularities that do underlie scientific laws, and exceptionless regularities that do not?

There is a fairly common response to this question, albeit a response that raises its own difficult issues. The response involves what are commonly referred to as "counterfactual conditionals," or just "counterfactuals." I might note that, besides their role in this context, counterfactuals play roles in a variety of other contexts, scientific and otherwise. Our next task, then, will be to get clear on what counterfactuals are.

Counterfactuals

Counterfactuals are a common feature of everyday speech and thinking. So it is almost certain that you are already familiar with the key ideas behind counterfactuals, even if you have not before come across the terms "counterfactual" or "counterfactual conditionals."

As usual, let's begin with some examples. Imagine yourself saying these statements: "If only I had studied harder for that test, I would have done better." Or "If I had not stayed out so late, I would not have overslept this morning." Or "If I had remembered my cell phone needed recharging, then the battery would not now be dead." Or "If I had gotten to the ticket counter earlier, I would have got tickets before they sold out." And so on.

There are all examples of counterfactual conditionals, or what are often just called counterfactuals. Notice first that each of these is a conditional sentence, that is, an if–then sort of sentence. This explains the "conditional" part of the phrase "counterfactual conditional."

Notice also that in every example, the "if" part reflects something that did not happen and that you know did not happen. You did not in fact study as hard as you might have for the test. You did not in fact get back from your night out as early as you might have. And so on. In each case, the "if" part of the statement reflects something that is not correct, that is contrary to fact. Or in other words, the "if" parts of each statement reflect something that is counterfactual, and this is where the "counterfactual" part of the phrase "counterfactual conditional" comes from.

Counterfactuals play an important linguistic role in everyday life and everyday thinking, in that they allow us to express how we think things would have turned out had conditions been different. If, counter to fact, you had studied harder for the test, then you would have done better. If, counter to fact, you had remembered to charge the battery in your cell phone, then that battery would not now be dead. And so on. These sorts of expressions are extremely common, and again, they play an important role in enabling us to express how we think things would have been had circumstances been different from the way they in fact were.

As indicated, counterfactuals are commonly seen as a key to distinguishing exceptionless regularities that we tend to consider candidates for scientific laws, and exceptionless regularities that we do not want to consider as candidates for scientific laws. The usual account of how counterfactuals do this is as follows.

Consider again examples of exceptionless regularities that we are not inclined to think of as underlying scientific laws, say the one above about all English sentences having been fewer than a million words, or the one about the way I always put my pants on. Notice that these regularities, while accurate given the way things happen to be, would not have been true under a wide variety of alternative circumstances. For example, if there had been a contest with a sizable cash award for constructing the largest grammatically correct English sentence, then it is likely that someone would have concocted an English sentence of more than a million words. And so under this counterfactual scenario, the regularity about English sentences would not hold. Likewise, if a computer programmer, say for fun, had developed a program for constructing long English sentences, then again the regularity about English sentences would probably not still be accurate. Likewise for my pants behavior. If at some time in my past I had broken one of my legs, chances are this would have altered my behavior, and the regularity about my pants behavior would not have been an exceptionless regularity. Likewise, had someone offered me a sum of cash for changing my behavior, or under any number of other counterfactual scenarios, the regularity about the way I put on my pants would no longer be an exceptionless regularity. In short, these sorts of regularities would not have been true under a wide variety of alternative circumstances.

In contrast, the regularity underlying Kepler's second law of planetary motion seems to remain an exceptionless regularity under any number of counterfactual situations. For example, had Jupiter been somewhat closer to the sun or further, or more massive or less massive, or had been a rocky planet rather than a gas giant, or under any number of a wide variety of counterfactual situations, Jupiter would still have behaved according to Kepler's second law.

In short, the sorts of exceptionless regularities that we tend to view as captured by scientific laws, such as Kepler's second law of planetary motion, tend to be resistant, in a certain sense, to counterfactual situations. In particular, such regularities tend to remain true even had circumstances been different in any of a wide variety of ways.

This distinction, then, between regularities that hold true under a wide variety of counterfactual conditions and regularities that do not is often viewed as a key difference between exceptionless regularities that are candidates for scientific laws, and those that are not.

Will this appeal to counterfactuals suffice to resolve the problem of distinguishing between law-like regularities, and mere accidental, non law-like regularities? Unfortunately, not in any sort of an easy way. In particular, an appeal to counterfactuals will not solve the problem of distinguishing law-like regularities from non law-like regularities without raising equally problematic issues. Those issues involve two considerations, one having to do with context-dependence, and the other having to do with what are commonly termed *ceteris paribus* clauses. A brief discussion of each of these is in order.

Context dependence   Although the above appeal to counterfactuals seems, at least initially, to make substantial headway on the problem of adequately distinguishing regularities we want to view as underlying scientific laws from regularities that do not, as so often happens, it does not take much probing to uncover further problematic issues. The first has to do with the context dependence of counterfactuals.

Above, in the preliminary discussion of counterfactuals, I withheld an important feature of counterfactuals, namely, whether we tend to view a counterfactual as true or false depends very much on the context. Consider again the example from above, "If I had remembered my cell phone needed recharging, then the battery would not now be dead." In the discussion above, we were implicitly assuming a more or less normal context, in which you would presumably want your cell phone charged. In such a context, we're inclined to treat this counterfactual as true.

But now consider an alternative context. Say you have a big test tomorrow, and you decide you do not want your cell phone to be charged until the test is over, so you won't be tempted to lose time talking on it. In this context, the counterfactual "If I had remembered my cell phone needed recharging, then the battery would not now be dead" is false – in this context you presumably would have remembered you did not want your phone charged and would have let the battery remain dead.

Or perhaps you had a fight with a friend and want to remain out of contact for a while, and so you prefer leaving your cell phone uncharged as a convenient excuse for not returning calls. Or any number of other possibilities. In short, there are endless contexts in which the counterfactual in question would be true, and endless contexts in which it would be false. And likewise for almost every counterfactual.

In short, whether a counterfactual is true or false is notoriously context-dependent. And the problem this raises, as it pertains to scientific laws, is this: Generally (maybe always), when the truth or falsity of something is context-dependent, that means that the truth or falsity depends on the knowledge and interests of the parties involved, and this in turn means the truth or falsity depends on people, and the interests and knowledge of those people.

At this point you can probably see the looming problem. Recall one of the main features of the typical view of scientific laws, discussed at the outset of this section, namely, that scientific laws are generally viewed as reflecting objective features of the world, that is, features that are independent of humans. But now we seem to have painted ourselves into a corner. We seem to need to appeal to counterfactuals in order to characterize what counts as a scientific law, in particular, in order to distinguish exceptionless regularities that are candidates for scientific laws from exceptionless regularities that are mere accidental regularities. But the appeal to counterfactuals brings with it context dependence. So if characterizing scientific laws requires counterfactuals, and counterfactuals are context-dependent, then counterfactuals are dependent on humans (or more precisely, the truth or falsity of counterfactuals is dependent on humans). So appealing to counterfactuals undermines the seeming objectivity of scientific laws.

Ceteris paribus *clauses*   There is another basic issue that arises in considering scientific laws as reflecting exceptionless regularities. Consider again Kepler's second law of planetary motion. If we look more closely at how planets actually orbit, we'll notice something interesting, namely, that Kepler's second law does not, strictly speaking, reflect an exceptionless regularity about planetary orbits.

The basic problem was mentioned earlier, and is easy to see. All sorts of factors can influence the orbit of a planet. For example, planets are occasionally struck by asteroids and comets, and such impacts influence the orbit of a planet. A particularly spectacular collision occurred as recently as the 1990s, when a massive comet collided with Jupiter, and while this collision did not throw Jupiter into anything like an entirely new orbit, it certainly had noticeable effects on Jupiter's orbit, including the effect that, for the period right round the collision, Jupiter did not quite obey Kepler's second law. That collision was a particularly spectacular one, but less spectacular ones occur all the time. Even more recently Jupiter was again struck by a quite large object, leaving a disturbance in Jupiter's atmosphere the size of the earth, and again altering the orbit of Jupiter.

Although collisions such as these provide somewhat dramatic examples, less dramatic events are happening all the time. Planets are constantly under all sorts of influences, from the gravitational influences of other planets, to passing comets and asteroids. Even the passing spacecraft we occasionally send will have an influence. And these effects, though small, are such that planets never orbit exactly as characterized by Kepler's second law.

Such events, which seemingly prevent a law from applying where it would otherwise apply, are present in probably every situation involving an alleged scientific law. Or to put the point another way, there is probably no scientific law that is straightforwardly, strictly followed.

The usual tact for attempting to avoid this issue is to employ what are termed *ceteris paribus* clauses, where *ceteris paribus* is a phrase roughly meaning "all else being equal." So, for example, we can say that if Jupiter is a planet, then *ceteris paribus*, that is, all else being equal (for example, there are no asteroid or comet impacts, no influences from other planets, and so on), it will obey Kepler's second law of planetary motion.

Not surprisingly, this account raises issues, of which I will mention two. First, you may already have noticed the connection between the discussion of *ceteris paribus* clauses and the discussion above of counterfactuals. The two are related.

Reading Kepler's second law as one that is accompanied by *ceteris paribus* clauses amounts to saying, in the case of Jupiter, that *if* it were the case that Jupiter was experiencing no additional forces, *then* it would orbit in accordance with Kepler's law. But, we know at the outset that Jupiter will in fact experience such additional forces. So the statement just made is itself a counterfactual, and so this account seems to inherit the problems we found above with counterfactuals.

In addition, note that there is no way to specify all the possible *ceteris paribus* clauses, simply because there are too many possibilities. We mentioned asteroid impacts, comet impacts, the influence of passing spacecraft, and the like above. But there are indefinitely many such influences, and we could not possibly list them all. The best we could do is to list influences such as comets, asteroids, and passing spacecraft, and then say something like "and other similar influences." But notions of similarity are tied to human interests. For example, whether two things are or are not considered similar depends crucially on the interests of those making the judgment. And so we again run into a problem related to what we saw above. If characterizing scientific laws requires appeals to *ceteris paribus* clauses, and such appeals require a notion of similarity which is itself dependent on human judgments, then our account of scientific laws again seems to lose the notion that scientific laws are objective.

Concluding Remarks

In concluding, I want to return to a point made at the outset of this chapter, namely, that the issues surrounding scientific laws have been the subject of extensive discussion and debate, especially over the past 40 or so years. That discussion and debate, while including the issues raised in this chapter, goes well beyond these issues.

My main goal in this chapter was not to summarize all the discussion concerning scientific laws that has taken place in recent decades. Rather, the main goal was to illustrate how quickly one runs into difficult issues, once one begins probing a bit into the question of what a scientific law is. The material above will, I hope, convey a flavor of some of the difficult issues that arise. In many ways, while certainly not an exceptionless regularity, this is a recurring pattern: once we probe into what seem to be relatively straightforward issues and concepts involved in science, we quickly run into difficult and puzzling problems.

Chapter Twenty-Two

The Development of the Newtonian Worldview, 1700–1900

As with the Aristotelian worldview, the Newtonian worldview is not a static set of beliefs. It developed and changed in the centuries following the 1600s, but throughout the changes, the core elements of the worldview remained steady. Our goal in this chapter is to illustrate some of the developments that took place between roughly 1700 and 1900.

In general, our approach will be to try to convey a flavor of how promising the development of the Newtonian view was during this period, such that, by 1900, it appeared that most of the major questions about the world had been answered within a Newtonian framework. We will look specifically at the development of several of the major branches of science during this period, and end with a discussion of some of the issues still unresolved at the beginning of the 1900s.

Remarks on the Development of the Major Branches of Science, 1700–1900

Our first task will be to make a few brief observations about some of the major branches of science, such as chemistry and biology, and how they developed during the period in question. These remarks should help to illustrate the ways in which the various branches of science became "Newtonized," that is, developed within the broad Newtonian framework. We will begin with the development of modern chemistry.

Worldviews: An Introduction to the History and Philosophy of Science.   Richard DeWitt
© 2010 Richard DeWitt

Chemistry

The beginning of modern chemistry is generally dated to the late 1700s, with the work of Antoine Lavoisier (1743–94). To understand why this is, it is helpful to look at chemistry before the 1600s.

When we think of chemistry today, we tend to think of a largely quantitative discipline. If you have taken a high-school or college-level chemistry class with a laboratory component, you will no doubt have experience with this quantitative aspect of chemistry. Lab work today generally involves careful measurements of weight, volume, temperature, and so on. In short, chemistry today is largely a quantitative discipline.

Not so before the 1600s, when, in contrast, chemistry was viewed largely as a qualitative discipline. That is, a chemist was concerned primarily with qualitative changes – changes in color, for example. The well-known goal of the alchemists, of turning lead into gold, will serve to illustrate. Lead and gold are, qualitatively speaking, fairly similar. Both are dense, highly malleable metals. In fact, the main qualitative difference between lead and gold is color – lead is dull gray, whereas gold is a shiny yellowish color.

If one can make a relatively small qualitative change in lead – in particular, introduce into lead the qualitative yellowish color of gold – then the result (at least given the existing view) should be gold. And since it is reasonable to think the elements involved in combustion are associated with yellowish qualities (fire is itself, for example, yellowish-colored), one idea was to use fire to transfer the yellow quality to lead.

This is a very simplified account of some of the activities of the alchemists, but note the qualitative emphasis. Incidentally, the alchemist's approach may seem a rather primitive approach to chemistry, at least by modern standards. But given the state of chemistry at the time, their work was not at all unreasonable (Newton, among many others, did a fair amount of alchemical research). Our best current science will probably also seem primitive by the standards in place, say, 500 years from now. You do the best you can with the state of knowledge at the time in which you live.

At any rate, the qualitative approach to chemistry changed dramatically in the late 1700s. Antoine Lavoisier began to do extensive chemical investigations using the weight balance as a central laboratory tool. In doing so, he proposed new views that were better at explanation and prediction than the existing theories, and soon his quantitative approach began to dominate chemistry.

By the early 1800s, chemists were able to specify a number of quantitative laws. For example, at this time John Dalton (1766–1844) formulated his atomic theory, which was very much within the Newtonian framework. Dalton argued that the behavior of gases could best be understood as resulting from particles interacting under a repulsive force. Note the similarities between this approach and the Newtonian approach. Newton explained the behavior of, for example, the planets, as resulting from those bodies being under the influence of forces. Dalton, likewise, is explaining the behavior of gases as being, fundamentally, a matter of objects and forces acting on those objects.

These interactions could be (and were) characterized by quantitative laws, and eventually these laws could be presented mathematically. Here we see chemistry being subsumed under the distinctive Newtonian approach, involving bodies under the influence of forces, with the forces characterized mathematically. Eventually, during the 1800s and into the 1900s, the Newtonian approach to chemistry would be so productive that branches of chemistry would shade off into branches of physics, such that physics and chemistry are no longer entirely separate disciplines, but rather, different levels at which to investigate the world. And the world being investigated, whether through chemistry or physics, was conceived of very much as a Newtonian world, one of bodies under the influence of forces that were describable through precise mathematical laws.

Biology

Biology also achieved its modern form during this period. Biology is a somewhat broader subject, and, notably, very important work was done in the 1500s and 1600s. But it was in the 1700s and 1800s that it became clear that biological phenomena were not isolated from the Newtonian view of the universe.

One way to illustrate this concisely is by a brief consideration of the issues at stake between biological vitalists, on the one hand, and biological mechanists, on the other. The vitalist view was that living substances were different from nonliving substances, and hence the laws (such as Newtonian laws) that apply to nonliving objects do not necessarily apply to living objects. On an intuitive level, the vitalist view is easy to understand. Look, for example, at your arm, and compare your arm to, say, a rock. On the face of it, the two seem very different. And in general, living things seem very different from nonliving things, and so it is far from clear whether life could be explained with the same laws that explain nonliving things.

However, work in the 1700s, and continuing into the 1800s and 1900s, made it clear that the vitalist view was mistaken. This work involved a number of areas and a number of researchers. Below we will look at only two such areas, but they should be sufficient to give a good idea of the kinds of results that would establish that biological phenomena are not different in kind from phenomena outside of biology.

First, consider some discoveries concerning the structure and function of nerves. Investigations into nerves, including dissection of nerve fibers and the recognition of the distinction between motor and sensory nerves, dates back to at least 500 BC. Nerve fibers were long thought to be pipes or channels for the vital fluid or vital force that was believed necessary for life, and this view of nerve fibers fit in well with the vitalist position. In the late 1700s, Luigi Galvani (1737–98) began a series of experiments showing that electrical currents would cause contractions in the muscles of a frog's legs. Galvani's work was soon continued and extended by Alessandro Volta (1745–1827). As a result of the work of Galvani and Volta (and a number of others as well), it was soon established that nerve conduction is an electrical phenomenon, which is a quite different view of nerves than the old view of nerves as pipes or channels for the vital fluid or force.

As work continued in the 1800s, the physical and chemical basis of the electrical activity associated with nerves would come to be well understood. For our purposes, the key point is that what was originally viewed as a purely biological phenomena, and one that fit well within the vitalist view, is now understood to be, at bottom, an electrical phenomenon resulting from physical and chemical processes not different in kind from those found outside of biology. And so this area of biology is seen, at bottom, to fit in well with the overall mechanistic, Newtonian understanding of physical and chemical processes.

As a second brief example, consider the early years of organic chemistry. Before the early 1800s, the standard view was that what would come to be called "organic" compounds could be produced only by living organisms. Moreover, organic chemistry was originally viewed as closely tied to vitalism, in that the vital fluid or vital force believed necessary for life was generally viewed as required to produce organic compounds. And for a number of years, this seemed to be a reasonable view, supported in large part by the fact that no one had been successful in producing organic compounds other than by means involving living organisms.

However, in 1828 Friedrich Wohler (1800–82) managed to produce urea, which is clearly an organic compound, from an inorganic compound. Chemists were soon able to produce other organic compounds from inorganic ones, including increasingly complex organic compounds. By the mid-1850s this feat became routine, and this too substantially undermined the vitalist view of the sharp distinction between living and nonliving things.

A final example involves work in evolutionary theory, largely in the early to mid-1800s. The end result would be that life in general, for example the variety of species, would be seen to be the result of natural processes operating according to natural laws. This will be taken up in detail in Part III of this book, so for now, we will leave it at that.

This is again a very brief sketch of these developments, but it illustrates the sorts of major advances that were made in biology in the period from roughly 1700 to 1900. And, importantly, these examples illustrate how biological phenomena came to be viewed, at bottom, as no different from nonbiological phenomena. Although there were still a few defenders of vitalism even into the early 1900s, by this time it was pretty clear that the mechanist view was correct. Discoveries in the 1900s, for example in genetics, sealed the case, and provided a good understanding of how living phenomena arise from molecular-level events. In general, by the early 1900s, biology, chemistry, and physics were united, and came to be viewed as investigating the same Newtonian world, although at different levels.

Electromagnetic theory

One final example will suffice to illustrate the ways in which various phenomena were brought within the Newtonian framework. Phenomena involving electricity and magnetism had been studied at least since the ancient Greeks, but the 1700s and 1800s saw the most dramatic advances in our understanding of such phenomena.

In the mid-1700s, Benjamin Franklin (1706–90) demonstrated that lightning was an electrical phenomenon, and demonstrated as well a number of interesting connections between electrical and magnetic phenomena. Then in the late 1700s and early 1800s, researchers such as Charles Coulomb (1736–1806) and Michael Faraday (1791–1867), and of course a number of others, made major advances in our understanding of electricity and magnetism. Coulomb, for example, discovered an inverse square law governing magnetic and electrical repulsions and attractions, in which the strength of electrical or magnetic attraction or repulsion between two bodies is inversely proportional to the square of the distance between the bodies.

It is worth noting that the "inverse square" nature of Coulomb's law is quite similar to the inverse square nature of Newton's account of gravity. Recall that, on Newton's account of gravity, the strength of the gravitational attraction between two bodies is inversely proportional to the square of the distances between the bodies. So too with Coulomb's law, and in this way, Coulomb's law is very much in the Newtonian spirit. More generally, notice the change in the overall approach to electrical and magnetic phenomena. For most of our history, going back at least to the ancient Greeks, electrical and magnetic phenomena were described in a qualitative way. Now, though, such phenomena are coming to be viewed as governed by precise mathematical laws, much in the spirit of Newton's approach.

In the first half of the 1800s, Faraday discovered yet more connections between electrical and magnetic phenomena. By far his most influential discovery, in practical terms, was that a magnetic field could induce an electric current. This is the basic principle that still underlies the production of electricity today – that is, the considerable amount of electricity we all use each day results, essentially, from Faraday's discovery.

Although this may have been, in practical terms, his most influential discovery, Faraday's suggestion that electricity, magnetism, and light might be aspects of the same underlying source came to have greater theoretical influence (though this idea soon came to have countless practical applications as well). This suggestion of Faraday's – that electricity, magnetism, and light are in some sense different aspects of the same underlying phenomena – would soon be developed into the electromagnetic theory put forth by James Clerk Maxwell (1831–79) in the mid-1800s. Faraday's discoveries had been described in a largely qualitative way, but Maxwell would discover the fundamental mathematical equations underlying and uniting phenomena involving light, electricity, and magnetism. These equations

– generally termed "Maxwell's equations" – are widely considered to be the most important discoveries of this period.

Once again, this is a very brief and selective overview of some of the key developments concerning electricity, magnetism, and light. But note again the same general pattern: this was a period of extraordinary advances in these areas, and phenomena that were once seen as distinct, and that were approached qualitatively, have now been unified and treated in the quantitative, mathematical manner fundamental to the Newtonian approach.

General comments

Although we have looked specifically at only three areas of science, these should give an idea of the ways in which various branches of science came under the Newtonian umbrella during the period from 1700 to 1900. Notably, this was a 200-year period of impressive accomplishments and discoveries in a wide range of sciences. In general, by 1900 the various branches of science were making rapid advances. The overall Newtonian approach was proving extraordinarily productive. By about 1900, the sense was that we were close to a complete understanding of nature, and that only a few reasonably minor problems remained to be worked out. Let's turn now to some of these problems.

Minor Clouds

In an oft-quoted phrase from a talk given in 1900, Lord Kelvin, one of the leading British physicists, noted that there were only a few "minor clouds" in the otherwise brilliant sky of modern science. The two main clouds Kelvin was referring to were, on the one hand, the results of the Michelson–Morley experiment, and on the other hand, some problems in understanding black body radiation. We will look at both of these problems below, as well as looking briefly at some additional areas that were problematic around the turn of the 1900s.

As it turns out, understanding the results of the Michelson–Morley experiment would require the development of Einstein's theory of relativity, whereas understanding the issues involved in black body radiation, as well as the other issues discussed below, would require the development of quantum theory. These theories are two of the most important branches of modern physics, and both have nontrivial implications for Newtonian science and aspects of the Newtonian worldview. In light of this, Kelvin's minor clouds turned out to be anything but minor. In the remainder of this chapter, we will look briefly at the Michelson–Morley experiment, issues involved in black body radiation, and other seemingly minor issues. In subsequent chapters we will explore relativity theory and quantum theory, with an eye toward the implications these theories have for the Newtonian worldview.

The Michelson–Morley experiment

The Michelson–Morley experiment involved the speed of light, and the means by which light moves. Albert Michelson (1852–1931) and Edward Morley (1838–1923) performed a number of experiments involving these issues, the most crucial of which were in the late 1880s. Some background information will help.

Consider the movement of water waves. Such waves are the result of the mechanical interactions of the medium through which such waves move, namely water. And of course, one cannot have movement of water waves without the presence of the underlying medium, namely water.

Likewise for sound waves. Sounds waves result from the mechanical interaction of the medium through which such waves travel. Air is the typical medium, though sound waves will travel through a variety of other mediums as well. And again, the underlying medium is necessary for the transmission of sound waves: no underlying medium, no waves.

In general, in keeping with the general Newtonian outlook, the movement of any wave was believed to require the mechanical interactions of an underlying medium. So if light is a wave – as there was good evidence to think it was – then given this Newtonian framework, the movement of light required the presence of an underlying medium. The situation is nicely summed up in the following passage on light from Deschanel's *Natural Philosophy*, which was a standard physics textbook from the late 1800s. (Before the word "science" became standard, "natural philosophy" was the term used for what we would call science. Incidentally, the publication date of Deschanel's text is just prior to the Michelson–Morley experiment, which would raise substantial problems for this Newtonian view of the transmission of light.)

> Light, like sound, is believed to consist in vibration; but it does not, like sound, require the presence of air or other gross matter to enable its vibrations to be propagated from the source to the percipient. … It seems necessary to assume the existence of a medium far more subtle than ordinary matter … [that is] capable of transmitting vibrations with a velocity enormously transcending that of sound. … This hypothetical medium is called *aether*. (Deschanel, 1885, p. 947)

The origin of the name "ether" is tied to the ether of old, that is, the ether from the Aristotelian worldview that was thought to be the element found in the superlunar region. But besides the name, there is little similarity between the Aristotelian ether and the medium thought to underlie the transmission of light. (Incidentally, I will continue to spell it "ether"; Deschanel's spelling, with the "a," is another common spelling.)

Note how this view of the transmission of light fits into the Newtonian picture of a mechanical universe. Light, like other phenomena such as sound and water waves, was thought to need the mechanical interactions of an underlying medium. The Michelson–Morley experiment was designed to provide more direct evidence for the existence of the ether. The idea was to send two beams of light out and

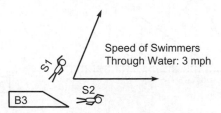

Figure 22.1 Boat and swimmer analogy

back (reflected from a mirror) in two directions, at 90-degree angles to each other. If light travels through a medium such as the ether, then given the fact that the Earth is presumably moving through this ether something like the way a boat moves through water, then we would expect the two beams to return at slightly different times. The key idea is similar to sending two swimmers out from a moving boat. Consider the illustration in Figure 22.1 of this analogy with boats and swimmers. Suppose all three boats are traveling through the water at the same rate, and that B1 and B2 are the same distance away from B3. One swimmer's job (S1 in the figure) is to swim out to the top boat (B1), and then return to the original boat (B3). Because of the movement of the boats through the water, this swimmer will have to swim at an angle toward the top boat, and again swim at an angle on the return to the original boat. The other swimmer's task (S2 in the figure) is to catch up to the boat (B2) that is ahead of the original boat, and then return to the original boat.

Relative to the original boat B3, both swimmers are swimming the same distance out and back. But note that, relative to the medium through which they are swimming (that is, the water), they are swimming different distances. In particular, S1 will swim a somewhat shorter distance, with respect to the water, than S2. (If you are interested, you can use the Pythagorean theorem, together with some algebra, to figure out the exact distance each person swims.) Since the two swimmers are swimming different distances relative to the medium through which they are moving, they will return to the original boat at different times, as illustrated in Figure 22.2.

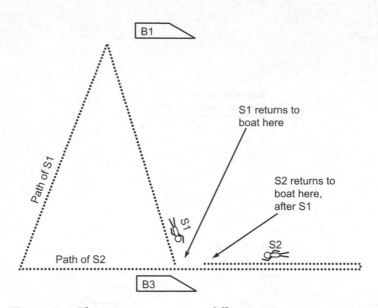

Figure 22.2 The swimmers return at different times

The idea behind the Michelson–Morley experiment is exactly analogous to the boat and swimmer analogy. In the experiment, Michelson and Morley sent two beams of light out from a fixed location, and this location is analogous to the boat B3 from which the swimmers were sent out. The two beams of light were sent out at right angles to one another, and these light beams are analogous to the swimmers S1 and S2. The beams of light were reflected off two mirrors at equal distances from the light source, and these mirrors are analogous to the boats B1 and B2.

If the Newtonian, mechanical view of the transmission of light is correct, in which light is propagated through the medium of the ether, then the light source and mirrors should be moving through the ether. This is because the light source and mirrors are themselves attached to the Earth, and the Earth, in its movement around the sun, will be moving through the ether. The ether, then, is analogous to the water in the boat and swimmer analogy. Although the light beams are moving the same distance relative to the light source, since the light source and mirrors are moving through the ether, the two beams of light would have to travel different distances through the ether (for the exact same reason the two swimmers travel different distances through the medium through which they are swimming). So, we would expect the two beams of light to return to the light source at slightly different times.

However, contrary to everyone's expectations, the light always returned at exactly the same time. This result was very surprising, and, as is proper in such cases, the experiment was repeated and verified numerous times. But the result was always the same – the light always returned at exactly the same time.

Notice how this fits the pattern of disconfirmation reasoning discussed in Chapter 4: if the Newtonian, mechanical picture of the transmission of light is correct, then the beams of light should return at different times. But they do not. So something is amiss.

Given the success of the Newtonian framework, it would not have been reasonable for scientists to reject the Newtonian view as a result of this one experiment. But something was not quite right, and as Lord Kelvin noted, the results of the Michelson–Morley experiment seemed to be one of the few clouds in an otherwise clear sky. As it turns out, these sorts of issues involving light were not minor at all, and, in fact, Einstein's relativity would eventually be required to solve these problems. And as we will see, relativity theory itself will have some interesting implications for our usual view of the universe. We will discuss that in the next chapter, but before doing so, we will briefly look at a few of the other seemingly minor clouds.

Black body radiation

Without going into too much detail, let me give a sketch of the problem with black body radiation. "Black body" is a technical term used in physics for an idealized object that absorbs all electromagnetic radiation directed toward it. For example, light is a form of electromagnetic radiation, so if we shine a light on a black body, it will absorb all the light, and so appear black (hence the name "black body"). In everyday life we do not come across such idealized objects, but we do have regular experience with black bodies that, while not quite the idealized black bodies of physics, will help illustrate the problem.

For example, consider a black coil burner on an electric stove. Such an object absorbs most of the light that hits it, and so it appears (mostly) black. Moreover, when heated, such an object will give off radiation. A coil burner on an electric stove, for example, will give off radiation both in the form of heat and also in the form of light (for example, when hot enough, the burner will glow red). And we can, of course, measure the pattern of radiation emitted by the coil.

An idealized black body, when heated, should also emit radiation. And given the developments in the Newtonian picture, as that picture developed during the 1700s and 1800s, one would expect a heated black body to give off a particular pattern of radiation. In fact, there were well-established equations, much in the quantitative Newtonian tradition, that should predict quite accurately the pattern of radiation we would expect from a heated black body. At the very end of the 1800s, and in the early 1900s, physicists were able to construct devices that should have emitted the pattern of radiation one would expect of a heated black body. However, the observed pattern of radiation emitted differed in important ways from that predicted from the Newtonian-based picture of radiation. In a nutshell, the situation was this: when looking just at the longer wavelengths of radiation emitted, the observed pattern of radiation matched closely the predicted pattern.

But at short wavelengths, the observed pattern of radiation was nowhere close to what was predicted. (Incidentally, these problematic shorter wavelengths are in the ultraviolet end of the electromagnetic spectrum, and hence this problem is sometimes referred to as the "ultraviolet catastrophe.")

The situation is analogous to that of the Michelson–Morley experiment, and is again a case of disconfirming evidence: given the existing view of radiation, a view very much in the Newtonian framework, one would expect certain experimental results. But in the case of black body radiation, those expected results are not observed. So in this case, as with the Michelson–Morley experiment, something seems to be amiss with the Newtonian picture.

Again, one certainly does not throw out an otherwise successful theory, much less the otherwise quite successful Newtonian framework, because of a relatively small number of problems. Nonetheless, the problem of black body radiation was another of the seemingly minor clouds.

It would eventually take quantum theory to explain the problem of black body radiation. As we will see in subsequent chapters, quantum theory has substantial implications for many of our assumptions about the world, and, in particular, it has substantial implications for important pieces of the Newtonian worldview. So as with the case of the Michelson–Morley experiment, the problem of black body radiation turns out to be more than merely a minor cloud.

Other issues

Although the results of the Michelson–Morley experiment, and the problem of black body radiation, were the two most substantial of the minor clouds referred to by Lord Kelvin, there were a few other puzzling issues at the beginning of the 1900s. Before closing this chapter, a brief mention of some of these issues is in order.

At the beginning of the 1900s, physicists realized that the light emitted from heated elements had unexpected patterns. For example, suppose we heat a sample of sodium. We will notice that it gives off a yellowish light (table salt contains sodium, and you can get a sense of this effect simply by heating a small amount of salt). This by itself is not problematic – it has long been known that some substances emit characteristic types of light when heated. What was surprising was that, in the early 1900s, physicists realized that the light emitted by, say, heated sodium, consists of only certain very specific wavelengths of light. (As it turns out, the pattern of wavelengths emitted by an element is unique to that element, and this fact has proved to be enormously useful in determining the makeup of unknown substances.) This very specific pattern of wavelengths emitted by heated elements, and, in particular, the fact that the light consists of only these specific wavelengths, was surprising. On the Newtonian-based view, one would expect the light emitted to consist of a very wide range of wavelengths, rather than being restricted to a few discrete wavelengths.

So again we have a case of disconfirming evidence for the Newtonian picture, although again, this seemed at the time to be a relatively minor problem. However, this too turned out to be a case that would be resolved only with the development of quantum theory.

During this same time period, at the end of the 1800s and the beginning of the 1900s, a variety of investigations were producing results that did not fit cleanly into any existing theory. These results were not necessarily direct problems for the existing theories, but neither was there an existing framework into which to place these results. A few examples will serve to illustrate these sorts of issues.

At the end of the 1800s and early 1900s, there was no shortage of physicists, some quite prominent, investigating what were typically called "cathode rays." We now understand cathode rays to be essentially a stream of electrons, but, at the time, the results from these researchers were often conflicting and, as mentioned, there was no overall framework into which to place them. So again, these results did not directly conflict with the existing general views, but they were somewhat puzzling.

At about the same time, what came to be called X-rays were discovered. We now view X-rays as a type of electromagnetic radiation, similar to visible light but of much shorter wavelength. As with cathode rays, many of the early results were puzzling. To cite just one example, some experiments suggested that X-rays had to be particles, but other experiments suggested equally strongly that they could not be particles, but instead had to be waves. So at the time, although many of the properties of X-rays were being discovered, there was no good understanding of what X-rays were, and, like cathode rays, they did not fit into any overall picture.

To take just one more example, this was also the time of the discovery of radioactivity, with research including, among other work, the important discoveries of Marie Curie (1867–1935) and Pierre Curie (1859–1906). (Marie Curie was the first woman to win a Nobel Prize in science, as well as the first person to win two Nobel Prizes.) Once again, the properties of radioactive elements proved puzzling. And as with the examples mentioned above, although radioactivity did not directly conflict with the generally accepted Newtonian view, the discoveries surrounding radioactivity did not fit cleanly into this framework.

The cases discussed above are by no means a complete list of the active areas of research in physics at the beginning of the 1900s, nor is it a complete account of areas with puzzling results. But these examples should be sufficient to give a sense of the types of results that, while not directly conflicting with the overall Newtonian view, nonetheless did not fit easily into that view.

Concluding Remarks

The period from 1700 to 1900 was a remarkably productive one, in which the framework supplied by those working in the 1600s, most notably Newton's contribution, was filled out in a very promising way. All the pieces seemed to be falling together beautifully, and the result was a Newtonian view of the universe in which almost everything seemed to be explained, or at least was expected to be explained.

In the section immediately above, we discussed two of the most prominent problems at the end of the 1800s, the problem of the Michelson–Morley results and the problem of black body radiation. We also looked briefly at a few other puzzling results from this time period. At the time, it was expected that the results of the Michelson–Morley experiment, and the problem of black body radiation, would eventually be resolved within an overall Newtonian framework. Likewise for the puzzling results discussed at the very end of the preceding section.

But as it happened, these results turned out to be more than just minor problems for the Newtonian view. In the remaining chapters we will look developments in recent times, largely from the 1900s. Two of these major developments – relativity theory and quantum theory – would eventually be able to account for the curious results of the Michelson–Morley experiment and the problem of black body radiation. The other major recent development is evolutionary theory, and we will explore it as well in the next part of the book. As we will see, all of these new developments have substantial implications for views we have had since at least the beginning of the Newtonian period.

Part III

Recent Developments in Science and Worldviews

In Part III, we explore Einstein's theory of relativity, quantum theory, and evolutionary theory. As we will see, all of these theories require substantial changes in our worldview. Having seen, in Part II, the changes required by new discoveries in the 1600s, we now see that recent developments require changes in our own view of the world. And as occurred in the 1600s, we see that some of the beliefs we have long taken as obvious empirical facts turn out, in light of recent developments, to be incorrect philosophical/conceptual facts.

Chapter Twenty-Three

The Special Theory of Relativity

In the previous chapter, we saw how the Newtonian view developed quite nicely during the two centuries following the publication of Newton's *Principia*, and that, by 1900, there were only a few seemingly minor issues that did not quite fit the Newtonian picture. One of these issues, the results of the Michelson–Morley experiment, would not be resolved until the development of relativity theory in the early 1900s. Our main goal in this chapter will be to understand the main aspects of the special theory of relativity. In the next chapter we will explore the general theory of relativity.

Albert Einstein (1879–1955) published the special theory of relativity in 1905 and, as the name suggests, the theory is not completely general, but rather is applicable so long as certain special circumstances apply. In 1916, he published the general theory of relativity, which, again as the name suggests, is a completely general theory, not restricted to the circumstances required in order to use the (much simpler) special theory.

Earlier in the book, we spent some time exploring some of the mistaken philosophical/conceptual facts of the Aristotelian worldview, especially the perfect circle and uniform motion facts. In this chapter, we look for the first time at some of our own beliefs that we have long taken as obvious empirical facts but that turn out, in light of relativity theory, to be mistaken philosophical/conceptual facts. We will begin with a brief discussion of two such beliefs.

Absolute Space and Absolute Time

The two main philosophical/conceptual facts in question involve what are commonly referred to as *absolute space* and *absolute time*. These beliefs about space and time are very commonsensical – most people take them as obvious empirical facts about space and time. We will begin with an example to help illustrate some issues involved with the notion of absolute space.

Suppose on a table before us we have some medium-sized object, say, a solid metal rod. And suppose we get out a trustworthy ruler, lay it alongside the rod, and determine that the rod is exactly one meter in length. This fact, that the rod is one meter long, is as clear an example of an empirical fact as we can have. We have direct, straightforward empirical evidence that the rod is indeed one meter long. So far, so good.

Now suppose we get a second more or less identical rod, and lay the second rod next to the first, and assure ourselves that the two rods are the same length, that is, one meter long. Now suppose I tie a string to this second rod, and begin twirling it rapidly over my head. So here I am, string in hand, twirling this metal rod over my head as fast as I can. Suppose I now ask you this question: As I am twirling the rod about, how long is the rod?

I suspect your natural inclination is to say that the rod is the same length as before, one meter long. This is a perfectly reasonable response, but notice that this belief, that the rod is still the same length as the rod on the table, one meter long, is not a straightforward empirical fact. Notice you have no way of directly measuring the rod as it is spinning about my head. So whatever the source of your belief that the rod is one meter long, it cannot be based on direct empirical evidence.

I suspect your belief that the rod that is in motion continues to be one meter long is based on these two factors: (a) your belief from a moment ago, based on direct empirical evidence, that the rod was one meter long, and (b) your belief that space is absolute, that is, that space does not shrink or enlarge as a result of motion (and so the distance between the ends of a rigid object such as the metal rod does not shrink or enlarge simply because that object happens to be in motion).

This latter belief, that distances do not change as a result of motion, is one way of phrasing the concept of absolute space. The key idea is that space is space – one meter is one meter, regardless of one's point of view or whether one is sitting at a desk or moving about the Earth at high speeds on a mission in space. And again, this notion – that space, for example, the amount of space between the ends of the metal rod, does not change simply as a result of motion – is what I will have in mind when I speak of absolute space. (It is worth mentioning that "absolute space," and "absolute time" as well, are sometimes used in a somewhat different sense, to refer to what are often called *substantivalist* versus *relational* views of space and time. I do not discuss the debate between advocates of substantivalist and relational views in the body of this chapter, but a brief discussion of the distinction can be found in the Chapter Notes at the end of the book.)

As we will see below, relativity theory challenges this commonsensical belief about space. Before discussing relativity, however, we will first consider absolute time. And again, we will begin with an example.

Suppose we know John and Joe are identical twins, and suppose they were born at the exact same moment. Having twins born at the exact same moment would be difficult (but not impossible, for example, if they were born by Caesarean section), but for now suppose we ignore this difficulty and accept that the two were born at the exact same time. Suppose also that you have never met Joe. Now suppose I tell you that John and Joe are normal, healthy individuals, and that John is 20 years old. And this time my question to you is this: Given that John and Joe are twins, and John is 20 years old, how old is Joe?

Again, I suspect your natural inclination is to say that Joe is also 20 years old. Again this is perfectly reasonable, but again note that your belief cannot be based entirely on direct empirical evidence. After all, you have never even met Joe, so you have little if any direct observational evidence about him. I suspect your belief that Joe is 20 years old is based on (a) your belief that John and Joe are twins and John is 20 years old, and (b) your belief that time is absolute, that is, that the amount of time that passes is the same everywhere (and so when 20 years have passed for John, 20 years must also have passed for Joe).

This latter belief, that the amount of time that passes is the same everywhere, is one way of phrasing the idea of absolute time. The idea is that time is time: the amount of time that passes is absolute, the same for everyone and everywhere.

As with the idea of absolute space, relativity theory also challenges this notion of absolute time. With this background material in mind, we will now move on to look at the special theory of relativity, and then later return to make clear the implications relativity theory has for our concepts of space and time.

Overview of the Special Theory of Relativity

Some aspects of special relativity are unusual, but the theory itself is not particularly difficult. In contrast, the general theory of relativity is appreciably more difficult. In this chapter, unless otherwise specified, when I speak of relativity theory I will always have in mind the special theory of relativity. Then as noted, we will turn to the general theory in the next chapter.

As is often the case, it is helpful to begin with a picture. Suppose Joe is on the ground, and (speaking from Joe's perspective), Sara is flying by overhead. The consequences of relativity are most dramatic when high speeds are involved, so let's imagine the speed involved is quite high, say, 180,000 kilometers per second (this is much faster than any person can travel with current technology). To illustrate the consequences to space and time, it will be helpful to have Sara and Joe each have clocks separated by some particular distance. Let's suppose Sara has two clocks, which we will call SC1 (for "Sara's Clock 1") and SC2. Suppose that, when Sara measures the distance between those clocks, she finds them to be separated by 50 kilometers. We will write this as "50 (s) kilometers," using "(s)" to indicate that this is the distance Sara measures. We will suppose also that Joe has two clocks, JC1 and JC2, which he measures as 1,000 kilometers apart. Figure 23.1 summarizes the situation.

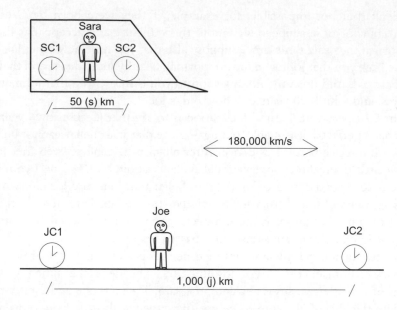

Figure 23.1   Illustration for special relativity

In a moment we will use this illustration to explore the implications of special relativity. But first, it will help to understand the two basic principles that lead to these implications. The first is the principle of the constancy of the velocity of light.

> *Principle of the Constancy of the Velocity of Light (PCVL)*: The speed of light in a vacuum will always be measured to be the same.

For example, if Joe and Sara measure the speed of light in a vacuum, they will both arrive at the same result. The speed of light in a vacuum is approximately 3.0×10^8 meters per second, or 300,000 kilometers per second. Incidentally, the speed of light is generally represented by "c." So Sara and Joe, if they measure c, the speed of light, will both measure light to be traveling at 300,000 kilometers per second.

Notice that, if the PCVL is correct, then light behaves very differently than typical objects. For example, suppose Sara and Joe measure the speed of a moving baseball rather than the speed of light. Suppose Sara shoots a baseball forward, that is, in Figure 23.1, she shoots the ball straight ahead of herself toward the right side of the illustration. Suppose that, from her point of view, she shoots the ball at 100 kilometers per second. In this case, when Sara measures the speed of the baseball, she will measure it to be moving at 100 kilometers per second.

But from Joe's perspective, the speed of the baseball is a combination of the speed at which it is shot forward (100 km/s) plus the speed at which Sara is moving

(180,000 km/s). So when he measures the speed of the moving ball, he will measure the ball to be moving at 180,100 kilometers per second.

But if the PCVL is correct, then light does not behave like the baseball. For example, if Sara turns on a flashlight while she is above Joe, then when Sara and Joe measure the speed at which the leading edge of the beam of light is traveling, they will both measure it to be going exactly the same speed – 300,000 kilometers per second. In short, if the PCVL is correct, then light behaves very differently than, say, a moving baseball.

Again, the PCVL is one of the basic principles on which the special theory of relativity is based. The other is generally called the principle of relativity (be careful not to confuse the principle of relativity with the theory of relativity). Roughly speaking, the principle of relativity is as follows.

Principle of Relativity (rough version): There is no privileged point of view from which to say who is in motion and who is at rest.

For example, in the scenario pictured in Figure 23.1, Joe is perfectly entitled to consider himself to be at rest and Sara to be in motion. But if the principle of relativity is correct (and there is good reason to think it is), then Sara is equally entitled to consider herself at rest and Joe to be in motion.

As noted, this is a somewhat rough version of the principle of relativity. A more careful formulation is as follows.

Principle of Relativity (more careful version): If two observers are in identical laboratories, except that the labs are moving in straight lines with uniform speed (that is, neither speeding up nor slowing down) with respect to each another, and if identical experiments are carried out in the two labs, then the results of the experiments will be identical.

For example, in the situation pictured in Figure 23.1, Sara and Joe are moving in a straight line with uniform velocity relative to one another. So if the principle of relativity is correct, any experimental result Sara achieves can be matched by Joe, and any result Joe gets can be matched by Sara.

Importantly, if the principle of relativity is correct, then there is no experimental difference between two labs moving relative to one another in a straight line at constant speed. So in the case of Sara and Joe above, there can be no empirical grounds for saying that one of them is "really" at rest while the other is "really" in motion. Take a moment to appreciate the importance of this fact: if the principle of relativity is correct, then there can be no empirical grounds – none whatsoever – for saying that either Sara or Joe is really at rest and the other is really in motion.

In Einstein's 1905 paper "On the Electrodynamics of Moving Bodies," in which he first presented what is now commonly referred to as the special theory of relativity, he put forth the principle of relativity and the principle of the constancy of the velocity of light as "postulates." That is, he essentially took these principles as givens, and showed that a consistent theory (the special theory of relativity) followed from them. Einstein then showed that this new theory could account for certain problematic issues that arose when applying Maxwell's electromagnetic theory (discussed in the previous chapter) to moving bodies. (This is why the title of the paper focuses on the electrodynamics of moving bodies, rather than making any mention of a new theory of relativity.)

But although the principle of relativity and the PCVL are treated as postulates in his original paper, they are quite plausible principles. For example, experiments such as the Michelson–Morley experiment (discussed in the previous chapter), as well as a number of similar experiments in which the speed of light always turned out to be the same no matter what the circumstances, suggest a principle such as the PCVL. In fact, Einstein makes note of such experiments when introducing the PCVL (it is not clear whether Einstein was specifically familiar with the Michelson–Morley experiment itself, though he was certainly familiar with other similar experiments).

Likewise, the principle of relativity is a plausible principle. Consider again the example of Joe and Sara in Figure 23.1. Given the picture, Joe is on the ground (presumably, Joe is on the surface of the Earth) and Sara seems to be on some sort of a ship. Our initial inclination is probably to say that Joe is the one at rest and Sara is in motion. But notice that this preference for the ground-based point of view is no doubt a result of the fact that we spend most of our existence on the ground. Thus, while it may be understandable that we naturally take the ground-based point of view, there surely is nothing special about this point of view. If we spent most of our time on the surface of Mars, we would naturally take that as our usual point of view. If we had been born and raised on the moon, we would naturally take that as our usual point of view. And if we spent most of our lives on ships such as the one Sara is on, we would naturally take that perspective.

The basic moral is that none of these points of view is special, or in other words, there is no privileged point of view. There are no grounds for saying that Joe is really at rest and Sara is really moving, or for saying that Sara is really at rest and Joe is really moving. All that can be said is that, from Joe's point of view, Sara is in motion, and that from Sara's point of view, Joe is in motion. In short, both the principle of relativity and the PCVL, although treated as postulates in Einstein's original paper, are plausible principles.

If we accept the principle of relativity, then when we speak of motion it must be understood as relative motion, that is, as motion relative to a point of view. This is an important point to keep in mind – it is not a problem to speak, say, of people and objects being in motion, but this must be understood not as absolute motion, but rather as relative motion, that is, motion from a particular point of view.

To briefly review: The basis of the special theory of relativity is the PCVL and the principle of relativity. Moreover, both the PCVL and the principle of relativity seem like plausible principles.

However, accepting the PCVL and the principle of relativity requires accepting that, for objects in motion, something surprising happens to space and time. In particular, the PCVL and the principle of relativity together entail the following.

(1) *Time dilation*: Time passes more slowly for people and objects in motion. In particular, when in motion, time moves more slowly by a factor of

$$\sqrt{1 - \left(\frac{v}{c}\right)^2}.$$

This equation is referred to as the *Lorentz–Fitzgerald equation*.

(2) *Length contraction*: Distances shrink for people and objects in motion. In particular, when in motion, distances shrink by a factor of

$$\sqrt{1 - \left(\frac{v}{c}\right)^2}.$$

(Note that this is the same equation as in (1), that is, the Lorentz–Fitzgerald equation.)

(3) *The relativity of simultaneity*: Events that are simultaneous from a moving point of view are not simultaneous from a stationary point of view. For example, suppose that from Sara's point of view, her two clocks, SC1 and SC2, are synchronized. From Joe's point of view, the two clocks are not synchronized. In particular, SC1 will be ahead of SC2 by the amount

$$\frac{\left(\dfrac{lv}{\sqrt{1 - \left(\frac{v}{c}\right)^2}}\right)}{c^2}.$$

In this equation, *l* refers to the length between the two clocks. The equation can be simplified to the equation $\dfrac{l^{\star}v}{c^2}$, where $l^{\star}$ refers to the length between the two clocks from the moving point of view. In what follows, I will use this simpler equation. Also, note that it is the clock that is in the rear, with respect to the direction of motion, that will be ahead with respect to time.

It is worth emphasizing that (1), (2), and (3) are deductive consequences of the PCVL and the principle of relativity. That is, using nothing more than high-school algebra, it is possible to mathematically derive (1), (2), and (3) from the PCVL and the principle of relativity. So if the PCVL and the principle of relativity are correct, then so long as basic mathematics is trustworthy, the effects summarized in (1), (2), and (3) must also be correct.

To see how (1), (2), and (3) apply, it will be easiest to consider a particular scenario. So consider again the situation pictured in Figure 23.1. Let's begin with what the situation looks like from Joe's point of view.

From Joe's point of view, Sara is in motion. Again, time moves more slowly for people and objects in motion. In particular, from Joe's point of view, time for Sara will be moving more slowly by the factor noted in (1) above, that is, by the factor specified by the Lorentz–Fitzgerald equation. So, for example, for every 15 minutes that pass on Joe's clocks, only 15 minutes multiplied by

$$\sqrt{1-\left(\frac{v}{c}\right)^2},$$

that is, 12 minutes, pass on Sara's clocks. Note that this is not a result of any sort of problem with Sara's clocks, nor is it any sort of illusion for Joe. When in motion, time moves more slowly. Since, from Joe's point of view, Sara is in motion, the amount of time that passes for Sara, and the amount of time that is measured by her clocks, will be less than the amount of time that passes for Joe and the amount of time that is measured by his clocks.

Likewise, distances shrink for people and objects in motion. For example, although Sara measures the distance between her two clocks as 50 kilometers, from Joe's point of view, the two clocks are separated by only 50 kilometers multiplied by

$$\sqrt{1-\left(\frac{v}{c}\right)^2},$$

that is, 40 kilometers. In short, from Joe's point of view, Sara's distances are shrinking.

Incidentally, as a detail not yet noted and one I will mention only briefly, it is worth pointing out that distances shrink by this factor only in the direction of motion. In this example, the direction of motion is horizontal, so to speak, and so (from Joe's point of view) Sara's distances, in the horizontal direction, shrink by the factor specified by the Lorentz–Fitzgerald equation. In the vertical direction, Sara's distances do not shrink at all. So in this case, for example, from Joe's point of view Sara will be skinnier but not shorter. Here are the particulars for those who are curious: where $\theta = 0°$ represents the direction of motion and $\theta = 90°$ represents the direction perpendicular to the direction of motion, distances at an angle θ between $0°$ and $90°$ shrink by a factor of

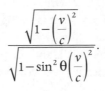

$$\frac{\sqrt{1-\left(\frac{v}{c}\right)^2}}{\sqrt{1-\sin^2\theta\left(\frac{v}{c}\right)^2}}.$$

Notice that, where $\theta = 0°$ (that is, when dealing with distances in the direction of motion), this equation reduces to the Lorentz–Fitzgerald equation (thus this equation, when dealing with distances in the direction of motion, gives the same result as applying the Lorentz–Fitzgerald equation). Where $\theta = 90°$ (that is, when dealing with distances perpendicular to the direction of motion), this equation reduces to 1, and hence there is no shrinkage of space when dealing with distances perpendicular to the direction of motion. Hereafter, we will consider only examples in which the distances are in the direction of motion, and hence this detail need not concern us further.

Finally, let us consider (3), the relativity of simultaneity. As noted in (3), events that are simultaneous from Sara's point of view will not be simultaneous from Joe's point of view. For example, suppose that from Sara's point of view, her two clocks are synchronized. Then from Joe's point of view they will not be synchronized. As noted in (3) above, SC1 will be ahead of SC2 by $\dfrac{l^{\star}v}{c^2} = 0.0001$ seconds. So from Sara's point of view, the two clocks read 12:00 noon at the exact same time. But from Joe's point of view, this is not the case. SC1 is 0.0001 seconds ahead of SC2, so when SC1 reads 12:00 noon, SC2 reads 0.0001 seconds before noon.

Thus far we have described the situation from Joe's point of view. But recall the principle of relativity. From Sara's point of view, she is at rest and Joe is in motion. So let us look at things from her perspective. From Sara's point of view, since Joe is in motion, time is moving more slowly for him by the factor specified by the Lorentz–Fitzgerald equation. For example, for every 15 minutes that pass for Sara, only 12 minutes pass for Joe. Likewise, since Joe is in motion, distances shrink, so that Joe's clocks are only 800 kilometers apart. And supposing that, from Joe's point of view, JC1 and JC2 are synchronized, they will not be synchronized from Sara's point of view. In particular, JC2 will be ahead of JC1 by $\dfrac{l^{\star}v}{c^2} = 0.002$ seconds.

So as mentioned above, if the PCVL and the principle of relativity are correct, strange things happen to objects in motion. Distances shrink, time moves more slowly, and events that are simultaneous from one point of view are not simultaneous from another point of view.

The Irresistible Why Question

It is worth emphasizing that the effects of motion on distance, time, and simultaneity, described above, have been well confirmed by countless experiments. So there is little question that motion has these surprising effects. Given this, an almost irresistible question at this point is the *why?* question. Why do distances and time contract for people and objects in motion? Why does whether two events occur simultaneously depend upon the motion of those observing the events? What we have just discussed above seems to run counter to some basic and deeply held convictions most people have about how space and time behave. If what we have just discussed is correct – and almost without question it is, as these effects have been empirically confirmed countless times – why does motion have these seemingly strange effects on lengths and time? The most accurate answer to this question is one that, unfortunately, I think most people do not find immediately satisfying. The answer does, though (or at least I think so), grow on you after a while. The best and most accurate answer to the question of why motion has these effects on space and time is this: That is simply the type of universe we live in. Our predecessors discovered, much to their surprise, that they did not inhabit the type of universe they always thought they had – the universe turned out, for example, not to be teleological and essentialistic, with heavenly bodies revolving in perfect circles at uniform speeds. Likewise, we have discovered, much to our surprise, that we live in a universe in which space and time are not the way most of us have always thought they were. In other words, much as our predecessors discovered that what had long been taken to be empirical facts turned out to be mistaken philosophical/conceptual facts, so too we have discovered that certain common-sense aspects of our beliefs about space and time, which most of us have long taken to be obvious empirical facts, have turned out to be mistaken philosophical/conceptual facts.

Is Special Relativity Self-Contradictory?

It might seem that some sort of contradiction must be lurking in relativity theory. For example, since from Joe's point of view, time is moving more slowly for Sara, Sara is aging more slowly than Joe. And from Sara's point of view, Joe is aging more slowly. Intuitively, it seems that they cannot both be right. For example, suppose Sara and Joe are twins. Then from Joe's point of view he is the older twin, and from Sara's point of view she is the older twin. According to relativity theory, they are both right. How can they *both* be the older twin, unless there is something contradictory about relativity theory?

The goal of this section is to convince you there is no contradiction in relativity theory. I have seen no better way to illustrate this point than that used by David Mermin in his *Space and Time in Special Relativity*. What follows owes much to his means of illustrating how there is no contradiction in special relativity.

Consider again the scenario with Sara and Joe. For this illustration, we will need Sara to have only one clock, so we will omit her second clock. Also, it will be easier to imagine that all the clocks are digital clocks (so we can speak, for example, of a clock reading "0.00" rather than reading 12 noon). I will present two snapshots in time, which I will simply label A and B. As preliminaries, let us suppose the following are true.

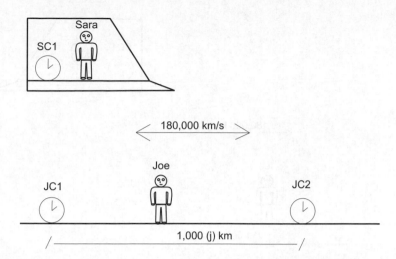

Figure 23.2 Snapshot A

- From Joe's point of view, his two clocks are separated by 1,000 (j) kilometers.
- From Joe's point of view, his two clocks are synchronized.
- The speed involved is 180,000 kilometers per second.
- At the moment Sara's clock is directly over Joe's first clock (JC1); both clocks read 0.00.

The first snapshot in time, snapshot A, will be as Sara's clock is directly over Joe's first clock (JC1). Given the background information above, this snapshot will look as illustrated in Figure 23.2. A fraction of a second later, Sara's clock will be directly over Joe's second clock, and this will be snapshot B, which will look as illustrated in Figure 23.3.

Given the scenario just described, here is what the situation will look like from Joe's and Sara's points of view.

From Joe's point of view:

(J1) Sara is in motion, left to right, at 180,000 kilometers per second.
(J2) JC1 and JC2 are 1,000 kilometers apart.
(J3) JC1 and JC2 are synchronized.
(J4) At snapshot A, when SC1 is directly over JC1, all three clocks read the same time, in particular, SC1, JC1, and JC2 all read 0:00.
(J5) At snapshot B, a fraction of a second later, SC1 passes over JC2. Between snapshot A and snapshot B, Sara has traveled 1,000 kilometers at 180,000 kilometers per second, so $\dfrac{1,000}{180,000} = 0.005555$ seconds have elapsed

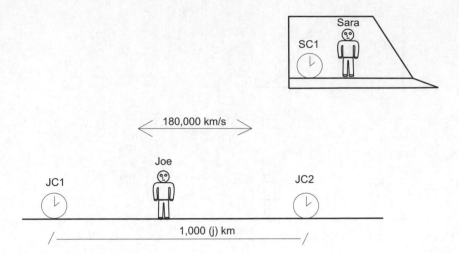

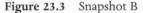

Figure 23.3   Snapshot B

between snapshot A and snapshot B. So, at snapshot B, when SC1 is directly over JC2, JC2 reads 0.005555. And since JC1 is synchronized with JC2, JC1 also reads 0.005555.

(J6)   Since Sara is in motion, time is moving more slowly for her and her clock. In particular, although 0.005555 seconds have passed for Joe between snapshot A and snapshot B, only $0.005555 \times \sqrt{1 - \left(\dfrac{v}{c}\right)^2}$ seconds, that is, 0.004444 seconds, have passed for Sara. In other words, at snapshot B, SC1 reads 0.004444.

Now let's look at the situation from Sara's point of view.

From Sara's point of view:

(S1)   Joe is in motion, right to left, at 180,000 kilometers per second.

(S2)   JC1 and JC2 are only $1{,}000 \times \sqrt{1 - \left(\dfrac{v}{c}\right)^2} = 800$ kilometers apart.

(S3)   JC1 and JC2 are not synchronized. JC2 is ahead of JC1 by $\dfrac{l^* v}{c^2} = 0.002000$ seconds.

(S4)   At snapshot A, when JC1 is directly under SC1, SC1 and JC1 both read 0.00. But JC2, since it is not synchronized with JC1 (see (S3) immediately above), reads 0.002000.

(S5)   At snapshot B, a fraction of a second later, JC2 passes under SC1. Joe has traveled 800 kilometers at 180,000 kilometers per second, so

$\dfrac{800}{180,000} = 0.004444$ seconds have elapsed. So at snapshot B, when JC2 is directly under SC1, SC1 reads 0.004444.

(S6) Since Joe is in motion, time is moving more slowly for him and his clocks. Between snapshot A and snapshot B, although 0.004444 seconds have passed for Sara, only $0.004444 \times \sqrt{1 - \left(\dfrac{v}{c}\right)^{2}} = 0.003555$ seconds have passed for Joe and his clocks. But remember that at snapshot A, JC2 read 0.002000 (see (S4) above). Since at snapshot A, JC2 read 0.002000, and between snapshot A and snapshot B 0.003555 seconds have passed for Joe, at snapshot B, when JC2 is directly under SC1, JC2 will read 0.002000 + 0.003555 = 0.005555.

Note an important point about all the facts Joe and Sara can jointly verify. For example, at snapshot A, their two clocks are adjacent to one another, so they can jointly verify the time on those clocks as the clocks are right next to one another. We can imagine they each take a photograph that show both SC1 and JC1 at snapshot A. Their photographs had better look the same, since they are photos of the same moment in space and time. And indeed, their photos will both show SC1 and JC1 reading 0.00 at snapshot A.

Likewise at snapshot B. Here, JC2 and SC1 are directly next to each other, and so Sara and Joe can independently take photos of the two clocks. And again, the two photos had best look the same, unless the universe is far stranger than we think it is. And indeed, the two photos will both show SC1 reading 0.004444, and JC1 reading 0.005555.

But although Sara and Joe agree on these facts they can jointly verify, they will disagree substantially on what has occurred. From Joe's point of view, 0.005555 seconds have elapsed for him between snapshot A and snapshot B, whereas only 0.004444 seconds have elapsed for Sara. Hence, from Joe's point of view, he is the older twin.

But from Sara's point of view, 0.004444 seconds have elapsed for her between snapshot A and snapshot B, whereas only 0.0035555 seconds have elapsed for Joe. Hence, from her point of view, she is the older twin.

In short, Sara and Joe both say that they are the older twin. And from their respective points of view, each is equally right.

What about their disagreements on what the other clocks read?

Before closing this section, it may help to consider some of the disagreements between Sara and Joe. As we saw above, they agree on what the clocks read in the situations (snapshot A and snapshot B) where they can jointly verify what certain of the clocks read. What about their disagreements? In particular, in snapshot A, Sara and Joe disagree on what Joe's distant clock JC2 reads. Likewise at snapshot B, they disagree on what JC1 reads.

We considered above the idea that Sara and Joe could take photos of the clocks as they were directly next to each other. And we saw that their photos would appear the same. What if they took long-distance photos of the other clocks involved? Recall that, at snapshot A, Sara and Joe disagree on what Joe's second clock, JC2, reads. Joe thinks that at snapshot A, JC2 reads 0.000, whereas Sara thinks that at snapshot A, JC2 reads 0.002. So suppose at snapshot A, Sara and Joe both take a long-distance photo of Joe's second clock, JC2. Such a long distance photo would be technically difficult, but not impossible. So would such a photo reveal a contradiction between Sara and Joe's points of view?

The answer will be no, but to understand this situation, we need to remind ourselves of some facts about light and long-distance photography. First, remind yourself that the speed of light is fast, but still finite. That is, it takes time for light to travel from an object to your eye, or from an object to a camera. To illustrate this fact, consider the light from the sun. The sun is about 93,000,000 miles, or 150,000,000 kilometers, from Earth. At 300,000 kilometers per second, light takes about eight minutes to travel from the sun to the Earth. So when you see the sun, the light hitting your eye (that is, the light that is causing you to have your visual image of the sun) left the sun about eight minutes earlier. In other words, what you are seeing is not the sun as it is at this moment; rather, you are seeing the sun as it was about eight minutes ago.

This point is much more dramatic when you think about light from stars. For example, the Andromeda Galaxy is the most distant object you can see with the naked eye, and this galaxy is 2,200,000 light years away. When you look at this galaxy, what you are seeing is the Andromeda Galaxy as it was 2.2 million years ago, not as it is now, and if you take a photograph of it, your photograph will be a photo of the Andromeda Galaxy as it was 2.2 million years ago.

So if at snapshot A, Sara and Joe take photographs of Joe's distant clock, that is, JC2, they must take into account the fact that it took some time for the light that caused the photograph to travel from the clock to their cameras. When they take this fact into account, here is what the situation looks like.

First point: When Sara and Joe take photos of JC2, their photos will both show the same time on JC2. Specifically, their photos will both show JC2 reading −0.003333. Don't miss the negative sign – the clock is reading 0.003333 seconds before it reaches 0.00. When the two of them take into account the time it takes the light to travel from JC2 to the camera, the situation is as follows.

From Joe's point of view:
Joe's work is easy. From his point of view, the light traveled 1,000 (j) kilometers from JC2 to his camera. At 300,000 kilometers per second, the light will have taken 0.003333 seconds to reach his camera. Joe deduces, correctly (for him), that at the moment the photo was taken, that is, at snapshot A, JC2 read −0.003333 + 0.003333 = 0.000. He concludes, correctly (for him),

that at snapshot A, both of his clocks read 0.000, and that his two clocks are synchronized.

From Sara's point of view:

Sara's calculations are a bit more difficult. Still, they require nothing more than basic algebra. If you want to skip the details, you can trust my figures and skip to the end of this section. But if you are curious, here is how Sara reasons.

From Sara's point of view, Joe is in motion toward her at 180,000 kilometers per second. So the light that produced her photograph had to have left JC2 when JC2 was much further away than the 800 (s) kilometers separating JC1 and JC2. That is, the light from JC2 – the light that results in her photo – is traveling toward her at 300,000 kilometers per second. Joe, on the other hand, is traveling toward her at 180,000 kilometers per second. So for the light and Joe to reach her at the same time, that is, at snapshot A, the light that results in her photo left JC2 when JC2 was much further away than the mere 800 (s) kilometers separating JC1 and JC2.

To figure out how far this light has traveled, Sara reasons as follows. Suppose d represents the distance the light traveled (again, the light that resulted in Sara's photo of JC2) and t represents the time it took this light to travel, at 300,000 kilometers per second, from JC2 to her camera. Then Sara knows that

$$t = \frac{d}{300,000}.$$

Keeping in mind that Joe is 800 (s) kilometers closer to her than is JC2, and that Joe is traveling at 180,000 kilometers per second, Sara also knows that

$$t = \frac{d - 800}{180,000}.$$

With a little algebra, from these facts Sara can deduce that the light that produced the picture left JC2 when JC2 was 2,000 (s) kilometers away. So the light traveled 2,000 (s) kilometers at 300,000 kilometers per second, which would take 0.00667 (s) seconds. Since JC2 is moving, its time is moving more slowly. In particular, when 0.00667 (s) seconds have elapsed, only 0.005333 (j) seconds will have elapsed on JC2. So when the light reached her camera (that is, at snapshot A, when Joe and his first clock are directly below her), 0.005333 (j) seconds have passed on JC2.

So Sara deduces, correctly (for her), that at the moment the photo was taken, JC2 read −0.003333 + 0.005333 = 0.00200. She concludes, correctly (for her), that Joe's two clocks are not synchronized. Rather, Joe's second clock (JC2) is 0.00200 seconds ahead of his first clock (JC1).

Once again, notice that Sara and Joe agree on what their respective photos read – both show, at snapshot A, that JC2 reads −0.003333. But they disagree on issues involving how much time has elapsed, how far apart objects are, and whether events are simultaneous.

Spacetime, Invariants, and the Geometrical Approach to Relativity

Shortly after Einstein published the special theory of relativity, one of his earlier mathematics teachers, Hermann Minkowski (1864–1909), recognized that what is called the *spacetime interval* is an *invariant* property of special relativity. Understanding the spacetime interval will allow us to get an idea of one of the central notions associated with relativity, namely the notion of spacetime, as well as allow us to understand the concept of variant and invariant properties. It will also allow us to see a brief sketch of a common alternative approach to relativity, that being the geometrical approach.

Occasionally, one hears claims that, according to Einstein's relativity theory, "everything is relative," or sentiments to that effect. As we saw above, length, time, and simultaneity do indeed differ between moving and stationary points of view, and thus such properties are relative to a point of view. But it is far from correct to think that all properties are relative.

We have already seen one property that is not relative – the speed of light. According to the principle of the constancy of the velocity of light, the speed of light (in a vacuum) will always be measured to be the same, regardless of what point of view is involved. According to relativity, then, the speed of light is not relative. Properties that are the same from every point of view, such as the speed of light in relativity theory, are considered invariant properties.

Note that different theories often treat different properties as variant and invariant. For example, length, time (that is, how much time has passed between two events), and simultaneity (that is, whether two events are or are not simultaneous) were considered invariant on the Newtonian view, but as we saw earlier in this chapter, these properties are not invariant according to relativity theory. On the other hand, the speed of light is invariant on relativity theory, but was not considered invariant on the Newtonian view. (Recall the Michelson–Morley experiment discussed in the previous chapter, which was designed to detect differences in the speed of light. As noted in that chapter, the expectation, given the usual Newtonian view on how light moved, was that the speed of light should be different in different circumstances. In other words, the speed of light was thought to be a variant property on the Newtonian view.)

Although the amount of time that has passed, and the distance between locations, can differ from one point of view to another according to relativity, Minkowski recognized that a certain property associated with a "combination" of space and time was the same from every point of view. That is, the pro-

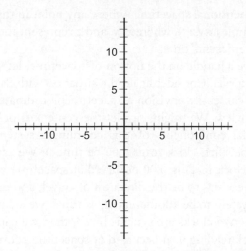

Figure 23.4 A typical Cartesian coordinate system

perty that Minkowski introduced, which is termed the "spacetime interval," is invariant on relativity theory. To understand the spacetime interval, we need to first understand the notion of spacetime. Although "spacetime" and the "space-time continuum" can sound like mysterious entities, the basic idea is quite straightforward.

To understand the notion of spacetime, let's begin by considering a typical two-dimensional Cartesian coordinate system, of the sort illustrated in Figure 23.4. We often (though not always) associate the horizontal and vertical axes as representing positions in space. For example, suppose we take the point (0,0) to be the center of a soccer field. If we take our units to be meters, then a point such as (8,11) might represent a point that is eight meters from the center of the field in one direction, let's say in the direction of one of the sidelines, and 11 meters from the center of the field in another direction, say in the direction of one of the goals.

Next, suppose we again let the horizontal axis represent positions in space. In particular, suppose this axis again represents distance in the direction of the side-lines on our soccer field. But now instead of letting the vertical axis represent another spatial dimension, we will instead let the vertical axis represent time. Suppose we have an individual who is at the center of the field, and who at time 0 begins walking toward the sideline, moving two meters every second. Then (0,0), (2,1), (4,2), (6,3), and so on, would represent this individual's position, at one-second intervals, in space and time. That is, (0,0) represents that the individual is located at point 0 at time 0, (2,1) represents that the individual is located at point 2 at time 1, at point 4 at time 2, and so on.

At bottom, this is all there is to the notion of spacetime. It is simply a way of representing positions in both space and time. A situation such as we just described, in which there is only one spatial dimension along with time, would be a two-dimensional spacetime. Including all three spatial dimensions, along with time, results in a four-dimensional spacetime, where any point in this spacetime can be represented as a 4-tuple (x,y,z,t), where x, y, and z represent the usual three spatial dimensions, and t represents time.

Now that we have a handle on the notion of spacetime, let's turn to the notion of the spacetime interval. Consider again the situation with Sara and Joe pictured in Figure 23.1. We can easily envision a spacetime coordinate system associated with Joe's point of view. We might, say, let the center of Joe's first clock be the origin of the spatial components of this coordinate system, and we can let the moment that Joe's first clock reads 0.00 be time 0. We can then say that the event of Joe's first clock reading 0.00 occurred at spacetime coordinate (0,0,0,0). Suppose we take the x axis to be the direction of travel, and take the spatial units of the coordinate system to be kilometers. If, as usual, we assume that (from Joe's point of view) Joe's two clocks are synchronized, then we can say that the event of Joe's second clock reading 0.00 occurred at spacetime coordinate (1000,0,0,0).

Now let's consider the spacetime interval between these two events, that is, the spacetime interval between the event of Joe's first clock reading 0.00 and the event of Joe's second clock reading 0.00. We can see that the space interval along the $\times$ axis is 1000, the space intervals along the y and z axes are 0, and the time interval is 0. If we let Δx, Δy, and Δz represent the space intervals between the two events along the x, y, and z axes, and let Δt represent the time interval between the two events, then the spacetime interval s between the two events is given by the following formula:

$$s^2 = c^2 \Delta t^2 - \Delta x^2 - \Delta y^2 - \Delta z^2$$

So in this case, the spacetime interval between the two events is $\sqrt{c^2 0^2 - 1000^2 - 0^2 - 0^2}$. (Incidentally, in this case the result will be an imaginary number, that is, one involving the square root of negative 1. Imaginary numbers are not as widely known as, say, the natural numbers or the rational numbers, but imaginary numbers are a well-understood and commonly used type of number in mathematics.)

In a sense, the spacetime interval is a sort of "distance" between events. Not the distance the events are separated in space, nor the distance they are separated by in time, but rather the "distance" they are separated by using a measure involving both space and time.

As mentioned above, the spacetime interval is an invariant property in relativity. To illustrate this, let's bring Sara back into the picture. We saw above that we could specify a spacetime coordinate system associated with Joe's point of view. We can, of course, also associate a spacetime coordinate system with Sara's point of view. For convenience (we need not do this, but it makes the discussion easier), we will assume that the origin point of Sara's coordinate system is the same as the origin point of Joe's coordinate system. Note that, from Joe's point of view, the coordinate system associated with Sara is a moving coordinate system. And since the coordinate system is moving, we know from our discussion earlier in this chapter that times, distances, and simultaneity will be affected.

For example, we saw above that in Joe's coordinate system, the event of his first clock reading 0.00, and the event of his second clock reading 0.00, occurred at coordinates (0,0,0,0) and (1000,0,0,0) respectively. But in Sara's coordinate system, those events will not occur 1000 kilometers apart, nor will they be simultaneous. In general, events will not have the same spacetime coordinates in Sara's coordinate system as they have in Joe's.

However, there are straightforward equations, known as the *Lorentz transformations*, that allow one to transform coordinates from a stationary spacetime coordinate system into a moving spacetime coordinate system. (I do not give these equations in the body of this chapter, but if you are curious, they can be found in the Chapter Notes at the end of the book. Also, here, as always in this chapter, it is assumed that the coordinate systems are moving relative to one another in a straight line with uniform motion.) When the coordinates (0,0,0,0) and (1000,0,0,0) from Joe's coordinate system are translated, using the Lorentz transformations, into the corresponding coordinates in Sara's coordinate system, they come out to be (0,0,0,0) and (1250,0,0, −0.0025) respectively.

If we use the formula above to calculate the spacetime interval s between these events in Sara's coordinate system, we will see that the value comes out the same as it did in Joe's coordinate system. And in general, the spacetime interval between any events in coordinate systems that are in straight-line, uniform motion relative to one another will be the same. So although the space interval and time interval between events will vary from coordinate system to coordinate system, the spacetime interval will not. Again, this is just to say that the spacetime interval is an invariant property in relativity theory.

In this section, then, we have gotten an idea of the notion of spacetime, and we have seen one of the more important invariant properties associated with spacetime, that being the spacetime interval. As a final note in this section, it is worth noting that this "geometrical" approach, that is, thinking of points of view as four-dimensional spacetime coordinate systems that are moving relative to one another, and using the Lorentz transformations to translate coordinates from one coordinate system to another, is a common way of approaching relativity. For many purposes, this geometrical approach provides a convenient way of picturing issues involving relativity. And of course, on this geometrical approach, one finds the same relativistic effects we discussed earlier, that is, time dilation, space contraction, and the relativity of simultaneity.

Concluding Remarks

In this chapter, we have looked at Einstein's special theory of relativity, and we have seen that the theory has non-trivial implications for our common-sense

beliefs about space, time, and simultaneity. With Einstein's special theory of relativity, we can see that some of the beliefs we have long had, and that most of us
took as obvious empirical facts, turn out to be mistaken. In this way, relativity
theory forces us to rethink some of these long-held beliefs. In the next chapter,
we look briefly at the general theory of relativity, noting that it too has interesting
implications for common views, especially common views on the nature of
gravity.

Chapter Twenty-Four

The General Theory of Relativity

Between 1907 and 1916, Einstein spent considerable time and effort developing the general theory of relativity. As mentioned earlier, general relativity is substantially more complex than special relativity. Our main goal in this chapter is to get a sense of the main aspects of general relativity, and to see some of the main implications of the theory. We will begin by looking at the basic principles on which general relativity is based.

Basic Principles

In the previous chapter, we saw that the special theory of relativity is based on two basic principles, the principle of relativity and the principle of the constancy of the velocity of light. In an important sense, the general theory of relativity is likewise based on two basic principles, generally termed the *principle of general covariance* and the *principle of equivalence*.

The principle of general covariance is usually summarized by the statement that the laws of physics are the same in all reference frames. This principle is most easily explained by comparing it with the principle of relativity from the previous chapter. Recall that the principle of relativity states that, if two labs differ only in that they are moving relative to one another in a straight line with uniform velocity, then any experiment carried out in one lab will have an identical result if carried out in the other lab. So, for example, in Figure 23.1, if the only difference between Sara and Joe is that they are in straight-line, uniform motion relative to one another, then if they carry out identical experiments, they will get identical results.

In the chapter on special relativity I tended to speak of "points of view," for example, describing a situation first from Joe's point of view and then from Sara's. Such points of view are often referred to as "frames of reference," or just

Worldviews: An Introduction to the History and Philosophy of Science.   Richard DeWitt
© 2010 Richard DeWitt

"reference frames." Reference frames such as those of Sara and Joe, which involve only uniform, straight-line motion, are referred to as *inertial reference frames* (or just "inertial frames"). Using the notion of an inertial reference frame, the principle of relativity can be phrased more succinctly to say that in all inertial reference frames, identical experiments will have identical results. Or to say the same thing in another way, the laws of physics are the same in all inertial reference frames. And indeed, the principle of relativity is often phrased in just this way.

Note that, when rephrased this way, we can see that the principle of general covariance is a generalized version of the principle of relativity (in fact, although the principle is now usually referred to as the principle of general covariance, Einstein often referred to it as the "general principle of relativity"). So whereas the principle of relativity says that, in essence, the laws of physics are the same in all inertial frames, the principle of general covariance says that the laws of physics are the same in all reference frames, regardless of how those reference frames might be moving relative to one another. This is exactly the sense in which general relativity is a general theory. Whereas special relativity applies so long as special circumstances are met – in particular, so long as we are dealing with inertial reference frames – general relativity removes this restriction, and applies to all frames of reference.

Let us now turn to the other basic principle of general relativity, that being the principle of equivalence. This principle states that effects due to acceleration, on the one hand, and effects due to gravity, on the other, are indistinguishable. There is probably no better way to illustrate this principle than by borrowing Einstein's usual example, which is the following.

Suppose you are in an enclosed room, about the size and shape of an elevator, which is such that you have no way to see outside the room. In the first case, suppose (unknown to you) that this "elevator" is sitting on the surface of the Earth, so that you are feeling the effects of the gravitational field of the Earth. What effects will you feel? Most notably, you will notice effects such as the feeling of being pulled toward the floor of the elevator, and you will notice also that dropped objects accelerate toward the floor at 9.8 meters per second per second.

Now suppose (again unknown to you) that you and the elevator are in a more or less empty region of space (and so not being influenced in any substantial way by any gravitational fields), but that the elevator is being accelerated "up" (that is, in the direction of a line drawn from the floor through the ceiling) at 9.8 meters per second per second. What effects will you feel as a result of this acceleration? Once again, you will notice the feeling of being "pulled" toward the floor of the elevator, and you will notice that dropped objects accelerate toward the floor at 9.8 meters per second per second.

The key point to note is that the effects due to gravity, in the first case, and the effects due to acceleration, in the second case, are indistinguishable from one another. This close relationship between effects due to gravity and effects due to acceleration was recognized as far back as Newton. Nonetheless, in Newton's physics they are treated as separate phenomena, and the close relationship between the two appears to be essentially a coincidence. But with general relativity, the principle of equivalence states that, in essence, there is no difference between the two types of effects – the effects are indistinguishable from one another.

In summary, then, the general theory of relativity, as with the special theory of relativity, is in an important sense based on two basic principles. With this in mind, we will move on to a brief discussion of the equations that are at the heart of general relativity, as well as look at some of the confirming evidence for that theory.

The Einstein Field Equations and Predictions of General Relativity

As noted, the general theory of relativity is, in an important sense, based on the principle of general covariance and the principle of equivalence. We saw in the previous chapter that the special theory of relativity is likewise based on two basic principles. And as we saw in that chapter, the mathematical "picture" that accompanies the two basic principles of special relativity involved some surprising effects for length, time, and simultaneity. Also as noted, the mathematical picture that accompanies the basic principles was not terribly difficult to work out (in that chapter we did not actually derive the effects on length, time, and simultaneity from the basic principles, but I noted that the derivation could be accomplished with no more than high-school algebra, and while the derivation is not trivial, it is not particularly difficult).

The situation with general relativity is quite different. Although the two basic principles of general relativity are not difficult to state, it turned out to be quite difficult to formulate the desired mathematical equations that would respect these principles. Einstein spent several years working on these equations, with many preliminary results that later had to be taken back or substantially revised. In short, the mathematical account required by the two basic principles took an enormous amount of work to figure out. And the equations themselves, unlike the equations that accompany special relativity, are quite complex.

By 1916, however, Einstein had worked out the equations he sought, and he presented them in a 1916 paper titled "The Foundation of the General Theory of Relativity." These equations are now referred to as Einstein's field equations, and they are the mathematical core of general relativity. The general idea is that solutions to these equations indicate how space, time, and matter influence each other. For example, one solution shows the way space and time should be affected by the presence of a body such as our sun. Another solution shows how space and time should be affected when a massive star collapses to form an extraordinarily dense remnant (for example, the effects on space and time of so-called black holes – the presumed remnants of very massive collapsed stars – are given by solutions to Einstein's field equations).

As with special relativity, general relativity also makes some unique predictions. I will mention a few of these very briefly, and then spend a bit more time on another consequence of general relativity, that being the curvature of spacetime.

At the beginning of Chapter 8, we briefly discussed the fact that for some decades, some peculiarities about the orbit of Mercury had been observed. Recall that planets orbit in ellipses, and this is all in keeping with the predictions of Newtonian science. Now envision the point of Mercury's orbit that is closest to the sun. This closest point of an orbit is termed the *perihelion*, and what had been observed, for some decades during the mid- to late 1800s, was that the perihelion of Mercury moved slightly during each orbit, such that the point was very, very slowly rotating around the sun. The amount the perihelion of Mercury moved each year was extremely small, but it was measurable, and moreover it was not what one would expect from the Newtonian account of planetary motion. In his 1916 paper, however, Einstein showed that his equations predicted that the perihelion of Mercury should advance each year, and the amount predicted by Einstein's general relativity was just the amount that had been observed. This, then, was a reasonably straightforward example of confirming evidence for general relativity, of the sort discussed in Chapter 4.

Likewise, in the 1916 paper Einstein showed that if general relativity is correct, then the wavelength of light moving away from a strong gravitational field should be shifted toward the red end of the spectrum. This effect is known as *gravitational redshift*. Since a star has a strong gravitational field, light leaving a star, for example, light from the sun, should be redshifted. The redshift predicted by general relativity is not easy to test, but in the experiments that have been conducted, the observed redshift matches well with the predications of general relativity, again providing confirming evidence for the theory. Light moving away from a less massive body such as the Earth should also be redshifted, and although the predicted effect is quite small, this too has been measured, and it also fits well with the predictions of general relativity.

In the previous chapter, in discussing the special theory of relativity, we saw that motion affects both space and time. In general relativity, motion also has similar effects on space and time. Moreover, gravitational forces (or equivalently, the effects of acceleration and deceleration) also affect space and time. For example, in the presence of a strong gravitational field, or, equivalently, in an accelerating reference frame, time moves more slowly. Importantly, unlike special relativity, there is a sense in which these effects are not symmetric. For example, suppose Joe remains on the surface of the Earth, and Sara leaves on a high-speed space mission. Suppose she accelerates for a period of time, then decelerates when she reaches her destination, turns around and accelerates again for a period of time as she begins the return trip to Earth, and then again decelerates when she approaches the Earth. During her trip she will experience the effects of acceleration and deceleration, and these will be effects not experienced by Joe. In this situation, general relativity predicts that less time will have passed for Sara than for Joe, and both Sara and Joe will agree on this.

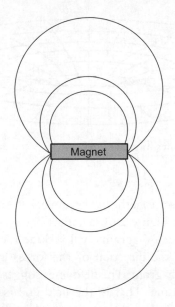

Figure 24.1   Magnetic field lines

Given the extremely accurate timekeeping devices we currently have, these sorts of effects on time are not difficult to test, and the effects predicted by general relativity have been well confirmed. Even the very small gravitational differences between the ground floor and top floor of a tall building should, according to general relativity, produce small differences (very small) in the rate that time passes on the top floor of a tall building compared to the rate that time passes on the ground floor. Even these very small differences in time have been measured, and are what general relativity predicts. In short, there are numerous cases of confirming evidence for general relativity.

As mentioned at the beginning of Chapter 4, the observed bending of starlight during the solar eclipse of 1919 provided the first confirming evidence for general relativity. This observation takes us into one of the more interesting aspects of general relativity, that being the curvature of spacetime. Hence this is worth going through a bit more slowly.

To get a sense of the curvature of spacetime predicted by general relativity, it might help to begin with an example that has little to do with relativity, but that will provide a convenient example with which to contrast relativity. Suppose we have a bar magnet, and suppose we put a piece of paper over the magnet. We then put iron filings on the paper, and shake the filings around a bit. In this situation, the iron filings will arrange themselves in a particular pattern, with this pattern reflecting the magnetic field surrounding the magnet. The magnetic field itself is often pictured as it appears in Figure 24.1. The lines in such a picture are commonly referred to as field lines, and in this case, they represent the strength and direction of the magnetic field. For example, where the field is

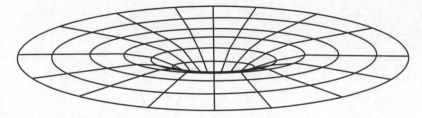

Figure 24.2   Typical field lines in general relativity

stronger, the lines are closer together; where weaker, the lines are further apart. (For ease of presentation, Figure 24.1 is somewhat simpler than most diagrams of field lines. Typically, such diagrams will include more detailed field lines, as well as arrows indicating the direction of the forces involved.) Now, notice an important feature of the diagram. The field lines suggest forces that exist in space and, presumably, also in time. That is, the field lines suggest that, say, in a particular region of space near the magnet, an iron filing would be under the influence of certain magnetic forces and, perhaps, would tend to move a particular way through space and time. In short, space and time typically provide a backdrop for such field lines, or to put it another way, such field lines appear to exist *in* space and time.

Now consider Figure 24.2, the sort of diagram often found in discussions of general relativity. On the face of it, the field lines in such a diagram look similar to the field lines in Figure 24.1. But there is a crucial difference: these field lines do not represent a field *in* space and time; rather, the field lines represent a curvature of spacetime itself. (Incidentally, as we saw in the previous chapter, spacetime is a four-dimensional continuum involving the usual three spatial dimensions together with a dimension representing time. Figures such as 24.2 typically represent two-dimensional "slices" through the four-dimensional spacetime.)

According to general relativity, the presence of a massive body results in a curvature of spacetime. The field lines in diagrams like Figure 24.2 represent such a curvature of spacetime in the presence of a body such as the sun. In this sort of diagram, if an object travels along the surface of the "slice," the shortest path between two points will be a curved line (such a shortest path is termed a *geodesic*). Since light will take the shortest path, light passing near a large body such as the sun should follow what appears to be a curved path. In short, if general relativity is correct, a massive body such as the sun should result in a curvature of spacetime, and hence we should be able to observe the bending of starlight as it passes near a body such as the sun.

In his paper of 1916, Einstein provided the amounts by which one would expect starlight to be bent near the sun. And as discussed in Chapter 4, the bending of starlight observed during the 1919 solar eclipse was in good agreement with Einstein's predictions, again providing confirming evidence for general relativity.

In summary, general relativity makes a number of unusual predictions, most notably those discussed above. Observations have been in agreement with these predictions, and, overall, general relativity is widely considered to be a well-confirmed theory.

Philosophical Reflections: General Relativity and Gravity

Before closing this chapter, one final issue closely tied to general relativity is worth discussing, that being the account of gravity provided by this theory. To say that the account of gravity is an issue tied to general relativity is perhaps a bit of an understatement, in that general relativity is often construed primarily as a theory of gravity.

As we saw in the discussion above, light travels along the shortest path available. In general relativity, bodies not under the influence of any forces will also move along the shortest path, that is, bodies generally move on geodesics. Importantly – and this is a crucial difference from the way gravity is typically conceived of on the Newtonian account – a body such as a planet does not move as it does because of any attractive force. Mars, for example, does not move in an elliptical orbit around the sun because of anything to do with a mutually attractive force – a gravitational force – between Mars and the sun. Rather, Mars, like any body in motion, moves in a straight line.

But a "straight line" in a curved space is a geodesic. As we have seen, according to general relativity, a body such as the sun will result in a curvature of spacetime. And according to the equations of general relativity, this curvature is such that the geodesic on which Mars should move would be what appears to be an ellipse about the sun. In other words, on general relativity, there is no attractive "force" between bodies such as Mars and the sun. Rather, Mars is simply moving in a straight line, but due to the curvature of spacetime, that "straight line" appears to be an ellipse about the sun.

Note that this is a very different treatment of gravity in general relativity, as opposed to the treatment gravity receives in Newtonian science. On the Newtonian picture, gravity is typically taken as an attractive force between objects. As we noted at the end of Chapter 20, such a force, if taken with a realist attitude, seems to be a case of action at a distance. And as discussed in Chapter 20, this suggestion of action at a distance was troublesome for Newton, and as a result, he professed to take an instrumentalist attitude toward gravity.

In spite of Newton's own claims on the matter, most people raised in the Newtonian worldview tend to take a realist attitude toward gravity. To use an example we saw earlier, if I drop a pen, and ask "Why did the pen fall?" the standard response will be that it fell because of the force of gravity. And if asked whether this force is real, the usual response is that of course it is. That is, people by and large tend to think of gravity as a really existing, attractive force between objects.

In short, gravity is typically taken, in the Newtonian view, with a realist attitude.

But now note an interesting consequence of the treatment of gravity on general relativity. As noted above, general relativity is a well-confirmed theory. And if we take general relativity with a realist attitude, it essentially forces us to adopt an instrumentalist attitude toward the Newtonian concept of gravity. That is, if objects fall toward the Earth, or planets move about the sun in elliptical orbits because of the curvature of spacetime and not because of any attractive forces between objects, then talk of gravity as an attractive force is at best a convenient, but not literally correct, way of speaking.

In summary, general relativity is a well-confirmed theory. Notably, with respect to prediction and explanation (of the movement of the perihelion of Mercury, or the bending of starlight, to name just two), general relativity does a better job than Newtonian theory. Newtonian theory is still a very useful theory, but if we want to consider which theory more accurately accounts for the known data, there is not much question that relativity does the better job.

As a consequence of this, if we are inclined to take a realist attitude toward theories in physics, then we ought to be realists about relativity theory and instrumentalists about Newtonian theory (Newtonian physics is, after all, still useful, although not, strictly speaking, a correct picture). But note that this forces us to adopt an instrumentalist attitude toward the notion of gravity as an attractive force – in other words, general relativity forces us to reassess an attitude (that is, a realist conception of gravity as an attractive force) that most tend to take for granted. In short, general relativity, as with special relativity, forces us to rethink some commonly held views.

Concluding Remarks

In this and the previous chapter, we have seen that the special and general theories of relativity have interesting implications for beliefs most of us have long held as basic, commonsensical beliefs. These include beliefs about length, time intervals, and simultaneity, as well as common beliefs about the nature of gravity. In particular, with respect to gravity, general relativity forces us to adopt an instrumentalist attitude toward the usual notion of gravity as an attractive force.

Recall that in the 1600s, new discoveries forced changes in the usual views of the world. We are likewise seeing that new discoveries are forcing us to reassess some of our common beliefs about the world. In the final chapter, we will return to a discussion of some of the implications Einstein's theory of relativity has for the Newtonian worldview. First, though, we will explore the other major branch of twentieth-century physics, that being quantum theory.

Chapter Twenty-Five

Overview of the Empirical Facts, Mathematics, and Interpretations of Quantum Theory

In the previous chapters, we explored the special and general theories of relativity, and saw that these theories have interesting implications, primarily for views on the nature of space and time and the nature of gravity. In this chapter we turn to another important branch of modern physics, that being quantum theory. We will soon see that recent discoveries involving quantum theory likewise have substantial implications.

Quantum theory is a slippery topic, and one that requires careful going if we are to get an accurate overview. Our strategy will be to first explain a key distinction between (a) empirical facts involving "quantum entities," (b) quantum theory itself, that is, the mathematical core of quantum theory, and (c) issues related to interpretations of quantum theory. After these three issues are clarified, we will then go on, in subsequent sections of this chapter, to explore each of these more fully.

Facts, Theory, and Interpretation

As noted, in any nontechnical discussion of quantum theory, there are at least three separate issues that should be kept distinct, these being (a) the quantum facts, that is, empirical facts involving quantum entities, (b) quantum theory itself, by which I will mean the mathematical core of quantum theory, and (c) interpretations of quantum theory, which involve philosophical questions about what sort of reality could give rise to the quantum facts and that, presumably, would be consistent with quantum theory itself. Unfortunately, in popular, nontechnical discussions of quantum theory, as often as not these issues come all mushed together. For example, it is not unusual to see claims that quantum theory shows that western science and certain eastern philosophies are converging on the same

Worldviews: An Introduction to the History and Philosophy of Science.   Richard DeWitt
© 2010 Richard DeWitt

view of the universe. But this is a mistake, or at best, is substantially misleading. Certain *interpretations* of quantum theory suggest such a convergence, but the interpretation of a theory, and the theory itself, are issues that should be kept separate. As one other example, it is likewise not unusual to see claims that quantum theory shows that the universe is constantly splitting into multiple, parallel universes. But again, certain interpretations of quantum theory suggest this, but quantum theory itself does not.

Issues surrounding quantum theory are admittedly complex. But if we approach the subject slowly and carefully, we can arrive at a good – and importantly, a reasonably accurate – overview of quantum theory and related issues. Our first step will be to describe briefly the distinction between the three issues mentioned above.

The quantum facts

When I speak of quantum facts, I mean simply empirical facts involving quantum entities. Such facts would include the outcome of experiments involving electrons, neutrons, protons, and other subatomic particles; experiments involving photons, that is, "units" of light; experiments involving particles (for example, alpha and beta particles) emitted during radioactive decay, and so on.

As we will see later in this chapter, these facts are surprising, but they are not controversial. In particular, there is no disagreement on what these facts are. There is substantial disagreement on how to interpret the facts, for example, on what sort of reality could produce such unusual facts. But issues involving interpretation need to be kept separate from accounts of the quantum facts themselves.

It is worth noting that I am leaving intentionally vague what precisely counts as a quantum entity. The entities mentioned above – subatomic particles such as electrons, protons and the like, photons, and particles involved in radioactive decay – are clearly quantum entities. So in most of the upcoming discussion, the quantum facts we discuss will be facts about such entities. But keep in mind that all objects – you, me, our desks and chairs, and so on – are composed of these smaller entities. But whether ordinary-sized objects should be considered quantum entities is a point of some dispute. Hence, in what follows I will emphasize mainly quantum facts that involve uncontroversial quantum entities, such as those mentioned above.

Quantum theory itself

Quantum theory, like much of physics since the work of Newton and others in the late 1600s, is a mathematically based theory, and when I speak of "quantum theory itself," I primarily have in mind the mathematics at the heart of quantum theory. The key mathematical components of quantum theory were discovered in the late 1920s, and that mathematics is used much as the mathematics in other branches of physics is used. Most notably, the mathematics of quantum theory is used to predict and explain the sorts of quantum facts mentioned above.

As a final brief point, the mathematics of quantum theory has been enormously successful. The mathematics of quantum theory has been around in roughly the same form for about 70 years, and it has yet to make an incorrect prediction. With respect to prediction and explanation, quantum theory is arguably the most successful theory we have ever had.

Interpretations of quantum theory

The interpretation of quantum theory is essentially a philosophical topic centered around the nature of reality. In particular, the various interpretations of quantum theory center around the question of what sort of reality is consistent with both the quantum facts and quantum theory itself. That is, presumably the quantum facts are the result of some sort of reality "out there," and given the success of the mathematics of quantum theory in predicting and explaining these facts, it is not unreasonable to think that the mathematics is in some way touching base with that reality. The interpretation question, then, is what sort of reality, consistent with the known facts and with the mathematics of quantum theory, could be producing those facts.

In the following sections, we will look more closely at the facts, the mathematics, and various interpretations of quantum theory. As noted, these items are easy to conflate, and doing so can (and does) lead to serious confusions and misunderstandings about issues surrounding quantum theory, as well as to confusions and misunderstandings about the potential implications of quantum theory. In what follows, I cannot encourage you too much to keep clear the distinction between the quantum facts, quantum theory itself, and the interpretation of quantum theory.

Some Quantum Facts

In this section, we look at some reasonably straightforward empirical results of experiments involving quantum entities such as electrons, photons, and the like. The experiments described below, or ones similar to them, are commonly used to convey some of the oddness associated with quantum facts. These experiments primarily involve electrons and photons. Electrons are components of atoms, and can easily be produced by an electron gun. Electron guns are devices that produce a stream of electrons, and such devices are common. For example, older-style televisions (not flat-screen models) have an electron gun in the back, and the television image is produced by guiding the electrons produced by this gun to appropriate locations on the screen. Photons, on the other hand, are "units" of light, and of course can be produced by any number of means, such as flashlights.

For ease of discussion, I want to move away, for a moment, from consideration of just quantum facts, that is, mere outcomes of experiments, and take a brief excursion into a reality discussion. Note this will briefly take us into issues involving interpretation, but doing so will make the ensuing discussion easier. We will quickly return to a straightforward consideration of just the facts.

A brief excursion into a reality issue

As we will see in a minute, some experimental outcomes involving quantum entities are most consistent with the view that quantum entities are waves, while some experimental outcomes are most consistent with the view that quantum entities are particles. Suppose we consider the reality question: Are electrons, photons, and the like really particles, or are they really waves?

First, take a moment to consider the fact that waves and particles are quite different. Consider particles first. A baseball is a good example. Particles are discrete objects, with reasonably well-defined locations in space and time. Particles interact with one another in typical particle-like ways, for example, by bouncing off one another, or perhaps by breaking into smaller particles.

A wave, on the other hand, is better viewed as a phenomenon rather than as a discrete object, and waves are typically spread out over a fairly large region, rather than being confined to a relatively small, well-defined location in space and time. A wave at a beach, for example, is not located in one particular place, but rather is spread out over a large area. Moreover, waves interact with one another in quite different ways than particles. Two waves will sometimes interact with one another so as to form a larger wave; sometimes two waves will interact in such a way as to essentially cancel one another out; and sometimes two waves will pass through one another, emerging unchanged from the interaction.

In keeping with the different nature of particles and waves, the two produce very different experimental effects. So one would think that determining whether electrons are particles or waves would be relatively easy. For example, suppose we use a source that provides a steady stream of particles, say, a paintball gun (these are guns that shoot small pellets containing paint, with the paint marking where the pellets hit). Suppose moreover that we shoot a steady stream of paintballs toward two small open windows. If we ask "What sort of pattern of hits will we expect?" the answer is easy. Many of the paintballs will hit the wall containing the windows, and those that pass through the windows will pile up on the inside wall, behind the windows. That is, we should expect to see a pattern of hits, on the inside wall, that correspond to where the windows are located.

Likewise, suppose for a moment that electrons are particles, and that we shoot thousands of electrons at a barrier with two slits, with a piece of photographic paper on the far side of the slits. Just as when we shot paintballs toward the two

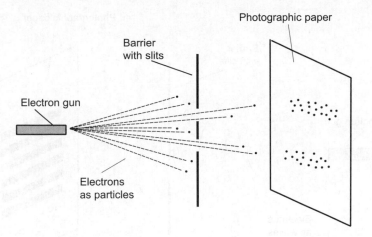

Figure 25.1   Electrons as particles

windows, if electrons are particles, then many of the electrons will hit the barrier, but those that pass through should hit the photographic paper in the areas behind the slits. Electrons can be recorded on photographic paper, so in this case we should expect to record what looks like thousands of individual particle hits, piled up on areas of the photographic paper behind the two slits. (Incidentally, photographic paper will not directly record electrons, but when coupled with a phosphor screen, which emits light when struck by electrons, photographic paper works well as an electron detector. For ease of discussion, we will continue to speak of the photographic paper itself as recording the electrons.)

Pictorially, this scenario would look like Figure 25.1. Importantly, do not overlook the fact that this picture is going well beyond the mere facts. In particular, since we cannot detect or otherwise observe electrons before they interact with some sort of measuring device, such as a piece of photographic paper, the depiction of the electrons between the electron gun and the paper is an interpretation, and not any sort of straightforward empirical fact. Again, this is a picture, an interpretation, of what reality is presumably like if electrons are particles. Keeping that in mind, Figure 25.1 is a picture of this scenario.

Notice the piling up pattern of electrons on the photographic paper. Again, this is what we would expect if electrons are particles, and this piling up pattern is what we will refer to as the "particle effect."

On the other hand, suppose electrons are waves, and suppose we again pass them through the same apparatus, with two slits and photographic paper. In this case, the two slits will split the wave into two waves. These two waves will then interact with one another, and the result will be a typical interference pattern produced by the interaction of two waves. In this particular case, we would expect the interaction of the waves to produce alternating light and dark bands on the photographic paper, with the light and dark bands indicating the areas where the waves reinforced one another, and areas where the waves canceled each other.

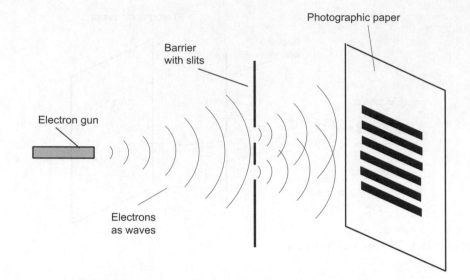

Figure 25.2   Electrons as waves

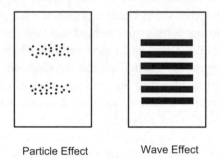

Particle Effect   Wave Effect

Figure 25.3   The particle effect and the wave effect

Such interference patterns are well known, having been studied as far back as the early decades of the 1800s.

So if electrons are waves, the two-slit apparatus should produce an effect similar to that shown in Figure 25.2. Once again, I emphasize that such waves cannot be directly observed, and so the picture is again an interpretation of what the underlying reality might be like. With that in mind, Figure 25.2, which I will refer to as the "wave effect," is what we should expect if electrons are waves.

In short, if electrons are particles, they should produce one sort of outcome, the particle effect, and if electrons are waves, they should produce a quite different sort of outcome, the wave effect. The particle effect and the wave effect are summarized in Figure 25.3.

We will now move on to describe a few experiments involving electrons. At this point, we are returning from our brief excursion into interpretative/reality issues, and will now simply describe the facts. That is, in the remainder of this chapter, we will describe some experimental setups involving quantum entities, and describe the results of those experiments.

Four experiments

The experiments described below are fairly standard examples, widely used to illustrate some of the puzzling features of quantum facts. The first experiment is that pictured in Figures 25.1 and 25.2 above. That is, we take an electron gun, shoot electrons at a barrier with two slits, and record the results on a piece of photographic paper.

Given this experimental setup, the outcome is a clear wave effect. That is, the photographic paper will have on it alternating dark and light bands. One more time, note again the sense in which this is a straightforward quantum fact. We are describing nothing more than straightforward observational results: if one sets up a basic two-slit apparatus, as described above, the result is a piece of photo-graphic paper with alternating dark and light bands.

For experiment 2, let's modify the first experiment slightly. In particular, suppose we have a setup exactly as for the first experiment, except that we include passive electron detectors to monitor each of the slits. That is, behind the top slit we put one electron detector, which we will call detector A, which will record any electron passing through that top slit. Next to the other slit we will place a second detector, which we will call detector B, which will monitor that bottom slit. The setup, then, looks as shown in Figure 25.4.

Here is the idea behind including the detectors: if electrons are waves, then the wave should pass through both slits simultaneously, and hence both detectors

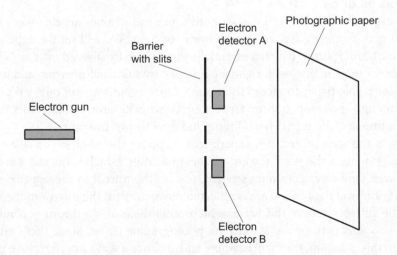

Figure 25.4 Two-slit experiment with electron detectors

should always fire simultaneously, and it should never happen that only one or the other detector fires by itself. If, on the other hand, electrons are particles, then each will at most pass through one or the other slit, and hence only one detector at a time should register the presence of an electron, and they should never fire simultaneously.

Recall that, in experiment 1, the result was a clear wave effect. The experiment we are considering now is exactly like experiment 1, except for the presence of the detectors. Since the detectors are passive detectors – that is, we would not expect them to interfere with the electron, but merely note whether an electron is present – our initial inclination would be to expect this experiment to also result in a wave effect. But instead, the result of this experiment is a clear particle effect. And in keeping with this, only one detector at a time fires. The detectors never register electrons in both slits simultaneously, as would be expected if electrons were waves. It is as if, in the presence of detectors, the electrons are acting as particles.

Moreover, suppose we hook up an on/off switch – say, something like a light switch – to the electron detectors, so that we can turn them on and off whenever we wish. We'll then be able to switch between the wave effect and the particle effect by simply moving the switch to the on or off position. When off, we will see the wave effect, and when on, the particle effect, and we can instantly change between the two, as often and as rapidly as we wish, by merely moving the switch back and forth.

These results are surprising, because it is difficult to imagine how such detectors could so substantially alter the outcome of the experiment. But again, this is just a fact about experiments involving quantum entities – if the experiment is set up as described in experiment 1, a wave effect is the result. And if we add electron detectors as in experiment 2, the outcome is a particle effect. And, as described above, we can switch between a wave and particle effect by merely turning the detectors off or on.

For experiment 3, we will use a photon gun rather than an electron gun. A photon gun is a device that produces "units" of light. We will set the experiment up with a beam splitter (which is essentially just a partially silvered mirror), a beam recombiner (essentially just a two-way mirror), two normal mirrors, and a piece of photographic paper to record the results. Once again, we will put two photon detectors into the setup, but for experiment 3, we will have the detectors turned off. In summary, the experimental setup looks as shown in Figure 25.5.

Here is the idea behind this experiment: Suppose the photon is a wave. The photon gun emits the wave toward the beam splitter, which splits the wave into two waves. One wave continues straight, toward the mirror in the upper right of the diagram, and the other wave is reflected down, toward the mirror in the lower left. The mirrors reflect the wave, which recombine at the beam recombiner, which also directs the wave toward the photographic paper. Since there are two waves in this situation, they will interfere and produce a wave interference pattern, that is, a wave effect, on the paper.

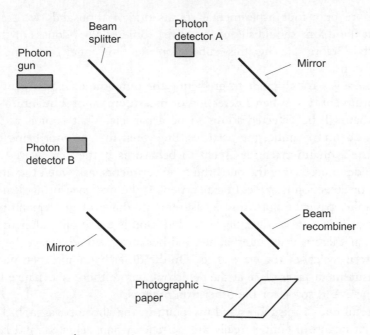

Figure 25.5 Beam splitter experiment

On the other hand, if photons are particles, then each photon will travel either through the upper/right route or through the left/lower route. Here there are no waves to interfere, and hence we would expect a particle effect.

Although we have included photon detectors in the diagram, for experiment 3 these detectors are turned off and hence play no role. And when we run this experiment, the result is a clear wave effect, as if photons were waves.

For experiment 4, we will use the exact same setup as experiment 3, except that we will turn the detectors on. By this point, you can probably guess, correctly, that something odd happens if we turn the detectors on. Again, these detectors are presumably playing a passive role as did the detectors in experiment 2. And again, if photons are waves, as the previous experiment suggests, then we would expect both detectors to fire simultaneously. After all, experiment 3 suggests that photons are waves, and hence a wave should be present at both detectors simultaneously.

But only one or the other detector goes off at a time, which is what we would expect if photons were particles, not waves. And the result on the photographic paper, in spite of the fact that this experiment is almost identical to the previous experiment, is a clear particle effect.

And exactly as with experiment 2, we can hook up an on/off switch to the detectors, and switch whenever we wish between the wave and particle effect by merely turning the detectors off or on. Let's take a minute to think about how odd this seems. It seems, in experiment 3, that we could get these results only if photons really are waves. And it seems, for experiment 4, that we could get these results only if photons really are particles.

These are only four experimental results out of thousands we could consider. But these four should suffice to convey some of the oddness of the quantum facts. Before closing this subsection, let me offer two more brief considerations.

First, here is a rough guide to predicting the outcome of experiments involving quantum entities. When a detection or measurement of a quantum entity is made, what will be detected seems to be a particle. That is, it is as if, when detected, quantum entities are particles. But when they are not being detected or measured, quantum entities seem to behave as if they are waves. So as a rough guide to predicting the outcome of experiments, ask when the first measurement or detection is made. In experiment 1, the first measuring device is the photographic paper. Before this measurement, think of the quantum entity behaving as if it were a wave. Since no detection is made until after the slits, it will be as if there is wave interference and hence we would expect the interference pattern typical of the wave effect. On the other hand, in experiment 2, the first measurement takes place at the detectors, before there is a chance for wave interference. And so on for the other experiments.

Be careful not to misunderstand my point in the above paragraph. I am not saying that quantum entities really are particles when detected, and really are waves when not detected. Rather, I am being agnostic on what is really going on, and merely offering a rough guide to predicting the outcome of such experiments. Treat quantum entities *as if* they are particles when detected, and *as if* they are waves when not being detected, and you will have a way of anticipating the outcomes of experiments such as those above.

As a second, related consideration, notice that acts of measurement or detection seem to play a curious role when quantum entities are involved. For example, the electron and photon detectors in the above experiments are measuring devices that measure the presence of an electron or photon. Such measuring devices seem to affect what happens, for example, whether we see a wave or a particle effect. And this is very puzzling. How could an electron, photon, or other quantum entity "know" whether there is a detector or other measurement device nearby? And for that matter, what exactly counts as a measurement? These are very difficult questions, and are part of what is often referred to as the "measurement problem." We will return to this problem later in the chapter, but for now, I just wanted to introduce these issues about measurements, and the curious role measurements play in quantum theory.

Overview of the Mathematics of Quantum Theory

The mathematics of quantum theory is sufficiently difficult as to make it impossible, in a text such as this, to present it in any detail, while at the same time remaining accurate and comprehensible. But while it may be impossible to explain here the details of the mathematics of quantum theory, it is not particularly difficult to give a good idea of what that mathematics is like, all the while remaining accurate and accessible.

My strategy will be to describe the mathematics in two separate, though overlapping, ways. I will begin with a section providing a very general descriptive overview of the mathematics of quantum theory. I will then provide a section that is still fairly general and descriptive, although with a sizable amount more detail. For those wishing more detail yet, there is a section providing such detail in the Chapter Notes at the end of this book.

Descriptive overview of the mathematics of quantum theory

Quantum theory is, at bottom, a "wave" sort of mathematics, as distinguished from a particle sort of math. A few notes on wave versus particle math may be helpful.

In physics, we find both "particle" mathematics and "wave" mathematics. By this I mean simply that a certain type of mathematics is used when the situation involves discrete objects ("particles"), and another type of mathematics is used when the situation involves waves. For example, if we drop a bowling ball off the roof of a building, the object involved is a discrete object (the ball) seemingly under the influence of various forces (gravitational forces, for example). The mathematics appropriate for this situation would be an example of what I am referring to as particle mathematics.

Waves, however, differ from particles (some of the key differences were discussed above), and hence the math appropriate for particles is not appropriate for situations involving waves. There are, however, well-established mathematics for dealing with waves. And in physics, wave mathematics is used just as particle mathematics is used. In particular, wave mathematics allows us to make predictions about what attributes of a system we will detect (for example, the amount of energy a wave is carrying), as well as how the system will evolve over time (for example, where the crest of the wave will be located at a future time).

Again, quantum theory is a wave sort of mathematics. But as mentioned, there is nothing unusual about this. Wave mathematics are everywhere in physics, and quantum theory is simply a particular version of a type of mathematics with which physicists are well acquainted.

There is one common question I want to address before ending this section, but otherwise, this concludes this very general descriptive overview of the mathematics of quantum theory. In summary, the math of quantum theory is a wave sort of mathematics, and the mathematics of quantum theory is used in the same way that other mathematics is used in physics. In particular, given the current state of a system, one can use the mathematics of quantum theory to make predictions about what attributes of the system might be observed, and to make predictions about what state the system will be in at a future time.

If the mathematics of quantum theory is a familiar sort of wave mathematics, why do we often hear that quantum theory is such an unusual theory? One can easily get the impression, from various writings on quantum theory, that quantum theory is somehow profoundly different from past theories in physics. Indeed, viewed a certain way, I think this sentiment is correct. For example, issues surrounding quantum theory force us to rethink some of the basic assumptions we have had about the world ever since the time of the ancient Greeks. But given that the mathematics of quantum theory is a familiar sort of wave mathematics, in what sense is quantum theory different from other theories in physics?

One minor difference, though one worth mentioning, is that the mathematics of quantum theory usually provides probabilistic predictions rather than definite predictions. For example, if we use the mathematics of quantum theory to predict, say, the position of an electron, the mathematics will provide us with the probability of detecting the electron at various positions. In contrast, in the case of the bowling ball dropped from the roof, the mathematics will provide us with a definite prediction. In short, the predictions of other branches of physics tend to be definite ("The ball will be detected in this location"), whereas the predictions of quantum theory are typically probabilistic ("There is such and such a probability of detecting the electron at this location").

But this is a fairly minor difference with other branches of physics. The major difference I have in mind concerns the *interpretation* of the mathematics. Since issues involving interpretation are a central topic of a later section, I will discuss it only briefly here.

The first point to note is this: The mathematics used in physics is, at bottom, just mathematics. As such, the mathematics has no necessary or inherent ties to the world. This is a point that is easy to overlook, but is central in understanding the sense in which quantum theory is an unusual theory. To understand this better, consider again a simple case such as the dropped bowling ball. In the mathematics used to make predictions about the falling ball, there is nothing that requires that chunk of mathematics to be interpreted as being about a falling object. The equation used is just an equation, just a piece of mathematics, just a collection of symbols joined together and manipulated according to the rules of the mathematics involved.

The fact is we *interpret* that chunk of mathematics in a certain way (for example, as being about a falling ball), and such interpretations have been extremely fruitful and useful (useful, for example, for making predictions). In addition, we tend to interpret the chunk of mathematics involving the falling ball in more or less the same way. For example, there is general agreement that this part of the equation represents the falling ball, this part of the equation represents time, this part represents the starting location of the ball, and so on. In short, there is general agreement on how that equation describes, or "paints a picture of," the situation involving the falling ball.

In cases such as the equation involved with the falling ball, it is precisely the fact that we tend to agree on the interpretation that disguises the fact that we are even making an interpretation. In other words, we are indeed interpreting the math, but we all interpret it the same way, and have done so for several hundred years. As a result, we tend not to recognize that using mathematics to make predictions about the world requires that we interpret the math as being about the world. But at bottom, the way we "tie up" the math with the world is not an inherent *part of* the mathematics; rather, it is an *interpretation of* the mathematics.

Immediately above, I tried to draw attention to the fact that there is general agreement on how to tie up the mathematics involved with the falling ball with the world. ("This part of the equation represents the ball," and so on.) This is exactly where a major difference with the mathematics of quantum theory arises: There is no consensus on how to tie up the mathematics involved in quantum theory with the world.

I need to be careful here so as not to be misunderstood. Consider an example where quantum theory is used to make a prediction about the location of an electron. Almost everyone agrees that the mathematics involved does tie up with the world, at least in a very general sense. For example, most agree that there are such things as electrons and measuring devices, that electrons can influence the measuring devices we use to record the position of electrons, and that the mathematics of quantum theory enables us to make predictions about how those measuring devices will behave in certain situations involving electrons. So in these very general terms, almost all agree that the mathematics of quantum theory does indeed tie up with things such as electrons and measuring devices.

If one tries to go beyond this general level, the picture of reality suggested by the mathematics of quantum theory is a thoroughly bizarre picture. The sense in which the picture is bizarre will be explored more fully in the section on the interpretation of quantum theory. So for now, I will simply leave it at that: the picture is a bizarre one, and that is why one often hears that quantum theory is such a strange theory.

In closing this section, it is worth emphasizing once again that the mathematics of quantum theory is not at all strange; rather, the strangeness arises in the interpretation of that mathematics. And, incidentally, this is a good place to recall a point made earlier in this text (from the chapter on instrumentalism and realism). One need not do this sort of interpretation at all. That is, taking an instrumentalist attitude toward a theory – in this case, quantum theory – is a common and respectable attitude. In the case of quantum theory, an instrumentalist attitude would amount to taking the following stance: we have the mathematics of quantum theory; we have experts highly proficient in the use of that mathematics; that mathematics allows us to make highly accurate and reliable predictions. Who could ask for anything more?

A somewhat more detailed, but still descriptive, overview of the mathematics of quantum theory

As described above, the mathematics of quantum theory is a wave sort of mathematics. Let's begin this section with a few facts about waves and wave mathematics.

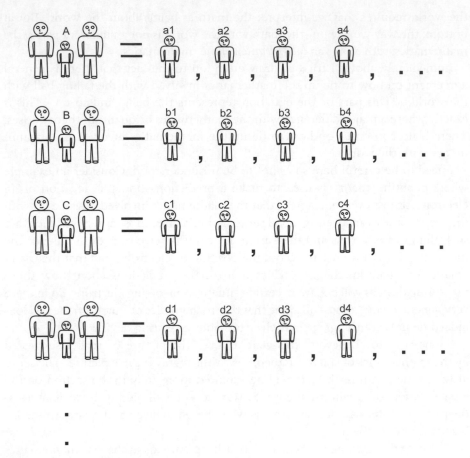

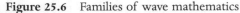

Figure 25.6 Families of wave mathematics

First, a fairly unremarkable fact, though one you may not have thought about explicitly: waves come in families. For example, the waves produced by stringed instruments (for example, guitars and banjos) have similarities not shared by waves produced by reed instruments (for example, clarinets and saxophones), which in turn have similarities not shared by waves produced by percussion instruments (for example, bass and bongo drums). This is not much different from the fact that you and other members of your family have similarities not shared by myself and other members of my family. In short, we can group waves into families.

Given that waves come in families, it is not surprising that the mathematics that apply to waves can likewise be grouped into families. Suppose, then, that Figure 25.6 represents the various families of wave mathematics.

In Figure 25.6, the family icons at the left represent the families of waves, which I will simply refer to as family A, B, C, D, and so on. This is similar to the way that your last name might represent your entire family. The icons to the right of the equal signs represent the individual members of each family, whom I will refer

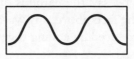

Figure 25.7   Representation of a wave equation

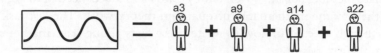

Figure 25.8   Adding family members to produce a particular wave

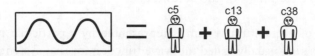

Figure 25.9   Members of another family can produce the same wave

to as a1, a2, a3, and so on. And again, this is similar to the way that your name, and the names of your parents, sisters, and brothers, represent the individual members of your family.

Now, in contrast to the unremarkable fact that waves come in families, here is a quite surprising fact: any particular wave can be produced by adding together appropriate members of any family of waves. For example, take a particular wave, say, one produced by plucking the top string of a guitar. Suppose Figure 25.7 represents the equation that characterizes this wave. Now, pick any family from the chart of families above, let's say family A. There will be members of family A such that, when those members are added together, they will produce the wave in question. That is, the wave in Figure 25.7 can be produced by adding together appropriate members of family A. Schematically, this would look as shown in Figure 25.8.

Notably, that same wave equation can be produced by adding together appropriate members of any other family. So, for example, there will be appropriate members of family C that can be added together to produce the same wave equation, as shown in Figure 25.9.

And in general, any given wave can be produced by adding together appropriate members of any given family of waves. As mentioned, this is a rather remarkable fact about waves, and one that has been extensively studied and has proven quite fruitful over the past century. It is, for example, this fact that makes it possible to reproduce music on electronic devices. (Think about it: isn't it rather astounding that the almost exact sounds of guitars, drums, horns, and so on, can be produced by small electronic devices that have almost nothing in common with guitars, drums, or horns? The key to how this works are the above facts about waves. In outline, the waves produced by, say, the earbuds of your iPod form a family of waves, and thus, because of the facts above, any wave of a guitar, trumpet, human voice, or whatever, can be produced by adding together appropriate members of the "earbud" family of waves.)

At bottom, this one unremarkable fact (that waves come in families), and the one remarkable fact (that any given wave can be produced by adding together appropriate members of any given family), are all the facts needed to get a better idea of the way the mathematics of quantum theory work. The last task of this section, then, is to explain how to relate these facts about wave mathematics to quantum theory.

The relation between these facts about wave mathematics and quantum theory can be stated quite briefly, although explaining this clearly will take a bit more work. Here is a summary of the bottom line:

(1) A state of a quantum system is represented by a particular piece of wave mathematics, usually called the wave function for that system.

(2) Each type of measurement one might perform on a quantum system is associated with a particular family of waves.

(3) Predictions about the outcome of a measurement on the quantum system are arrived at by finding the members of the family of waves (the family associated with that measurement) that, when added together, produce the wave function.

Let's work on unpacking what this means, beginning with an explanation of (1) above. Consider a quantum system, say a particular electron in a particular setting. In a way similar to the way a wave produced by a guitar string can be represented by a particular piece of wave mathematics, so too the electron in this particular setting can be represented by a piece of wave mathematics. Again, this piece of mathematics is called the wave function for this system.

Let's suppose the wave function for the electron is as shown in Figure 25.10. So understanding (1) is relatively straightforward. A quantum system, such as an electron in a particular setting, can be represented by a wave function.

Understanding (2) is a bit more involved, but still fairly straightforward. Recall there are families of waves, and that the mathematics associated with waves also come in families. In the mathematics of quantum theory, each such family is associated with a particular type of measurement. For example, the position of an electron is one measurement one might perform on the electron mentioned above. And among the various families of wave mathematics, one family will be associated with measurements of position. Another family will be associated with

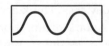

Figure 25.10   Representation of the wave function for an electron in a particular setting

Wave Family: Associated with measurement of:

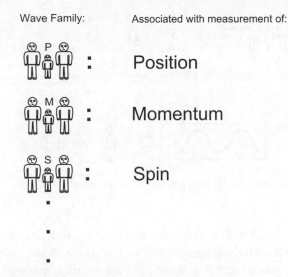

P : Position

M : Momentum

S : Spin

Figure 25.11   Families associated with measurements

measurements of the momentum of the electron, another family with the spin of the electron, and so on for the various possible measurements we might perform on the electron. In short, as stated in (2), each family of wave mathematics is associated with a possible measurement one might perform on a quantum system.

Pictorially, this would look as is shown in Figure 25.11. The figure shows only three types of measurements, but there are actually infinitely many types of measurements one might perform on a quantum system. And as suggested by Figure 25.11, and described in (2) above, each such measurement will be associated with a particular wave family. (Incidentally, I've chosen the easy to remember P, M, and S, but it should be noted that these are not the standard abbreviations used by physicists to represent position, momentum, and spin.)

Understanding (3) is probably the trickiest part of this story. Let's approach this by way of example. Suppose we have a particular electron in a particular situation, and we want to make predictions about what will happen if we perform a measurement on the position of the electron. For simplicity, suppose for this example, that there are only two possible locations where the electron might be found. Now, from (1) we know that there is a wave function associated with that electron (see Figure 25.12) and from (2) we know that, among the various families of

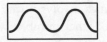

Figure 25.12   Wave function for electron

Figure 25.13   Family P associated with measurements of position

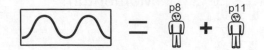

Figure 25.14   Wave function decomposed into members of Family P

mathematics associated with waves, there will be one family, let's say this is family P, that is associated with measurements of position (see Figure 25.13).

Now recall the remarkable fact above, that appropriate members of any wave family can be added together to produce any particular wave function. So in particular, there will be appropriate members of family P that, when added together, produce the wave function for the electron. Let's suppose that p8 and p11 are the members of family P that add together to produce the wave function for the electron. Schematically, this is shown in Figure 25.14.

With this information in mind, we are in a position to explain (3). Recall that we are interested in making a prediction about the position of the electron, that P is the family associated with measurements of position, and that p8 and p11 are the members of P that add together to produce the wave function that represents the electron.

These two family members, p8 and p11, will enable us to make the desired predictions. In this example, there were two areas where the electron might be found. As it turns out, some straightforward and standard mathematical manipulation of p8 will result in a number between 0 and 1, and this number represents the probability of finding the electron in the first area. Likewise, the same straightforward mathematical manipulation of p11 will also result in a number between 0 and 1, and this number represents the probability of finding the electron in the second area. This, then, is how p8 and p11 lead to the predictions about the position of the electron.

(It is not crucial for understanding this discussion, but if you are curious, here is an overview of the mathematical manipulation just mentioned: p8 is a piece of wave mathematics associated with a particular wave. All waves have amplitudes associated with them, and so p8 will likewise have an amplitude associated with it. The mathematical manipulation involves squaring the amplitude of p8. Because of the nature of the mathematics involved, this will always result in a number between 0 and 1. As mentioned, that number will represent the probability of finding the electron in the first area. Do likewise for p11 – that is, square the amplitude associated with p11 – and that number will be the probability of finding the electron in the second area.)

Thus, (1), (2), and (3) describe how the mathematics of quantum theory is used to predict the probabilities of observing the possible outcomes of measurements performed on quantum systems. In summary: (1) states of a quantum system are represented by the wave function for that system; (2) wave families are associated with types of measurements; and (3) the members of a family that add up to the wave function allow us to make predictions concerning the outcomes of measurements associated with that wave family.

The evolution of states over time   Before closing this section, a very brief note on the evolution of states over time is in order. Thus far, we have discussed only the outcomes of measurements on a quantum system in a given state. As noted, physics is generally concerned not only with the outcome of measurements at a particular time, but also with measurements we might make in the future. To return to the example of the ball dropped from the roof of a building, well-known mathematics allows us to do both, that is, we can use well-known equations not only to predict the outcome of measurements at the present time, but also to predict the outcome of measurements we might make at a future time.

In quantum theory, the evolution of a system over time is predicted by *Schrödinger's equation*. Recall from (1) above that the current state of a system is represented by the wave function for that system. In very brief outline, Schrödinger's equation allows us to calculate, from the wave function representing the current state of the system, what state the system will be in at future times.

Schrödinger's equation rounds out our overview of the mathematics of quantum theory. As noted above, the mathematics of quantum theory, at bottom, performs the same role as the mathematics we have had in physics for hundreds of years. In particular, the mathematics of quantum theory, as summarized in (1), (2), and (3), allow us to predict the outcome of measurements at a particular time, and via Schrödinger's equation, allow us to predict what state the system will be in at a future time.

Interpretations of Quantum Theory

The most controversial issues involved with quantum theory arise in the interpretation of quantum theory. Again, the interpretation issue is more of a philosophical issue, largely involving the question of what sort of reality could underlie the facts and the mathematics. Our main goal in this section is to explore this interpretation issue. We will begin with some background material.

Background discussion

Suppose we consider an experimental arrangement, not too different from some experiments discussed earlier. In particular, suppose we have a photon gun capable

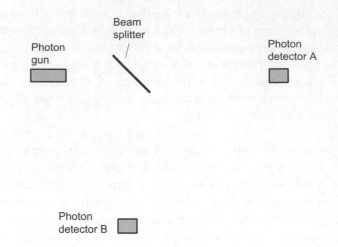

Figure 25.15 Beam splitter arrangement

of emitting single photons. We send the photons toward a beam splitter, that is, a partially silvered mirror. We will place two photon detectors after the beam splitter, and for ease of discussion, let's suppose the detectors beep whenever they detect a photon. The setup is as pictured in Figure 25.15.

Suppose we have a button, and each time we press the button, a photon is emitted toward the beam splitter. There is no question about what the quantum facts will be in this case: each time we press the button, either detector A or detector B will register a photon, but never both at the same time. And moreover, over a long run of button presses and photon detections, the photon will be detected at detector A 50 percent of the time, and at detector B 50 percent of the time. In other words, 50 percent of the time detector A will beep, and 50 percent of the time detector B will beep, but they will never beep simultaneously.

Likewise there is no issue about the predictions made by the mathematics of quantum theory. The mathematics predict that each time the button is pressed, there is a 50 percent chance of registering a photon at detector A, and a 50 percent chance of registering a photon at detector B. In short, the predictions are in keeping with the facts.

This again illustrates the point that there is little controversy about the facts, and little controversy about the mathematics themselves. But consider now some *interpretation* issues.

First, recall some points made earlier in this chapter about interpreting mathematics. In particular, consider again the case of the ball dropped from the roof of a building. This case involves an ordinary object, namely a bowling ball, and part of the relevant mathematics is naturally interpreted as representing that ball. The ball seems to us to fall through ordinary three-dimensional space over the course of a short length of time, and again, parts of the relevant mathematics are naturally interpreted as representing ordinary three-dimensional space and ordinary time. The ball seems to be falling through a medium that offers at least some resistance, that is, the air through which the ball falls slows the ball down somewhat, and again, part of the mathematics is naturally interpreted as representing the resistance of the air. And finally, the ball seems to be under the influence of forces, for example, gravitational forces, and once again, parts of the relevant mathematics are naturally interpreted as representing such forces.

In short, the mathematics that applies to the case of the ball dropped from the roof lends itself to a straightforward interpretation. And that interpretation coincides nicely with the sort of reality we think is involved in this case.

What happens if we try to interpret the mathematics of quantum theory in the same sort of straightforward way? Consider again the setup pictured in Figure 25.15 above. Suppose we push the button once, emitting a photon. What does the mathematics suggest?

The state of the system just after the button is pressed is represented by a wave function. As noted in the previous section, this state evolves over time as governed by Schrödinger's equation. Schrödinger's equation will provide us with the mathematical representation of the state of the system at any point. So suppose we look at the state of the system just before either of the detectors has registered.

Just before either detector registers a photon, the mathematics represents the electron in what is termed a *superposition* of states, with one state representing the electron as a wave traveling toward detector A, and the other state representing the photon as a wave traveling toward detector B. But recall that the detectors never fire simultaneously. Instead, either detector A will beep or detector B will beep. Suppose in this case detector A beeps, indicating that it has detected a photon. This itself is not particularly puzzling – after all, one of the states in the superposition of states seemed to represent a wave traveling toward that detector. But what about detector B? Recall that one of the states in the superposition of states represented a wave traveling toward B, so what happened to that state? Why didn't B beep as well? What happened to the wave that seemed, at least if we try to interpret the mathematics in as straightforward a way as possible, to be traveling toward B?

This example illustrates certain aspects of what is called the *measurement problem*. The measurement problem is one of the most perplexing issues involved in the interpretation of quantum theory, and it is an issue that will arise at several points in this section. In the next few paragraphs, I will try to outline some of the issues and questions involved in the measurement problem. But I would encourage you, throughout the rest of the section, to keep an eye out for the curious issues and questions surrounding measurement.

Two words of forewarning: first, the measurement problem typically does not appear to be much of an issue at first. In other words, it usually takes a while for the problem to soak in, so to speak. But the more you think about measurement issues, the more puzzling they should seem. Second, there are many somewhat different ways to view the measurement problem. These different ways are, in a sense, different faces of the same problem, but each emphasizes slightly different aspects of the puzzling issues involved.

In general, the measurement problem involves the fact that, if we try to interpret the mathematics of quantum theory in a reasonably straightforward way, then something very odd, and very counterintuitive, seems to be happening when measurements of quantum systems take place. The mathematics of quantum theory typically represent such systems as being in a superposition of states, but we never observe (or never seem to observe) such superpositions of states. For example, in the case above, when a measurement occurs we observe either detector A beeping, or detector B beeping. In either case, it leaves the question of what happened to the other state? Did the measurement "collapse" or "reduce" (to use two common terms) the superposition of states to a single state? If so, how could an act of measurement have this effect? And for that matter, what is a measuring device, and how does a measuring device differ from a nonmeasuring device? After all, the physical interactions that occur during what we consider measurements do not seem in any way fundamentally different from physical interactions that occur in situations that we do not consider measurements. So how can any principled distinction be drawn between measuring devices and nonmeasuring devices, and between measurements and nonmeasurements?

There are no agreed-upon answers to such questions. Again, something curious seems to happen when measurements are made, and how to interpret what happens is a matter of dispute. We will discuss alternative answers in a moment, but it will make the discussion easier if we first introduce the Schrödinger's cat thought experiment.

From what has been said thus far, one might be tempted to assume that quantum weirdness resides only in the micro level of entities such as photons, electrons, and the like, and not in the macro-level world of you, me, and our houses and cars, and so on. To indicate that the situation is not quite so simple, let's consider Schrödinger's cat, which is a well-known creature that can be found lurking around most discussions of quantum theory. Schrödinger's cat raises no substantially new issues than those raised by the superposition of photons in the experiment above, but his cat does serve to emphasize the oddness of the situation, and also that the weirdness is not necessarily confined to just the micro level. Schrödinger's cat will also help illustrate some features of the interpretations we will consider momentarily.

Schrödinger's cat   In the mid-1930s, when some of the oddness of the new quantum theory was becoming evident, Schrödinger presented a thought experiment which further illustrates some of the oddness of the quantum picture. Incidentally, a "thought experiment" is just what the name suggests, that is, an experiment we are asked to think through rather than actually perform.

Schrödinger asks us to imagine a cat placed in a sealed box, along with a mild source of radioactivity. The radioactive source is such that, over the course of one hour, there is a 50 percent chance that it will emit a radioactive particle. If such a particle is emitted, it will trigger a detector that in turn is rigged to break open a vial of poison, which would be fatal to the cat.

Incidentally, it is common to use the word "particle" to speak of quantum entities such as photons, electrons, radioactive particles, and the like. But as should be clear from the discussion of the past sections, neither "particle" nor "wave" is quite right when speaking of quantum entities. I will continue, as is standard, to use the word "particle," but nothing should be read into this with respect to the question of whether quantum entities really are particles rather than waves.

Schrödinger's point, of course, was not to abuse cats – again, this is a thought experiment, not one intended to actually be carried out (nor, if you think about it, would doing the experiment yield any interesting data). Rather, his point was to connect up the oddness of the micro level to macro-level events. In doing so, he was also attempting to provide an argument against a particular interpretation of quantum theory, which we will discuss below.

To get a better sense of the Schrödinger's cat thought experiment, consider a slightly altered version that involves a modification of the setup described in Figure 25.15 above. The implication of this version of Schrödinger's cat is the same as in Schrödinger's original paper, but using the setup above will simplify our discussion.

Imagine placing the experimental setup described in Figure 25.15 into a large opaque box. Along with the setup we will place a cat. We will also rig photon detector A, but not detector B, to a vial of poison, just as in Schrödinger's original thought experiment. That is, if detector A registers a photon, it will open the vial of poison, which would be fatal to the cat. On the other hand, if photon detector B registers a photon, then nothing happens. The setup would then look as shown in Figure 25.16.

Suppose the entire apparatus, cat and all, is in the sealed box, so that we cannot see what happens inside the box, nor can we tell whether one or the other of the photon detectors has registered a photon, nor can we hear anything that goes on inside the box. We do, however, keep the button that causes the photon gun to emit a photon outside the box.

Now, given this setup, suppose we push the button once. Suppose we consider the situation a few seconds later, after the photon has had plenty of time to reach the detectors. Recall that the evolution of states over time is governed by Schrödinger's equation, and that the Schrödinger equation represents the situation as being in a superposition of states. In this case, the two states would involve one state representing the photon as having been registered at detector A, with the other state representing the photon as having been registered at detector B. But recall also that the detection of a photon at A will result in the poison being released, and hence result in a dead cat. A detection of a photon at B, on the other hand, will leave the cat alive and well. So the superposition of states now would seem to involve two states, one in which the cat is dead and the other in which the cat is alive and well. In other words, if we ask what is really going on inside

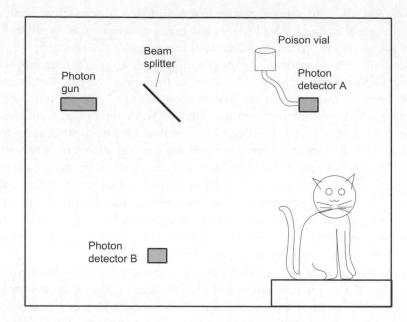

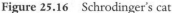

Figure 25.16   Schrodinger's cat

the box, it seems that quantum theory is representing the cat as existing in a superposition of cat-dead and cat-alive states.

Again, this is not in principle different from the superposition of the states of the photon discussed in regard to Figure 25.15 above. Schrödinger has, though, moved the oddness from the micro to the macro level.

Before closing this section, it is worth noting that Schrödinger himself was trying to show that something must be missing from the mathematics of quantum theory. He took it as a given that a cat cannot exist in a combined both dead-and-alive state, and hence the mathematics of quantum theory must be leaving something out.

The attempt to modify quantum theory, so as to supply the sort of missing element(s) Schrödinger thought the cat experiment pointed to, resulted in what are commonly termed *hidden variable* interpretations of quantum theory. That is, the idea behind such interpretations is that quantum theory, as described above, is not capturing what reality is really like, and what is needed is to add something (so-called hidden variables) to the theory to better bring it back into line with our intuitions about reality.

Hidden variable interpretations are in contrast to what is generally referred to as the *standard interpretation*, or the *Copenhagen interpretation*. Both names are rather misleading, because there is no one, single, well-defined interpretation that constitutes the standard, or Copenhagen, interpretation. Rather, the standard interpretation is more like a general approach to interpreting quantum theory, and within this approach several variations are found. Let's explore some of these variations.

Variations on the standard (Copenhagen) interpretation

In contrast to hidden variable interpretations, advocates of the various versions of the standard interpretation agree that quantum theory is a complete theory, and that no "hidden variables" or other additions are needed. In fact, though, even these advocates do add at least one thing to the mathematics described above. Typically, what is added to the mathematics described so far is what is commonly termed a *projection postulate*. A projection postulate governs what is generally referred to as the "collapse" or "reduction" of the wave function. The idea is that, before a measurement, a quantum entity is represented as existing in a superposition of states. For example, before measurement, the photon emitted in the experiment described above is represented by a wave function, where the wave function itself consists of a superposition of states. But upon measurement, the superposition of the two states "collapses" to a new state represented by a new wave function. The collapse is governed mathematically by the projection postulate, and the result is the new wave function mentioned. In this example, this new wave function represents either an electron at detector A or an electron at detector B. Thereafter, this new wave function evolves according to the Schrödinger equation until, perhaps, another measurement occurs.

There is no question that the mathematics described earlier, coupled with a collapse of the wave function or projection postulate, accords well with what we observe. That is, the predictions made are, and have been for over 70 years, dead on. But if we try to picture this in terms of what is "really" going on, the collapse of the wave function is certainly a difficult event to picture.

Suppose we ask advocates of the standard interpretation what is "really" going on. For example, in the experiment pictured in Figure 25.15, where is the electron really located just before the measurement takes place? Is the electron really a wave existing in both channels just before measurement? And does one of these waves then instantaneously disappear right at the moment of measurement?

Advocates of the standard interpretation generally take the view that there are no answers to questions such as these. For example, we simply cannot say where the electron is really located before measurement. Likewise, if you ask about other attributes of the electron – its momentum, spin, and so on – the same applies. In general, we cannot say what attributes the electron has before a measurement occurs.

Importantly – and this is very crucial to understanding the sense of what is being said here – according to the standard interpretation, the reason we cannot say what attributes a quantum entity has before measurement is *not* simply because we do not know what those attributes are. Rather, we cannot say what those attributes are because those attributes do not exist prior to measurement. There is no deep, independent reality consisting of objects with definite attributes existing prior to measurements of those attributes.

This is a very counterintuitive view, so it is worth taking a moment to clarify it. Suppose I tell you I have a number of coins in my pocket. You presumably do not know how many coins there are, but I am sure you are convinced that there is a definite number. Maybe two, or three, or eight, but whatever the actual number, you no doubt believe there is a definite, independent fact about how many coins are in my pocket.

This is a case of everyday lack of knowledge. You cannot say how many coins are in my pocket because you simply do not know how many there are. This is *not* the sort of lack of knowledge involved in the standard interpretation. On the standard interpretation, there is nothing to know. The electron has no definite position before measurement. It has no definite spin. And so on.

Advocates of the standard interpretation are not denying the existence of reality. That is, there is a reality, there are quantum entities, there is an electron "out there." But that electron, and quantum entities in general, do not possess definite attributes before they are measured.

(To be accurate, I should note that everyone, including advocates of the standard interpretation, agrees that quantum entities have a small number of attributes independent of measurement. These attributes, such as mass, are referred to as "static attributes." But aside from these few static attributes, according to the standard interpretation the remaining attributes of a quantum entity do not exist prior to a measurement.)

In short, although there is a reality, it is not a reality that consists of quantum entities with definite attributes before measurements. This is a much more radical lack of knowledge than the everyday lack of knowledge illustrated by the coin example.

Within this general description of the standard interpretation, there are a number of variations possible. We have noted that according to the standard interpretation, quantum entities do not have attributes until a measurement occurs. However, little has been said about these two questions: (a) what counts as a quantum entity? and (b) what counts as a measurement? Let's take a moment to consider these two questions.

With respect to what counts as a quantum entity, we have thus far used as examples entities such as electrons, photons, particles emitted during radioactive decay, and the like. Everyone agrees that these are quantum entities.

But are these elementary particles the only quantum entities? Keep in mind that all objects are presumably composed of nothing more than these basic entities. So there is a good case to be made that "quantum entities" really should refer to everything. That is, everything should be considered a quantum entity. At any rate, note that the answer to the question of what counts as a quantum entity is neither simple nor uncontroversial, and the question admits of more than one reasonable answer.

Likewise for the question of what counts as a measurement. This is a significant question, especially for advocates of the standard interpretation. Note that, on the standard interpretation, the collapse of the wave function occurs when a measurement is made. But what counts as a measurement? To see that there are a variety of ways to answer this question, consider the Schrödinger's cat scenario. When does the first event that can be considered a measurement take place? Is it when one or the other photon detectors becomes involved in a physical interaction involving the presence of an electron? One might initially be inclined to say that that is the first measurement. But keep in mind that the processes involved with those detectors are just physical processes, no different in kind from the other physical processes taking place inside the box. For example, the photon has a physical interaction with the beam splitter, but we generally do not count that physical interaction as a measurement. But the physical processes involved when the photon interacts with something we call a "detector" are not particularly different from the physical interactions it has with the beam splitter. So it is not at all clear why the photon's interaction with the detector should count as a measurement either. In other words, there seems nothing inherently special about those photon detectors, so why would they result in the collapse of the wave function? The processes undergone by the detectors happen to be of special interest to you and me, because we may well be interested in the presence or absence of a photon, but aside from their interest to us observers, there is nothing special about those photon detectors. So it is far from clear that the photon detectors should be considered the first measuring device.

What about the cat's auditory system? Perhaps the cat hears either detector A or detector B beep, and perhaps the cat's perception of the beep counts as the first measurement. On the other hand, the physical occurrences that lead to the cat's perception of the beep are not in principle any different from any of the other millions (and more) physical interactions taking place inside the box, so again, why should these particular interactions be considered a "measurement" when the others are not? Maybe they should not be, and odd as it sounds, perhaps the cat itself exists in a superposition of states, one in which it hears a beep from detector A and one in which it hears a beep from detector B, and soon after, a superposition in which one state involves a dead cat and the other a live cat.

Finally, does the first measurement occur when we open the box, see what the detectors read, and see whether the cat is dead or alive? Our initial inclination is to think that the wave function must have collapsed by this point, because we do not detect a cat in a superposition of alive and dead states (but as we will see, not all interpretations of quantum theory agree with this). But does the collapse occur only at this moment? Is human consciousness the key to measurement?

In general, the two questions of what counts as a quantum entity, and what counts as a measurement, do not admit of single, uncontroversial answers. And depending on how these questions are answered, one arrives at various versions of the standard interpretation. Since all the variations include the view that, in some sense, reality is dependent on measurements, I will term these variations "measurement-dependent realities," and distinguish them as mild, moderate, and radical. We will turn now to a discussion of these variations on the standard, or Copenhagen, interpretation.

Mild measurement-dependent reality According to this mild version, "quantum entity" refers only to the most basic elementary particles such as electrons, neutrons, protons and the wide variety of other subatomic particles, photons, particles emitted during radioactive decay, and the like. That is, the only quantum entities are the most basic "stuff" of the universe, and it is only this elementary level that lacks definite attributes until measurements are made.

Once again, let me repeat a point made above. Advocates of this variation, and of the other variations as well, are not claiming that nothing exists before measurement, and that measurements then bring a reality into existence out of nothing. Rather, the idea is that there is a reality independent of measurement, but as far as the most basic components of the universe go, it is a largely undefined reality. To return to the example of the coins in my pocket, it would be similar to saying yes, there are coins in my pocket, but there is no definite number of coins, nor are there definite shapes or sizes to the coins. That is, there is a reality consisting of coins, but it is a reality lacking any definite attributes.

To apply this analogy to the quantum level, the advocates of this mild version of measurement-dependent reality are saying that there is a reality consisting of electrons, photons, and the like, but that reality is largely an undefined reality. The electrons are neither in this location nor in that location nor in any other location, they have neither this spin nor that spin, and so on, until a measurement of such attributes takes place. Only with measurements do such entities acquire definite attributes.

With respect to the question of what counts as a measuring device, this version answers the question in a broad way. For example, in the Schrödinger's cat scenario, the measuring devices are taken to include the photon detectors, the cat's auditory system, and our looking into the box and observing the detectors and the cat.

With respect to the Schrödinger's cat example, note that, on this interpretation, the first measurement occurs when the photon reaches the detector. This then collapses the wave function, and either detector A or detector B registers a photon. And depending on which, either the vial is opened or it is not. So on this interpretation, the cat never exists in a superposition of dead and alive states.

In short, on the mild measurement-dependent reality interpretation, only the elementary particles such as electrons, photons, particles emitted during radioactive decay, and so on, can exist in a superposition of states. And almost any sort of measurement suffices to collapse the wave function. To put it roughly, there is still plenty of quantum weirdness, but the weirdness exists only at the micro level.

Moderate measurement-dependent reality   The mild version considered above took quantum entities to be only the elementary particles. But, presumably, everything is composed of such particles. The coffee cup on the desk in front of you, for example, is at bottom composed of just these elementary entities. If the coffee cup is composed of nothing but quantum entities, there is a good case to be made that the coffee cup itself should likewise be considered a quantum entity, albeit a larger and more complex one than the basic particles.

If we answer the question of what counts as a quantum entity in this broad sense, that is, we count all objects as quantum entities, then almost any object can in principle exist in a superposition of states. In this interpretation, then, some of the quantum weirdness bleeds up to the macro level.

However, the moderate measurement-dependent reality interpretation, as with the mild measurement-dependent interpretation, takes a broad view of what counts as a measurement. So, although almost any object can in principle exist in a superposition of states, measurements generally suffice to collapse such superpositions well before we or any other creatures could experience them. To use the example of Schrödinger's cat as a concrete case, on this interpretation, the photon detectors, the cat, and so on count as quantum entities, and so could in principle exist in a superposition of states. However, given the broad approach to what counts as a measurement, the photon detector suffices to collapse the wave function well before any superposition of cat-alive and cat-dead states could occur. So although macro-level objects such as you, me, the cat, and so on, could in principle exist in a superposition of states, in practice the wave function will collapse before we would ever experience such a superposition.

Radical measurement-dependent reality (consciousness-dependent reality)   Suppose, in keeping with the moderate variation above, we take a broad view of what counts as a quantum entity, that is, everything counts. But suppose we take a narrow view of what counts as a measurement. In particular, suppose we adopt the view that only human consciousness constitutes genuine measurement. So, for example, in the Schrödinger's cat scenario, the first measurement does not take place until we open the box and look at the detectors and cat.

This is a much more radical variation, for here, wave functions are not collapsed until human observation is involved. So unobserved situations exist in superpositions of states – cats can exist in a superposition of dead and alive states, and so on. In short, the world, when unobserved by humans, does not consist of definite states. Unobserved objects are not definitely located at any particular position, and so on.

What would motivate anyone to adopt this radical view of the world? The problem is exactly that of what counts as a measurement. Consider again the Schrödinger's cat scenario. As discussed above, the physical processes that occur in what we consider to be measuring devices do not seem different in kind than other physical processes. So it is difficult to see how such a process could result in a collapse of the wave function. If – and this is a big if – you view human consciousness as involving processes that are different in kind than other processes, then in the chain of events from the emission of a photon to our opening of the box and observing what is inside, the observation by human consciousness is the only sort of event that is different from the others. And so this, or at least so it has seemed to some advocates of this interpretation, is the natural place to put the collapse of the wave function.

Hidden variable interpretations

In general, hidden variable interpretations are those that maintain that the mathematics described thus far constitutes an incomplete theory. In particular, the mathematics leaves out "elements of reality," to use Einstein's phrase. What is needed, according to such interpretations, is to supplement the theory described so far with what have come to be called "hidden variables," that is, additional elements that fill in the perceived incomplete aspects of existing quantum theory. There are two main versions of hidden variable theories. We will begin with Einstein's version.

Einstein's realism   The interpretation of quantum theory that I will refer to as "Einstein's realism" is the common-sense interpretation, that is, the interpretation that attempts to reconcile quantum theory with our usual sense of what the world is like.

I should stress at the outset that Einstein's interpretation – or at least, the key elements of the interpretation that were most important to Einstein – is no longer compatible with certain newly discovered quantum facts. As mentioned previously, any interpretation of quantum theory has to respect the quantum facts, and new facts have emerged in recent years, such that Einstein's interpretation is no longer compatible with those facts. Einstein died before these facts emerged, so it is interesting to speculate on (though we will ultimately never know) what his reactions to these new facts would have been.

In short, Einstein's view was that there simply could not be the sort of lack of knowledge that is central to the standard interpretation. Recall that, on the standard interpretation, quantum entities do not have definite attributes until they are measured to have those attributes. Again, according to the standard interpretation, it is not merely that we do not know what those attributes are; rather, quantum entities do not *have* such attributes. Einstein's view was that quantum entities must have definite attributes before they are measured to have them. Again, this is the common-sense view. With respect to the coins in my pocket, common sense tells us that there must be a definite number of coins, even if we do not know what that number is. Likewise, common sense, so Einstein urged, tells us that quantum entities also must have their attributes all along, even before any measurements are performed.

However, the mathematics of quantum theory does not represent quantum entities as having attributes before a measurement takes place. This is the sense in which Einstein argued that the mathematics of quantum theory must be incomplete. There are "elements of reality" – for example, quantum entities with definite attributes before measurement – that quantum theory does not capture. Hence, Einstein's view was that a new theory would take the place of quantum theory – one that did all that quantum theory did, but included the "hidden variables," not present in existing quantum theory, that reflected these elements of reality. Einstein did not have specific proposals as to how quantum theory ought to be supplemented, but he was convinced that quantum theory needed to be supplemented.

We will discuss Einstein's views further, and the new facts that are so problematic for that view, in the next chapter. For now, we will simply repeat that, in summary, Einstein's is the common-sense interpretation. Moreover, certain key elements of that interpretation (which we will explore in the next chapter) are no longer compatible with the known facts. As such, Einstein's interpretation – and for that matter, any interpretation that fits with certain widely held views on how the universe works – is no longer viable.

Bohm's realism David Bohm, in the late 1940s and early 1950s, formulated a modification of the mathematics of quantum theory. Let's look briefly at some key aspects of Bohm's view.

Bohm treats quantum entities as particles, under the influence of what is commonly termed a *guidance wave*. Notably, Bohm's mathematics, and the standard mathematics, appear to make identical predictions. Thus there is no empirically testable difference between the two approaches.

The picture Bohm's version paints of the underlying reality, however, is different from that suggested by the standard interpretation of quantum theory. As we saw above, on the standard interpretation quantum entities do not have definite characteristics, such as position, before measurements. Very briefly, on Bohm's view, quantum entities do have definite positions independent of any measurements. And so the knowledge we lack, say, of the position of an electron, is not a matter of the electron having no such position before measurement; rather, our lack of knowledge is basically the same sort of lack of knowledge we have as to how many coins are in my pocket.

With respect to the Schrödinger's cat thought experiment, since the photon has a definite position before measurement, there will be a fact of the matter (although we will not know it until we open the box) as to whether detector A or B registered the photon. Hence there will be, on Bohm's interpretation, no superposition of a live and a dead cat. There is simply our lack of knowledge as to which is the case.

A likely question at this point is this: If Bohm's view has all the predictive power of the usual approach to quantum theory, and if, on Bohm's view, we can view the underlying reality as being the sort of well defined reality most of us are more comfortable with, why isn't Bohm's view the standard view? That is, why is his approach, and his interpretation, a minority view? Why hasn't his view taken over as the majority view?

Issues surrounding Bohm's interpretation are complex and contentious, and there are no easy answers to these questions. However, I will suggest two general (and partial) answers to such questions. First, note that Bohm's mathematics cannot be shown to be better than the existing mathematics. Again, Bohm's mathematics, and the standard mathematics of quantum theory, appear to make identical predictions. And recall that the standard mathematics for quantum theory had been in place for quite a few years before Bohm proposed his modifications. Physicists had become comfortable with the existing mathematics. But now Bohm suggests an alternative mathematics. But since his mathematics makes no new predictions, it cannot be shown to be any better than the mathematics already in place. Thus, from a practical point of view, there is no compelling reason to replace the existing mathematics with Bohm's new approach. And since his interpretation of quantum theory is tied to his mathematics for quantum theory, I suspect the lack of enthusiasm for his mathematics resulted in a lack of enthusiasm for his interpretation.

Second, although Bohm's mathematics can arrive at the same predictions as the standard mathematics, it is not clear that Bohm's mathematics lends itself to an interpretation that is any less problematic than the standard interpretation. In a nutshell, here is the perceived problem: the guidance wave mentioned must be taken as representing something in reality. (If it is not, then there is no reason at all to prefer Bohm's approach. That is, if one is going to take an instrumentalist attitude toward quantum theory, then one might as well stick with the standard mathematics that most physicists use.) Moreover, Bohm's guidance wave requires faster-than-light influences (*superluminal* influences is the standard way to describe such faster-than-light influences), and there is a common perception that the superluminal influences in Bohm's approach conflict with Einstein's theory of relativity. In short, there is a widespread perception that Bohm's approach conflicts with relativity, and if forced to choose between Bohm's interpretation and Einstein's relativity, there is not much question that relativity would win out.

As noted, there is a widespread perception that Bohm's interpretation and relativity theory conflict. Whether this perception is correct is another, and difficult, question. As we will see in the next chapter, recently discovered quantum facts rule out any interpretation that does not allow for some sort of superluminal influences. In light of these newly discovered facts, advocates of Bohm's interpretation argue that the superluminal influences required by Bohm's interpretation are no worse than those required by any viable interpretation. But some critics of Bohm's interpretation argue that his interpretation requires more "robust" superluminal influences than those required by other interpretations, and that, in particular, the superluminal influences in Bohm's interpretation at least violate the spirit, if not the letter, of relativity theory.

The issues involved in this debate are complex, and it is an open question as to whether the tension between Bohm's interpretation and relativity theory is genuinely problematic. It is safe to say, though, that Bohm's interpretation is often perceived as not fitting well with Einstein's relativity, and this is one of the main reasons why Bohm's interpretation is not more widely accepted.

The many-worlds interpretation

Before concluding this general section, it is worth describing one final interpretation, that being the many-worlds interpretation. The many-worlds interpretation is not a variation on the standard interpretation, nor is it a hidden variable interpretation.

It will be easiest to understand the many-worlds interpretation if we contrast it with the standard interpretation. As noted earlier, the collapse of the wave function is one of the most puzzling aspects of any of the versions of the standard interpretation. One virtue of the hidden variable interpretations discussed above is that they involve no collapse of the wave function (thus they are sometimes, along with the many-worlds interpretation, categorized as "no collapse interpretations"). It is likewise a virtue of the many-worlds interpretation that it too does not require a collapse of the wave function. But, as with the hidden variable views, this virtue comes at a cost.

Consider again the beam splitter experiment depicted in Figure 25.15. Suppose we push the button once. Recall that, after the photon passes through the beam splitter but before it reaches the detectors, the mathematics of quantum theory represents the photon as existing in a superposition of states. Again, one state represents the photon as a wave going toward detector A, and the other state represents the photon as a wave going toward detector B.

Now suppose, a fraction of a second later when the photon has had time to reach the detectors, that we hear detector A beep, indicating that it has detected a photon. According to all the versions of the standard interpretation, there has been a collapse of the wave function, that is, the superposition of states has collapsed to a single state, in this case, a state involving a photon at detector A.

What does the many-worlds interpretation say about this scenario? According to the many-worlds interpretation, the wave function never collapses. That is, the superposition of states continues, even after the photon reaches the detector. The superposition of states now consists of one state representing detector A as having detected a photon, with the other state representing detector B as having detected a photon. On the many-worlds interpretation, the wave function never collapses, and superpositions of states continue.

A similar story holds for the Schrödinger's cat scenario. Neither the photon detectors, nor the cat's auditory system, nor you and I looking in the box collapses the wave function. Again, on this interpretation there is no such thing as a collapse of the wave function.

An obvious question is this: If the wave function does not collapse, and there is the superposition of states just described, then why do you and I not observe such superpositions of states? Why, in the example depicted in Figure 25.15, do we observe only detector A registering a photon? And in the Schrödinger's cat scenario, why do we not observe the superposition of cat-alive and cat-dead states?

The answer is that you and I are part of one of the states making up the superposition of states. You and I inhabit, so to speak, one of the states involved in the superposition. You and I happen to inhabit (or, perhaps it would be better to say, we are part of) the state in which detector A registered the photon, or in the Schrödinger's cat scenario, the state in which the cat is dead. But since there has not been (and never is) a collapse of the wave function, the other states also exist. You and I have counterparts that inhabit the other states. (Incidentally, no word is quite right, but "counterparts" probably comes closest.) Whereas we heard detector A beep, our counterparts heard detector B beep. And whereas we are looking at a deceased cat, our counterparts are gazing at a cat that is perfectly alive and happy.

In short, there is no artificial and mysterious "collapse" of the wave function. The wave function representing the entire universe – that is, the wave function representing everything, including you and me and all our counterparts – evolves according to the Schrödinger equation. This wave function represents a universe consisting of superpositions of an enormous number of states, and the number of these states is constantly increasing. Exactly one of those unimaginably huge number of states is the state that you and I inhabit. But from a broad viewpoint, there is nothing special about this state – it is no more or less "real" than all the other states involved in the overall superposition.

The overall picture that goes along with the many-worlds interpretation is of a constantly branching tree, with each branch representing a state involved in the huge superposition of states. A new branch sprouts each time a quantum entity enters a situation leading to a superposition of states. Since this occurs very, very often, new branches of this tree are sprouting at a furious pace.

A few final observations on the interpretation of quantum theory

As we have emphasized, interpretations of quantum theory must be constrained by the known quantum facts and, presumably, they will at least be compatible with the mathematics of quantum theory. But as we have seen, the facts themselves are quite strange, and the mathematics does not lend itself to a common sense picture of reality. And largely for these reasons, the interpretations of quantum theory all tend to be quite contrary to common sense. The sole exception to this is Einstein's interpretation, which, as we have noted, is no longer a viable option.

We should note that, with respect to the interpretation question, an instrumentalist attitude is a common approach. That is, one taking a purely instrumentalist attitude toward quantum theory will not dabble in the interpretation issue. With this attitude, quantum theory is viewed as a convenient tool for making predictions, much as Ptolemaic epicycles were viewed as a convenient mathematical tool for making predictions. But when asked about the reality underlying the theory, an instrumentalist will take an agnostic attitude. And again, this is a perfectly respectable approach to take toward quantum theory. Indeed, there is much to be said for the view that this is the most practical attitude for working physicists to take, at least during working hours.

But for those of us who are not working physicists, and even for physicists after the lab has closed for the day, it is difficult to resist the question that has occupied so much of western thought since the time of the ancient Greeks: *What kind of universe do we live in?* Our answers to this question have been heavily influenced by our best science, and quantum theory is certainly one of the most important and successful theories in our history. Thus it is appropriate that quantum theory influence our current views on the sort of universe we inhabit. But notably, the available interpretations of quantum theory paint a picture of a universe very much unlike what we have always thought it was like.

The versions of the standard interpretation, Bohm's interpretation, and the many-worlds interpretation all have aspects that are attractive, and aspects that are not so attractive. It is worth taking a minute to summarize the virtues and vices of each of the interpretations.

As we saw, the standard interpretation has the virtue of being a reasonably "minimalist" interpretation, that is, an interpretation that sticks reasonably close to, and takes at pretty much face value, the standard mathematics described earlier in this chapter. If, for example, the mathematics suggests that an electron in a particular situation has no definite position, then so be it. We will just have to accept that the world does not consist of definite entities that always have deter-minate properties.

But, as we also saw, advocates of the standard interpretation do add the collapse of the wave function to the picture suggested by the usual mathematics of quantum theory. The collapse presumably occurs when a measurement takes place, and this leaves the advocates of the standard interpretation with no good answers to difficult questions surrounding the measurement problem. What in the world is happening during a measurement? Since a measurement process is just a physical process, not different in kind than processes we do not consider to be measure-ment processes, how could there be any real difference between a measurement process and a nonmeasurement process? Likewise, if everything is composed of quantum entities, how could there be any real difference between measuring devices and the quantum systems they are measuring? These questions are all variations of the measurement problem, or perhaps better, they are different angles from which to view the measurement problem. The collapse of the wave function raises difficult questions for advocates of the standard interpretation, and advocates of these interpretations lack good answers to such questions.

In contrast, Bohm's interpretation has the noteworthy virtue of avoiding the measurement problem. On his interpretation, there is no collapse of the wave function, and hence advocates of his interpretation do not have the difficult questions mentioned above. On Bohm's interpretation, there is no fundamental difference between measuring devices and the quantum systems they measure, no mysterious collapse of the wave function, and no measurement problem.

But Bohm's interpretation is widely seen as not fitting comfortably with Einstein's theory of relativity. Since Einstein's relativity is a key branch of modern physics, this is a potentially serious vice. So Bohm's interpretation, in spite of its virtues, carries with it some substantial baggage.

The many-worlds interpretation likewise has the substantial virtue of avoiding the measurement problem. Here, too, there is no collapse of the wave function, and hence none of the difficult questions that accompany the collapse. Moreover, this interpretation is a truly minimalist interpretation. That is, it accepts at face value the picture suggested by the standard mathematics discussed in the previous chapter. If the mathematics of quantum theory suggests that systems involving quantum entities exist in superpositions of states, then so be it. You, me, all the objects around us, and so on are merely part of one state among the enormously complex superposition of states involving everything there is. This is what the mathematics suggests, and the advocates of the many-worlds interpretation accept it at face value.

But the virtues of the many-worlds interpretation come only with the vice that this interpretation is probably the most counterintuitive of all. It is difficult to envision reality being as different as that painted by the many-worlds interpretation, consisting of unimaginably large numbers of counterparts of you and me and most of what we see around us. And it can be equally difficult to envision a reality that does not consist of the single, definite world we seem to see around us, but rather consists of a superposition of a huge number of states, of which that we see around us is but one. So in short, the many-worlds interpretation, as with the others, has both its virtues and its vices.

Perhaps, at this point, we can better understand why an instrumentalist attitude toward quantum theory is so common. Review for a moment some of the theories we considered earlier in the book. For most of western history (written history, that is), the Ptolemaic theory was our best theory in terms of accounting for the astronomical data. But as we saw, the theory requires epicycles. And it is difficult to imagine how planets could really move on small circles like that, that is, it is difficult to take epicycles with a realist attitude. And so, given such difficulties with viewing planets as really moving the way they were described on the Ptolemaic system, it was common for astronomers to take an instrumentalist attitude toward Ptolemaic epicycles.

With the acceptance of Kepler's view of planetary motion, with planets moving in elliptical orbits at varying speeds, we had a picture of planetary motion that easily could (and was, and is) taken realistically. But this view of planetary motion only tells us that planets move this way, with no fuller understanding of why they move as they do. Newton's physics seemed to provide an explanation for planetary motion: elliptical orbits are what one would expect from the principle of inertia and the principle of universal gravitation.

But as we saw at the end of Chapter 20, if taken realistically, Newton's concept of gravity seemed to involve mysterious "action at a distance." And again it was difficult to imagine how there could really be such mysterious, "occult" forces. And, also as we saw toward the end of Chapter 20, largely for these reasons Newton himself professed to take an instrumentalist attitude toward gravity. (Again, people raised in the Newtonian worldview tend to take a realist attitude toward gravity, but this is probably because they are raised with the notion of gravity, and hence tend not to recognize that there are some odd features of gravity.)

And as we are seeing, quantum theory likewise does not lend itself to a commonsensical picture of reality. And so the fact that an instrumentalist attitude toward quantum theory is common may be simply the latest version of a trend that has been around for some time, that of taking an instrumentalist attitude toward theories that do not lend themselves to unproblematic realistic interpretations.

But if we do want to consider the reality question – we are interested in the question of what sort of world we live in – it is worth noting that all the available interpretations have aspects that are not so appealing. And as mentioned at the outset of this section, not only is there is no consensus as to which interpretation, if any, is correct, there is not even a consensus on which interpretation is preferable. Which interpretation one favors tends to boil down to the largely aesthetic question of what flavor of oddness one prefers. Or, perhaps more accurately, what flavor of oddness one finds least bothersome.

Concluding Remarks

This has been, admittedly, a long chapter. But that is to be expected with a topic as complex as quantum theory. To return to the general point with which we began the chapter, it is important to keep in mind the distinction between quantum facts, the mathematics of quantum theory, and interpretations of quantum theory.

The major sections of this chapter dealt in turn with each of these topics. As we saw, the quantum facts are surprising, but there is no controversy over what those facts are. And the mathematics of quantum theory is, as noted, a variation on a wave sort of mathematics that is not at all unusual in physics.

The most controversial and difficult issues arise with the interpretation question, that is, the question of what sort of reality is consistent with the known facts and, presumably, consistent also with the mathematics of quantum theory (or in the case of Bohm's interpretation, an alternative to the standard mathematics of quantum theory). As we saw, none of the usual interpretations (other than Einstein's, which as noted is no longer a viable option) paint a picture of reality that is anything like the usual picture we have had for at least the past 2,500 years.

In the next chapter, we will explore some recent empirical results that have additional implications for the reality question. In particular, the results clarify what sorts of interpretations are viable options. As we will see, the new results do not make the situation any less odd, but they do help identify where the oddness lies.

Chapter Twenty-Six

Quantum Theory and Locality: EPR, Bell's Theorem, and the Aspect Experiments

In the previous chapter, we explored some quantum facts, saw an overview of the mathematics of quantum theory, and explored some of the interpretations of quantum theory. In recent years, certain new quantum facts have arisen, and these facts allegedly have implications for the interpretation of quantum theory. Our main goals in this chapter are (a) to understand these recent experiments involving quantum theory that are often claimed to have substantial implications for our view of reality, and (b) to analyze these alleged implications. In particular, we will look carefully at the claim that these recent experiments show that any "local" view of reality must be mistaken. As with previous chapters, let us begin with some introductory material.

Background Information

As emphasized in the previous chapter, it is important to keep distinct the quantum facts, quantum theory itself, and the interpretation of quantum theory. In this chapter we are primarily interested in recent experiments that are often claimed to have substantial implications for the interpretation of quantum theory. These experiments fall under the heading of quantum facts. In other words, the experiments in question, and the outcomes of those experiments, are nothing more or less than new quantum facts. On the other hand, the alleged implications of the experiments are part of the reality question involved in the interpretation of quantum theory. Again, quantum facts are important to the interpretation question because such facts constrain the types of answers we can give to that question.

Worldviews: An Introduction to the History and Philosophy of Science.   Richard DeWitt
© 2010 Richard DeWitt

In particular, whatever reality is like, it had better be the sort of reality that can produce the known facts.

Allegedly, certain recent quantum facts rule out any "local" view of reality. That is, recent quantum facts supposedly are such that only a nonlocal reality could produce them. We will at some point have to consider carefully what is meant by "locality" and "local" views of reality, but first I want to try to describe these new quantum facts in a way that is as accessible as possible.

These new facts are most easily understood by looking at what I will call the EPR/Bell/Aspect trilogy. This trilogy consists of the EPR thought experiment, presented in a 1935 paper by Einstein, Podolsky, and Rosen; John Bell's 1964 proof, generally referred to as Bell's theorem or Bell's inequality; and a series of experiments from Alain Aspect's lab, beginning in the mid-1970s and culminating in a series of important experiments in the early 1980s.

In what follows, I will be presenting slightly simplified versions of the EPR, Bell, and Aspect material. For example, I will present EPR, Bell, and Aspect as if all three were concerned with simple polarization attributes of photons, whereas in fact the story is a little more complicated. Also, I will present Bell's theorem as if it were the design for an experiment, whereas it actually is a mathematical proof rather than an experimental design. We will also somewhat simplify issues involved with the polarization of photons, and somewhat simplify the description of the relevant experiments. These minor simplifications will make this material substantially easier to follow, while in no way distorting the key ideas behind EPR, Bell's theorem, and the Aspect experiments. Let's begin with the EPR thought experiment.

The EPR Thought Experiment

The discussion to follow focuses largely on the "polarization" of photons. The discussion requires no knowledge of what polarization is. This is good, because no one knows what polarization "really" is anyway. For this discussion, all you need to know is that polarization is an attribute of photons (in something very roughly like the way orangeness is an attribute of pumpkins), and that polarization can be detected by polarization detectors.

Suppose our polarization detectors will register any photon as having "Up" polarization or "Down" polarization, and that for an individual photon, there is a 50:50 chance of being measured as having Up polarization and a 50:50 chance of being measured as having Down polarization.

Suppose we generate a special pair of photons, in particular, a pair of photons in the "twin state." By saying the photons are in the twin state, I mean simply that if we measure the polarization of the two photons, they will always be measured to have the same polarization. That is, in much the same way that, when we measure the sex of identical twins, the sex detectors will register both as male or

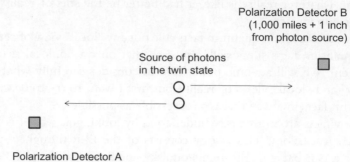

Figure 26.1   A typical EPR setup

both as female, when we measure the polarization of photons in the twin state, the polarization detectors will register both as Up or both as Down. (More precisely, if we use identical polarization detectors to measure the polarization, the detectors will both read Up or both read Down. For the remainder of this section we will assume we are using identical detectors for both measurements.)

Notably, aside from the results of such a measurement, nothing more or less is meant by saying the photons are in the twin state. In particular, nothing is being said or implied about what the photons are "really" like when they are not being measured. In short, I am simply describing some quantum facts: if you generate a pair of such photons and perform a polarization measurement on them, the polarization detectors will both read Up or both read Down. Again, that is all that is meant by saying the photons are in the twin state. Try to stick to just these facts, and resist the temptation to imagine a picture of what is "really" going on with the photons when they are not being measured.

Now suppose we take these photons that are in the twin state, separate them, and send them off in opposite directions toward two polarization detectors. Call the detectors A and B, and let's suppose that B is slightly further away from the photon source than A. The setup is shown in Figure 26.1.

Let us focus on the photon speeding toward A. Suppose that as the photon reaches A, the detector registers it as having Up polarization. At this point we know that, a fraction of a second later, when the other photon reaches detector B, it too will be registered as having Up polarization (we know this because the photons are in the twin state, and so will always be measured as having the same polarization). And indeed, a fraction of a second later, the photon at B is indeed registered as having Up polarization.

That is all there is to the EPR scenario. Given what I have said so far, nothing should seem remotely surprising or curious about this scenario. So, what's the big deal? What were Einstein, Podolsky, and Rosen up to in devising this scenario?

In the setup described above, Einstein, Podolsky, and Rosen were trying to convince us that quantum theory is *incomplete*, that is, that there were "elements of reality" (to use a phrase found in the EPR paper) that were not included in quantum theory. Specifically, Einstein, Podolsky, and Rosen are arguing that

(1)   the photons must have a definite polarization before they are measured as having that polarization,

but

(2)   quantum theory does not represent either photon as having any definite polarization before it is measured as having that polarization.

So, the EPR argument goes, quantum theory is an incomplete theory of reality, in that quantum theory fails to represent an element of reality, namely, the polarization of the photons before they are detected.

Claim (2) is correct, that is, this is indeed a feature of quantum theory (in other words, it is a feature of the mathematics of quantum theory). So if EPR can convince us that (1) is correct, they have a powerful argument for their conclusion that quantum theory is an incomplete theory. Our next task, then, is to look carefully at the reasons for believing claim (1). This can get complicated, but go slowly and what I say should become clear.

The argument for (1)

To understand the argument for (1), we need to understand what has come to be called the *locality assumption*. As with a lot of basic assumptions, it is remarkably difficult to put into words. Before attempting a definition, let me first illustrate the locality assumption with some examples.

Suppose I put some object, say a ballpoint pen, on a table in front of you. I ask you to move the pen, but without touching it, blowing on it, shaking the table, paying someone $5 to move the pen for you, using any "psychic powers of the mind" you may have, and in general, without in any way having any contact whatsoever (physical or otherwise) with the pen. Chances are, you think I am asking you to do something impossible. But *why* do you think it is impossible? Presumably because you think that one thing (in this case you) cannot influence or have any effect on another thing (in this case the pen) unless there is some sort of contact (for example, physical contact, or communication, or at least *some* sort of connection) between the two things.

Another example: Suppose we send Sara and Joe to get donuts every morning. We send Sara to the Donut King on the north side of town, and we send Joe to the Donut King located some distance in the opposite direction, on the south side of town. They leave at the same time, and we might even send someone with them, to make sure they go to their assigned donut store. Suppose every day that Sara chooses cream-filled donuts, Joe also chooses cream-filled donuts, and every day that Sara chooses chocolate-covered donuts, Joe also chooses chocolate-covered donuts. And in general, whatever type of donut Sara chooses, Joe chooses the same. This goes on day after day, week after week, month after month. In this scenario, our intuitions are that there must be some sort of connection, some sort of communication, between Sara and Joe. We tend to feel that what happens at one location – in this case, Sara's choice of donuts at her donut store – cannot influence what happens at another location – Joe's choice of donuts at his donut store – unless there is some sort of connection or communication between them.

The statement immediately above is one way of phrasing the locality assumption. In short:

> *The locality assumption (rough version)*: What happens at one location cannot influence what happens at another location unless there is some sort of connection or communication between the two locations.

What exactly is meant by "some sort of connection or communication," and what it means for one thing to "influence" another, can be understood in a variety of ways, and this fact has led to a great deal of misunderstanding and miscommunication concerning the locality assumption. Later in this section we will explore the various ways of understanding these notions, and thus various ways of understanding the locality assumption. For the sake of the EPR argument, this somewhat rough version of the locality assumption will suffice.

With the locality assumption on the table, EPR's argument for (1) can be finished quickly. From the locality assumption it follows that the measurement of the polarization of the photon at detector A cannot influence the measurement of the polarization of the photon at detector B. The reason for this is simply that the two detectors are too far apart, and hence there simply is not time for any sort of signal or communication or influence to travel from A to B. At least, there can be no such influence unless the influence travels faster than light, and since it is widely accepted (based on Einstein's relativity theory) that no such influence can travel faster than light, there seem to be excellent reasons for thinking that what happens at detector A cannot influence what happens at detector B. Thus the perfect correlation between the polarization of the photon at A and the photon at B can be explained only if the photons had a definite polarization before they were detected. In other words, if the locality assumption is correct, then (1) follows.

One can summarize the EPR argument in this way: Either the locality assumption is false or quantum theory is an incomplete theory. But (EPR continues) no one in their right mind would give up the locality assumption, so (EPR concludes) quantum theory must be an incomplete theory.

Bell's Theorem

Note that there would be no point in actually carrying out the EPR thought experiment. This is because the central issue is whether the photons have their polarization before they are detected, and carrying out the experiment can only tell us about their polarization when they are detected, not before.

In 1964, John Bell (1928–90) began to consider the question of whether there was any way to modify the EPR scenario in such a way that actually carrying out the modified experiment could tell us something interesting. And, in fact, he did find a way to modify the experiment in an interesting way. Bell's result is generally referred to as *Bell's theorem* or *Bell's inequality*. As the name suggests, Bell's result is essentially a mathematical proof. However, it is easier to understand if we think of it as a design for an experiment, and so that is the way I will describe it below.

Bell himself, along with David Mermin and Nick Herbert, have produced some very good nonmathematical expositions of Bell's theorem. In what follows, although the Coke machine analogy is mine, some key ideas are taken from what I think are the best aspects of their explanations. Bear with me here – this will take a few minutes, but when we are finished you will have done an informal deduction of Bell's theorem.

Let's begin with a Coke machine analogy. Consider the picture in Figure 26.2. In this setup, we have two more or less identical Coke machines, which we will call machine A and machine B. We also have a button, and each time we press the button, each machine produces a can of soda. Let's suppose, in fact, that each time we press the button, each machine produces either a diet Coke, which I will abbreviate with a D, or else a 7-Up, the Uncola (as they used to call it), which I will abbreviate to a U. Each machine also has a dial with three settings, labeled L (left), M (middle), and R (right).

Finally, suppose there is no apparent communication or connection between the two Coke machines. That is, we can find no wires, no radio links, nor any other sort of connection between A and B. With this in mind, we will describe four scenarios.

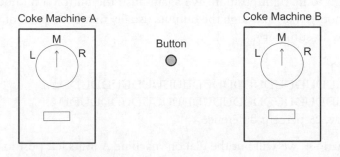

Figure 26.2 Coke machine analogy

For scenario 1, the dial on machine A is set to its middle position, and likewise the dial on B is set to its middle position. Suppose we push the button hundreds of times, and observe that every time we push the button, the machines produce identical soft drinks. That is, every time A produces a diet cola, B likewise produces a diet cola, and every time A produces a 7-Up (an Uncola), B likewise produces a 7-Up. Moreover, we notice that the machines produce a random mixture of diet colas and 7-Ups. That is, although they always produce identical sodas, 50 percent of the time it is a diet cola, and 50 percent of the time it is a 7-Up.

Suppose we use A:M to indicate that the dial on machine A is set to the middle position, and likewise B:M to indicate that the dial on machine B is set to its middle position. Suppose we use D and U to represent the two types of sodas, and record the output of each machine (for example, A:M DUDDUDUUUD represents the output of 10 button presses for machine A when set to its middle position). Then we can summarize scenario 1 as follows:

Scenario 1
A:M   DUDDUDUUUDUDDUUDUDDUDUUUDUDD …
B:M   DUDDUDUUUDUDDUUDUDDUDUUUDUDD …
Summary: identical outputs

For scenario 2, we will move the dial on A to its left position, and leave the dial on B in its middle position. Suppose that when we set the dials in this way, and press the button a number of times, we notice that although the machines usually produce identical sodas, occasionally they do not match. In particular, we notice that there is a 25 percent difference in the output of the two machines when the dials are set in this way. To summarize:

Scenario 2
A:L   DDUDUUDUDDUUDUDUUDUDDDUDUUDU …
B:M   DUUDUDDUDDUUDUUDUDUDUDUDDUUU …
Summary: 25 percent difference

For scenario 3, we will return the dial on A to its middle position, and move the dial on B to its right position. We again push the button a number of times, and again notice that, although the outputs usually match, 25 percent of the time they do not. In summary:

Scenario 3
A:M   UUDUDDUDUUDDDUDUUUDUDUUDDUDD …
B:R   UDDUDUUDDUDUDUDDUUUUDUDDDUDD …
Summary: 25 percent difference

For scenario 4, we will put the dial on machine A in its left position (as in scenario 2), and put the dial on machine B in its right position (as in scenario 3). Leaving out the results, here is the setup for scenario 4:

Scenario 4
A:L   ???
B:R   ???
Summary: ???

Now let's consider what we would expect as an outcome when the dials on A and B are set this way. In particular, let's assume the following:

(a)   There is no communication or connection between the two machines.
(b)   The locality assumption is correct.

If (a) and (b) are correct, then the 25 percent difference in scenario 2 has to be a result only of changes to machine A resulting from moving its dial. And likewise, the 25 percent difference in scenario 3 must be the result only of changes made to B resulting from moving B's dial. That is, if there is no communication or connection between the two machines, then moving A's dial can affect only the output of machine A, and moving B's dial can affect only the output of machine B.

So if moving the dial on A produces a 25 percent difference in A's output, and moving the dial on B produces a 25 percent difference in B's output, then ask this key question: When we move both dials, as in scenario 4, what is the *maximum* difference there can be between the outputs?

Before moving on, pause for a moment to understand and answer this question. If you see the answer, then you have done what is essentially an informal deduction of Bell's theorem. The correct answer is that, given (a) and (b), the maximum difference in scenario 4 is 50 percent. That is, if moving the dial on A produces a 25 percent difference in A's output but does not affect B's output, and moving the dial on B produces a 25 percent difference in B's output but does not affect A's output, then changing both dials can result in a combined difference of at most 50 percent. And this deduction, that in scenarios such as this there can be a maximum difference of 50 percent, is essentially Bell's theorem.

Of course, Bell was not concerned with the output of Coke machines, and indeed, the Coke machine situation is merely an analogy. To see how this relates to quantum theory, let's tie the Coke machine analogy up with quantum theory.

Suppose that, when we press the button, instead of producing cans of sodas in Coke machines, each time we press the button we produce a pair of photons in the twin state, exactly as described in the EPR scenario pictured in Figure 26.1. And instead of Coke machines, we have polarization detectors A and B, again just like in Figure 26.1. But unlike the basic EPR scenario of Figure 26.1, these detectors come equipped with dials like those on the Coke machines, with L, M, and R settings. This modified EPR scenario would appear as in Figure 26.3.

Polarization detectors can in fact have settings equivalent to the L, M, and R settings depicted in Figure 26.3. Now suppose we perform some experiments exactly analogous to the Coke machine scenarios. Suppose we set both

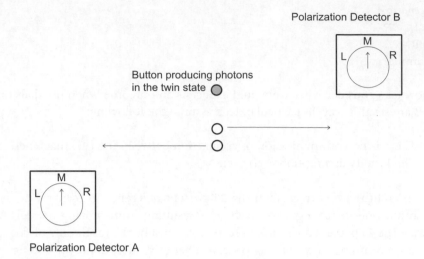

Figure 26.3   Modified EPR scenario

polarization detectors to their middle position, and push the button repeatedly, each time generating a pair of photons in the twin state which are then sent out to their respective detectors. Recall that for photons in the twin state, so long as the detectors are identical, the photons will both be detected as having Up polarization or having Down polarization. In particular, the photons will both be detected as having the same polarization so long as the detectors are set to identical settings. So in this scenario, with both detectors set to the middle position, and assuming we use D and U to represent Down and Up polarization, then the result of this experiment will be exactly as summarized above for scenario 1. That is, the output of the detectors will be identical.

Note that this is simply a quantum fact, that is, the outcome of a quantum experiment: set the detectors up so that both are in their middle position, send pairs of twin photons toward them, and the outcome will be that the detectors register every pair of photons as either both being Up or both being Down. Moreover, this result is exactly what quantum theory predicts.

Now change the polarization detectors as in scenario 2 above. The detectors are no longer both set to the same position, so we cannot expect to get identical readings for each pair of photons. But as a matter of experimental fact, the result in this scenario is exactly as depicted in the summary of scenario 2. Again this is simply a quantum fact, and again is exactly what quantum theory predicts.

And likewise, change the polarization detectors as in scenario 3, and the outcome is again as depicted in the summary of scenario 3, and once again, this is simply a quantum fact and is what is predicted by quantum theory.

So far, so good – nothing out of the ordinary. But now set the detectors as in scenario 4, with A set to its left position and B set to its right position, and consider what results we would expect. Using the same notation as used for the Coke machine analogy, these scenarios can be summarized as follows:

Scenario 1
A:M   DUDDUDUUUDUDDUUDUDDDUDUUUDUDD ...
B:M   DUDDUDUUUDUDDUUDUDDDUDUUUDUDD ...
Summary: identical outputs

Scenario 2
A:L   DDUDUUDUDDUUDUDUUDUDDDUDUUDU ...
B:M   DUUDUDDDUDDUUDUUDUDUDUDUDDUUU ...
Summary: 25 percent difference

Scenario 3
A:M   UUDUDDUDUUDDDUDUUUDUDUUDDUDD ...
B:R   UDDUDUUDDUDUDUDUDDUUUUDUDDDUDD ...
Summary: 25 percent difference

Scenario 4
A:L   ???
B:R   ???
Summary: ???

So we ask the same question we asked of the Coke machines: If the locality assumption is correct, and there is no communication or connection between the two polarization detectors, what is the maximum difference there can be between the output of the two detectors? And again, the answer (which is essentially Bell's theorem) is that the maximum difference there can be between the output of the two detectors will be 50 percent.

But here is the punch line: quantum theory does not agree with this 50 percent figure. Instead, if the detectors are set as in scenario 4, quantum theory predicts there should be an almost 75 percent difference between the output of the two detectors.

In other words, Bell discovered that predictions based on quantum theory, on the one hand, and predictions based on the locality assumption, on the other hand, do not agree. That is, the simple deduction illustrated by the Coke machine analogy shows that, if the locality assumption is correct, then when polarization detectors are set as in scenario 4 there can be at most a 50 percent difference between the outputs of the detectors. But a prediction based on the mathematics of quantum theory gives an expected difference of almost 75 percent.

In short, Bell showed that quantum theory and the locality assumption are incompatible with one another. They cannot both be correct.

Aspect's Experiments

Remember that Bell's theorem, as I have presented it via the Coke machine analogy, is essentially an experimental setup. As described, the experiment sounds relatively straightforward. In fact, the experiment is technologically quite difficult, and there was no way, when Bell produced this result in 1964, of actually carrying it out. In the ensuing decades, however, a number of physicists worked on actually carrying out Bell-type experiments. The best of these experiments were carried out in Alain Aspect's lab at the University of Paris during the late 1970s and early 1980s. (If you are curious, the main difficulty in carrying out this experimental design is in ensuring that there is no possibility of any connection or communication between the two detectors.)

To summarize the results of these experiments, Aspect's experimental results show that, in the conflict between the locality assumption and quantum theory, quantum theory wins. That is, Aspect's results indicate strongly that the locality assumption is wrong. Since Aspect's experiments in the 1970s and 1980s, these results have been replicated and verified a number of times by a number of different labs using a number of experimental setups.

With respect to views on the nature of reality, the implications of Bell's theorem and Aspect's results are substantial. Aspect's results are quantum facts, and any respectable view of reality must respect the facts. And these facts seem to dictate that the locality assumption must be rejected.

However, we should be careful at this point. Recall that the locality assumption was phrased pretty loosely in the discussion above. The next major issue, then, is to discuss the locality assumption with much greater care, thereby clarifying the implications of the Bell/Aspect results.

Locality, Nonlocality, and Spooky Action at a Distance

Recall that the two main goals of this chapter are (a) to explain recent experiments that are often claimed to have substantial philosophical implications, and (b) to analyze these alleged implications, especially with respect to the claim that these experiments show that any "local" view of reality must be mistaken. At this point we are finished with (a), and are ready to move on to (b). Let's begin with the rough statement of the locality assumption as it appeared earlier:

The locality assumption (rough version): What happens at one location cannot influence what happens at another location unless there is some sort of connection or communication between the two locations.

As mentioned earlier, there are various ways to understand the phrase "some sort of connection or communication," and the notion of "influence." The first task, then, is to make these issues more precise.

There are good reasons to believe the speed of light is a sort of universal speed limit. As such, we can exploit it to put constraints on the possibility of two events being connected: namely, two events can be connected only if the amount of time between the two events is at least as great as the amount of time it would take light to travel between them. For example, there can be (and is) a connection between the event of me dialing my home number on my office telephone, and the event of my phone in my home office ringing a few seconds later. The time between the two events is greater than the amount of time it would take light to travel between the two locations. Thus, there is the possibility of a connection between these two events. And, of course, the two events are in fact connected, and the nature of the connection is pretty well understood.

In contrast, it takes approximately eight minutes for light to travel from the sun to the Earth. So there can be no connection between an event happening on the sun (a solar flare, for example) and an event happening on the Earth (a disruption of radio communications, for example), unless those two events are separated by at least eight minutes.

Using this notion, we can put constraints on the notion of "some sort of connection or communication." In what follows, unless otherwise noted, we will understand "connection" to mean the *possibility* of a connection, where there is a possibility of a connection between two events if and only if the amount of time between the two events is equal to or greater than the time it would take light to travel between them. If there is no possibility of this sort of connection between two events, then we will say that the second event occurred "at a distance" or, equivalently, that the second event was "at a distant location."

This leads to a more refined version of the locality assumption. The emphasis on the speed of light stems from Einstein's theory of relativity, and this sort of influence seemed to be the sort of influence that most worried Einstein (Einstein once referred to this sort of influence as "spooky action at a distance"). Given this, let's call this Einstein Locality:

Einstein Locality: An event at one location cannot influence an event at a distant location.

In the Aspect experiments, the events in question were indeed "at a distance," that is, the events were such that there could be no connection between the events (again, unless that connection was faster than light). The details of how Aspect accomplished this were some of the most technologically challenging parts of the experimental setup, and basically he accomplished this by an arrangement that was equivalent to rapidly and randomly (or at least quasi-randomly) altering the positions of the detectors just before the photons reached them. In short, Aspect managed to arrange the experiment so that changes in the positions of the detectors occurred so quickly that there was no time for any signal to pass from one detector to the other (again, unless that signal traveled faster than light).

And in these experiments, there is some sort of influence between what happens at one detector and what happens at the other. That is, events at one detector (changing the position of the dial) seemed to influence the events at the distant detector (whether the detector registered Up or Down polarization). Thus, the

Bell/Aspect experiments show that Einstein Locality is mistaken. In short, it appears that events at one location can influence events at a distant location.

As mentioned above, the notion of "influence," as used in the definitions of the locality assumption and Einstein Locality, is not an entirely clear concept. As a final (and important) topic in this section, it is worth discussing what can and cannot be said about the "influence" pointed to by the Aspect experiments.

The word "influence" is often meant in the sense of causal influence, that is, in the sense of one event causing another. Let's take a minute to construe Einstein Locality in terms of causal influence. Let's call this Causal Locality:

> *Causal Locality:* An event at one location cannot causally influence an event at a distant location.

Do the Bell/Aspect experiments show Causal Locality is mistaken? This is a difficult question, with some arguing that the correct answer is "yes" and others arguing the opposite. The main difficulty lies with the notion of causation itself. Generally, when we speak of causes, we have in mind examples of, say, errant baseballs causing windows to break, broken glass causing flat tires on our cars, the impact of my fingers on the keyboard causing the keys to depress, in turn causing electrical signals to be passed from my keyboard to my computer, and so on.

These sorts of causal influences are fairly well understood in terms of current physics. If we restrict our use of "causal influence" to the sorts of causes that are well understood in terms of current physics, then we should be hesitant to speak of the sorts of influences found in the Aspect experiment as being causal influences. Again, this is simply because we do not have much of an idea – no idea, really – of the nature of the influences pointed to in Bell/Aspect experiments, and so such influences cannot fall under the heading of influences that are well understood in terms of current physics. In short, if we understand the notion of "causal influences" in this way, then the results of the Bell/Aspect experiments do not show Causal Locality is incorrect.

On the other hand, if we understand "causal influences" in a broader way, then a good case can be made that the Bell/Aspect experiments do indicate that Causal Locality is wrong. For example, suppose we understand "causal influences" to apply (speaking somewhat roughly) to cases where there is a strong correlation between events, and moreover, a strong correlation that cannot be explained by any appeal to any common causes. (By "common causes" I mean that the correlated events are correlated not because one causes the other, but because both are the result of another, common cause or causes. For example, my outside thermometer reading below 32 degrees Fahrenheit, and the water in the nearby pond being frozen, are closely correlated. But neither causes the other; rather, they are both the result of a separate, common cause, namely, sufficiently cold weather.) A reasonable case can be made that the Bell/Aspect results meet this criterion – that is, there certainly is a strong correlation between the settings of one detector and the readings at the other detector, and a good case can be made that this correlation cannot be explained as a result of any common cause. In short, if we construe "causal influence" along these lines, then a good case can be made that the Bell/Aspect results do show that Causal Locality is mistaken.

What are we to make of this, especially with respect to the question of whether the Bell/Aspect results show Causal Locality is mistaken? First, it is worth noting that there is no unequivocal answer to this question. Either answer to the question is reasonable, so long as it is made clear how the notion of "causal influence" is being used. As we have seen, much hinges on the complicated issue of how one construes causation. Notably, the question of Causal Locality illustrates just how complex issues surrounding the Bell/Aspect results can become, and how seemingly simple issues can quickly become complex.

Before closing, it is worth considering one other common way of understanding the notion of "influence." The influences I have in mind are influences we can use to send information. Cases of telephone calls, shouting out a window, sending signals with Morse code, and so on, are examples of this sort of influence. This leads to what we can call Informational Locality:

Informational Locality: An event at one location cannot be used to send information to a distant location.

Do the Bell/Aspect experiments show Informational Locality is incorrect? In other words, can we exploit the influence between the events at distant locations to send information? For example, could we set up a detector here on Earth, another on Mars, and use a Bell/Aspect situation to instantaneously send information between the two locations?

Nothing in the Bell/Aspect setup requires that the detectors be reasonably close to each other. So, in principle, we could have one detector on Earth, another on Mars (or in a galaxy thousands of light years away, for that matter), and we would still have the same result. That is, what happens at one detector will apparently instantaneously influence what happens at the distant detector. This, then, might tempt us to think we could exploit this influence to instantaneously send information to distant locations, thereby violating Informational Locality.

However, there appears to be no way to exploit the Bell/Aspect setup to allow for the sending of any information between the locations. Basically, the reason is because the series of Ds and Us is random. To see this, it will help to focus on a more concrete example.

Suppose you are at detector A, perhaps in Tulsa, Oklahoma, and I am at detector B, perhaps 2,000,000 light years away in the Andromeda Galaxy. Suppose I keep the dial on my detector in the R position. Looking back at scenarios 3 and 4 above, we know that, if you have the dial on your detector in its M position, there is only a 25 percent chance that what my detector registers will be different from what your detector registers. But if you put your dial in the L position, the odds of my detector reading differently than yours jump to 75 percent. In short, you can substantially influence the odds of whether our detectors match or not simply by switching the dial between M and L. Since you have so much influence on what happens at my tape, it seems you should be able to exploit this influence to send information to me instantaneously, thereby showing Informational Locality to be false.

However, you will not be able to use this influence to send information to me. Here is the problem. To send me a message, you need to be able to influence whether I receive a D or a U on my tape. Even if you could only influence the probability of my receiving a D or a U, this would be sufficient to send me a message. So no question about it: if you could manipulate your dial so as to influence whether I received a D or a U, then you could send me a message.

The problem is, you do not have this sort of influence. All you can influence is the probability of whether what my detector says matches what your detector says. So although you can influence the probabilities that our detectors will have the same readout, this does you no good in terms of sending a message. To send a message you need to be able to influence whether I receive a D or a U on my detector, but all you can do is influence the odds that the output of our detectors match.

In general, there seems to be no way to exploit the influence of EPR/Bell/Aspect setups to send information to a distant location. So contrary to how it initially seems, the Bell/Aspect experiments give us no reason to think that Informational Locality is false.

Concluding Remarks

In summary, the Bell/Aspect experiments clearly demonstrate that Einstein Locality is mistaken. That is, the experiments show that there can be some sort of influence (or connection, or correlation, or whatever word you prefer – none of the words is quite right) between events at distant locations. As we have seen, there is no unequivocal answer to the question of whether the Bell/Aspect experiments show that there is any sort of *causal* influence between events at distant locations – much here depends on how we construe the notion of causation. And as we just saw, although there is some sort of influence involved, it seems that the Bell/Aspect results do not give us any reason to think that the influence can be used to send information between distant locations.

This leaves us, then, with the question of what kind of influence it is. This is a question for which we can, at last, give a precise answer. The answer to the question "What kind of influence is it?" is this: no one has the foggiest idea.

Chapter Twenty-Seven

Overview of the Theory of Evolution

In the previous four chapters, we looked at some ways in which new developments, most notably involving relativity theory and quantum theory, are requiring us to rethink substantially some basic and long-held assumptions about the sort of universe we live in. In this chapter and the next we will be exploring relatively recent work (from the mid-1800s to the present) concerning evolutionary theory. As with the discoveries discussed in previous chapters, it will not take long to see that evolutionary theory likewise forces us to rethink some common and long-held views.

In one sense, these chapters are much easier than the previous ones, in that at least in outline, the theory of evolution is far easier to understand than relativity theory or quantum theory. But in another sense these chapters are more difficult. One difficulty stems from the fact that there are some extremely common, deeply rooted, and substantial misunderstandings about the most fundamental aspects of evolution. And replacing deep and widespread misunderstandings can be difficult. But a second difficulty arises because evolution has implications that many find difficult to accept.

We will hold off discussing the difficult implications of evolutionary theory until the next chapter, and in this chapter focus on two relatively straightforward goals. The first is to get a clear understanding of the basics of evolutionary theory, especially with respect to clarifying what evolutionary theory is and is not. The second main goal is to understand the ways in which evolutionary theory has developed from the 1800s to the present day, with an eye toward seeing how, at present, evolutionary theory provides a unifying framework for much of biology.

Worldviews: An Introduction to the History and Philosophy of Science.   Richard DeWitt
© 2010 Richard DeWitt

Overview of the Basics of Evolutionary Theory

The term "evolution" is, I think, often thought to mean something much more complex than it does. At bottom, evolution is nothing more than this: change over time. To be sure, there are other, sometimes more detailed, characterizations of evolution, especially within specialized fields within biology. And even this basic characterization of evolution needs to be understood in a certain way. For example, we do not want to think of changes in a landscape over time caused by erosion, or a photograph fading over time as a result of exposure to light, as instances of evolution. Rather, the sorts of changes over time that we have in mind are changes in populations over time, and in particular, changes in populations over successive generations. For simplicity, I will speak of evolution as change over time, with the understanding that we are really speaking of changes in populations over successive generations.

The main goal of this section is to clarify the basics of the explanation, most famously associated with Darwin and Wallace (more on them below), of evolutionary change. Let's begin with a description of the two basic ingredients for evolution.

The basic ingredients for evolution

Notice an important feature of the characterization of evolution given above: it is *not* specific to biological organisms. "Populations" and "generations" are terms often used in a biological context, but they need not be. We can speak of populations of cars (for example, model types) and generations of cars (model years, perhaps), populations (again models) and generations of personal computers, and likewise for a wide variety of nonbiological entities. In short, evolution need not be thought of as applying only to biological populations.

In fact, nonbiological cases provide better examples with which to start our exploration of the basic ingredients for evolution. Consider, for example, consumer goods such MP3 players, cell phones, personal computers, and the like. Recall that evolution basically is a matter of change over time, and notice that these sorts of products certainly undergo change over time, and a good bit of change at that.

Suppose we are looking for a general explanation – that is, general in the sense that it will apply to all of these examples – of why such products undergo change. The details will of course differ from one type of product to another. But there is a straightforward and general explanation we can give, and it goes as follows.

First, note that all the products just mentioned (MP3 players, cell phones, personal computers, and so on) show a substantial amount of variation from one model to another. For example, suppose we gather our friends and colleagues, and we all put our cell phones on a table and compare them. We cannot help but be struck by the wide variety of features. Large screens, small screens, smart phones, dumb phones, some with built-in cameras, some without, this service plan and that service plan, some with keyboards, some without, some with clever advertising campaigns, some not so clever, and so on. In short, we would find a wide range of variation. (Incidentally, you would also likely be struck by how primitive mine is. I still use my original cell phone, and although it is only six years old, it is striking how in only six years it has gone from being state of the art to being an absolute dinosaur. Change in these products takes place at a rapid pace.) The same story would apply if we compared our MP3 players, personal computers, or any of a large number of objects.

Notice also that the variations we are speaking of are variations that can be passed on to later generations. Should a particular type of variation prove attractive, the company responsible will no doubt ensure that the feature appears in later generations of its product, and other companies are likely to imitate that feature. So we are dealing not only with variations, but with variations that can be passed on to subsequent generations.

There is a second general observation we can make about cell phones, MP3 players, personal computers, and the like. These products are, in a certain sense, engaged in what we might call a "struggle for existence." This might sound a bit overdramatic, but it really is quite apt. Products that are not sufficiently successful compared to the products they compete with will cease to be produced, and soon will no longer exist.

And of course, a key factor in determining which products are successful in this struggle for existence, and which products are not, is exactly the variation we discussed above. Models with features consumers find appealing will be purchased in greater numbers than models with less appealing features, and companies will see to it that the appealing features continue to be represented in later models. In contrast, companies will generally tend to discontinue features of models that consumers did not find appealing. In this way, such features will stop being represented in later generations. In short, the struggle for existence results in a difference in what variations continue to be present in later generations.

These are very simple examples, but they illustrate an important point. In looking for a general explanation of the changes these sorts of products undergo over time, we have identified two basic features, namely:

(1) variations (of a sort that can be passed on to subsequent generations), and
(2) a struggle for existence (which affects what variations continue to be present in later generations).

These simple ingredients suffice to provide a very general, and very apt, explanation for how products such as MP3 players, cell phones, personal computers and the like undergo change over time. And if we think about it a minute, it does not take long to realize that *anytime* these ingredients are present, we are almost certain to find change over time, that is, we are almost certain to observe evolution. Put this way, this account of change over time is at its core a remark-ably simple and straightforward view.

As mentioned earlier, evolution is most closely associated with biology. The ties between the examples above and biological evolution are quite straightfor-ward. As with the examples above, biological organisms exhibit a wide range of variations, of a sort that can be passed down to subsequent generations. Biological organisms also tend to reproduce at a rate that exceeds the number of organisms that can survive. So biological organisms, as with the products above, are in a struggle for existence, and the variations that enhance an organism's ability to survive and reproduce in the environment in which it finds itself have a greater chance of being represented in later generations. In short, the wide range of vari-ations found among organisms, coupled with the struggle for existence among these organisms, results in differences in whether particular variations continue to be present in later generations.

Some clarification: What evolutionary theory is and is not

The theory of evolution is, at its most basic, a straightforward account of how change over time results from the basic ingredients (1) and (2) above. Although simple at its core, this account is prone to some easy misunderstandings.

In my view, the most serious of these misunderstandings, because it causes the most substantial confusions about what evolutionary theory is and is not, involves teleology. In particular, there is a widespread misconception that evolution is a goal-directed process, that is, that evolutionary theory paints a picture of change in which that change is directed toward certain goals. The first subsection below, then, focuses on this issue, and also discusses related issues such as whether it makes sense to speak of higher and lower species, more evolved and less evolved species, primitive and advanced species, and the like.

The second most pervasive confusion about evolutionary theory, I think, con-cerns the role of chance. Thus a brief subsection on this topic is provided below. The other two issues we explore below involve less fundamental misunderstand-ings about evolutionary theory. But they are common misunderstandings, and almost always, when they are expressed, it is done in a way that indicates confu-sion about the theory of evolution. As such, these too deserve mention. These are, respectively, the common claim that evolution is "only a theory," and the idea that according to evolutionary theory, humans are descended from apes.

Teleology Perhaps the most widespread and most misleading misperception about evolutionary theory is that certain specific traits, such as intelligence, lan-guage, tool use, and the like, are "higher" or "better" traits, and thus that the evolutionary process would be expected to produce organisms with such traits. For example, consider this not uncommon question: If evolutionary theory is correct, then why haven't other animals developed the sort of intelligence that humans have? Why haven't other animals evolved language, or the ability to use sophisticated tools, or upright posture?

It is difficult to overstate the extent of the confusion about evolution this question indicates. It is probably a safe bet that the majority of people, if asked, would say they found this question to be a sensible one. And if this is correct, then it indicates that a majority of people are confused about a fundamental aspect of evolution. As a brief aside, I should note that I do not blame people for having this misperception. I think that, for decades, people such as myself, and fellow scholars and educators, have done a woefully inadequate job of explaining key components of evolutionary theory in a manner that makes the key aspects of evolutionary theory clear to nonspecialist audiences.

By the end of this discussion, I hope it is clear why the question above makes no sense, or at least, makes no sense if one has a correct view of evolutionary theory. To begin, note that the question makes sense only if one thinks of evolution as a process that is biased toward producing certain types of characteristics, for example, characteristics such as language, upright posture, tool use, and in general features we consider "advanced" and that, not surprisingly, we associate with our own species. Or to put it more simply, the question assumes evolution is at bottom a goal-oriented, teleological process.

But evolution is no such thing. The traits that result from the struggle for existence that is part of the evolutionary process are the traits that help organisms survive successfully and reproduce in the environment in which they find themselves, *whatever those traits might be*. In this context, there are no "better" or "worse" traits, no advanced or less advanced traits, no species with traits that make them "higher" or "lower" species in any meaningful sense of the words. Nor are any species "more evolved" or "less evolved," again at least not in any interesting sense of the word. The only types of species there are are species that have survived. And they have survived in large part because of the traits – again, whatever those traits may have been – in them and their predecessors that had helped them survive and reproduce in their environment.

In short, the evolutionary process is not directed toward producing any particular type of traits, other than the traits that happen to have helped in terms of survival and reproduction. The evolutionary process is not a goal-directed process, that is, it is not a teleological process in any fundamental sense. (As a brief aside, some scholars in this area argue that at least some evolutionary explanations do involve teleology. But even among those who think evolutionary explanations involve teleology, there is broad consensus that the evolutionary process is not goal-directed, that is, not teleological, in the broad and misleading sense that is the focus of this discussion.)

With this point about evolutionary theory in mind, return now to the question with which we began this section, that is, the question of why, if evolution is correct, other organisms have not evolved the sorts of traits humans have. As noted earlier, and which I hope is now clearer, this question makes sense only if one presupposes that human traits are somehow better than other traits, and that somehow these traits would be ones that the evolutionary process would select for a wide range of other organisms. But that presupposition is confused. In short, evolutionary change is not a goal-directed, teleological process.

Chance   Another common misconception about evolutionary theory involves the role of chance. It is not uncommon to hear that the evolutionary account is one based on chance, but chance cannot produce the complex organisms we see any more than a hurricane blowing through a scrapyard of discarded airplane parts would be likely to produce a complex structure such as a modern airliner.

This view of evolutionary theory is likewise badly misguided. Chance does play a substantial role in evolution. For example, sexual reproduction (and asexual reproduction as well, although arguably to a lesser extent) involves a good deal of chance occurrence, including what genes the offspring wind up with. Chance astronomical events such as asteroid impacts that dramatically affected the environment are thought to have substantially influenced the evolutionary development of life on earth. A similar story goes for chance geological processes such as the opening of a massive rift valley that separated groups of organisms that had previously coexisted. And so on for countless other chance events that have had an impact on evolutionary change.

There is no question that chance plays a substantial role in the evolutionary account. But importantly, evolutionary theory in no way suggests that evolution occurs through chance alone. Rather, it is chance *coupled with a selection process*. In the discussion above of the basic ingredients for evolution, ingredient (2), the struggle for existence, is exactly this: a selection process. And there is no question that coupling a selection process with chance processes can produce complex forms.

To give just one quick and quite simple example of complexity arising from chance processes coupled with a selection process, consider the Game of Life. This is a computer simulation that, in its most basic form, starts with a randomly distributed collection of basic entities, usually referred to as live cells. To this random assortment of live cells is added an extremely simple selection process, which affects whether new live cells come about or whether existing live cells die out. Although the selection process is an exceptionally simple one, the result of this selection process coupled with the initial chance distribution of live cells generally results in ongoing, unpredictable processes in which a surprising amount of complexity arises. For example, it is common to find a variety of multicellular entities as a result of this combination of chance and a selection process. Some such multicellular entities "eat" other entities, some produce "offspring" at regular intervals, some "travel" (including traveling great distances), some shoot out entities at regular intervals that "pull" other entities toward them, some shoot out entities that push others away from them, and so on.

The Game of Life was first developed in 1940, and it has spawned an entire field of study involving the way complex structures and behavior can emerge from chance events coupled with a selection process. The field itself is a fascinating one, but for our purposes the main point is that it illustrates that complex structures and behavior can emerge from chance processes, when those chance processes are coupled, as they are in evolutionary theory, with a selection process. In short, the sort of claim we considered at the start of this section, that a chance process such as is found in evolutionary theory could not have produced complex organisms, indicates a fundamental misunderstanding of evolutionary theory.

Evolution as "only a theory"   It is not unusual to hear evolution characterized, generally dismissively, as "only a theory." In this brief subsection we will see how this is, in a sense, correct, but in this sense only trivially so. In a more important sense the claim, at least as it is generally intended, is misguided.

The sense in which the claim is correct is the trivial sense in which *everything* in science is only a theory. As we saw in earlier chapters, during the period of the Aristotelian worldview it was believed that one could arrive at scientific facts that were absolutely certain, that is, facts that, in our terminology, would not merely be a theory. But also as we saw, that view of science was replaced during the 1600s and the years following, and in particular, no one any longer thinks of any views within science as being absolutely certain. Instead, we now view our best theories as just that – the best accounts we have that handle the data and, if you take a realist approach, accounts that you also believe reflect, at least in part, the way things are. But in no case within science are views currently taken as absolutely certain. So evolution is indeed only a theory, but only in the trivial sense that everything within science is only a theory.

On a closely related subject, it is worth noting that, contrary to what is often taught in early science classes, there is no universally accepted, unequivocal use of scientific terms such as "hypothesis," "theory," "law," "principle," and the like. Rather, such terms are used in a broad variety of ways.

In particular, the word "theory" is commonly used in at least two very distinct ways. To see this, consider two examples, the first being string theory and the other relativity theory.

We have not discussed string theory in this book, but fortunately, only a little background is needed to make the point I want to make. String theory is a relatively recent proposal within physics, and as with all theories, it is designed to handle a certain body of data. Notably, string theory is highly speculative, in the sense that there is no direct experimental data to support it, and for that matter, even its leading proponents admit that it is difficult to imagine, at least with current technology, any experimental designs that would put the key tenets of string theory to an experimental test. Yet string theory is an interesting and viable area of research within physics. It may turn out to be a lasting and valuable contribution to physics, or like many theories, it may be a dead end. But it nicely illustrates one way in which the word "theory" is commonly used in science, namely, to indicate a proposed way of handling data, and a way that is at present speculative, and that has as yet not been subject to empirical confirmation or disconfirmation. In short, in this context the word "theory" is used to indicate an interesting idea but one that is still highly speculative and has not yet been empirically confirmed.

But now consider another common, and quite different, use of the word "theory." For example, consider the theory of general relativity discussed in earlier chapters. This is always referred to as "relativity theory," "the theory of relativity," or something similar. But in this context the use of the word "theory" is entirely different from the way the word is used in the context of string theory. In particular, unlike string theory, relativity theory is a highly confirmed theory. As discussed in earlier chapters, the empirical support for relativity theory is quite substantial. So relativity theory, unlike string theory, is anything but a speculative, empirically unconfirmed idea. In this usage, "theory" is used to indicate a view with so much empirical support that there is little question that it is our best view, and if perhaps not the final word on the subject, at least one which fits with the empirical data so well that its key elements are likely to be preserved in later accounts.

In the context of evolutionary theory, the word "theory" is used in this latter sense and not in the former sense. That is, the word is used in the way it is used when we speak of relativity theory. As such, it is misguided to describe evolution as "only a theory." In contrast, the empirical evidence supporting evolutionary theory is extraordinarily strong; in fact, the empirical support for evolutionary theory is stronger than for almost any other view within modern science.

Humans as descended from apes   The final issue I wish to discuss in this section involves the proper view from within evolutionary theory of the issue of ancestors and descendants. It is not uncommon to hear, usually in some way meant to reflect negatively on evolutionary theory, that according to the theory of evolution humans descended from apes. There is a sense in which this is entirely correct, but not, I think, in the sense usually suggested by the claim.

The first point to note is that humans are not descended from any species of ape currently in existence. Our closest currently existing relatives are the Common (Robust) Chimpanzee and the Pygmy (Bonobo) Chimpanzee. But we certainly did not descend from either of these species of chimpanzees, or from any other species of currently existing apes such as gorillas or orangutans. The reason is simple: four to six million years ago, when our ancestors began to acquire traits that we now associate with humans, none of these modern species of apes were in existence. And so it simply is not possible that we descended from any species of ape currently in existence.

Here is the correct way to think of our relationship to modern apes. By way of analogy, consider the relationship between you and some blood relative of yours (that is, related by family rather than by marriage). For the sake of a concrete example, suppose we consider the relationship between you and your cousin Sara. One would never say that you were descended from Sara. Rather, the correct picture is that you and Sara share a common ancestor from which you both descended. How far back you have to go to find this common ancestor depends on how closely related you and the relative in question are. If the relative is your brother or sister, then your last common ancestor would be your parents. If the relative in question is your cousin (more precisely, your first cousin), then your last common ancestor is one of your pairs of grandparents. If the relative in question is your great-great aunt, then you have to go back four generations to find your last common ancestor, which would be one of your sets of great-great-great grandparents.

In short, the correct view is not that we are descended from modern apes. Rather, the correct idea is that modern humans, and modern apes, share a common ancestor. But here, unlike going back a generation or two or three to find the last common ancestor, in this case we have to go back about a quarter of a million generations, or about five million years before we would find the last common ancestor between modern humans and any modern ape.

As a quick side note, there are numerous species to which we are much more closely related than we are to modern chimps. For example, numerous other species of humans, such as the Neanderthals, have existed. Most paleoanthropologists (those whose expertise is human origins) recognize at least four other species of humans, some many more than this. And we are much more closely related, in the sense of sharing a more recent last common ancestor, to these other species of humans than we are to any modern chimp. However, we are the last surviving species of humans – our closer relatives, the other species of humans, are now all extinct.

Before concluding this section on ancestors, it is worth clarifying a feature of evolutionary theory. You and I, and all humans, are related to *all* life on earth. That is, we share a common ancestor with every single living thing on the planet. As noted, the closest living relatives to modern humans are the Common Chimpanzee and Bonobo Chimpanzee, and this simply means that humans and these chimps share a last common ancestor that is more recent than any ancestor shared by humans and any other living species. Gorillas are humans' next closest surviving relative, but the common ancestor of humans and gorillas is more distant than our last common ancestor with chimpanzees. As we consider more and more distant relatives, we have to go further back in evolutionary history to find our last common ancestor. And this does not apply only to animal life. Humans and trees have a last common ancestor as well, but in this case, we would have to go *way* back in time to find our last common ancestor. Likewise for humans and bacteria and viruses and, for that matter, humans and any other species on earth. Every living organism is our relative, as is every organism (as far as we know) that has ever existed on earth.

The Development of Evolutionary Theory from the Early 1800s to the Present

The key ingredients behind evolutionary theory discussed in the first section of this chapter, namely, (1) variations (of a sort that can be passed on to subsequent generations) and (2) a struggle for existence resulting in differences in whether certain variations are or are not represented in subsequent populations, are most famously associated with Charles Darwin (1809–82) and Alfred Russel Wallace (1823–1913). The main goal of this section will be to provide a historical overview of the development of the theory of evolution.

We will begin by exploring the development of Darwin and Wallace's key insights, and then look at developments from roughly 1850 to 1900, 1900 to 1950, and 1950 to the present.

Darwin and Wallace's work

Darwin's and Wallace's key ideas were not entirely unprecedented, in that a few other individuals had suggested similar ideas before them. To name just two, Darwin's grandfather Erasmus Darwin had hinted (somewhat obscurely) at ideas similar to the key ingredients noted above. Likewise, 30 years before Darwin and Russell published their key works on evolution, an expert in wood products for shipbuilding named Patrick Matthews had articulated principles quite similar to (1) and (2) above. Darwin's grandfather, however, suggested his ideas mainly in some of his published poetry rather than in any sort of scientific publication. And Matthews mentioned these principles in the context of a book devoted to the best sources of wood for naval purposes, and beyond this, he never promoted or defended these key ideas (at least not until after the publication of Darwin's and Wallace's key works). In short, Darwin and Wallace deserve primary credit, if not for being the absolute first to develop the key ideas, then at least for being the individuals responsible for thoroughly articulating and defending them. And of these two, for reasons explained below, Darwin deserves the major share of the credit.

The development of Darwin's views   In this brief subsection we will take a look at the origins of Darwin's views, and in the next subsection the origins of Wallace's views. I should note that these subsections are intended as a quick sketch of some of the events leading up to their discoveries. Thorough and quite excellent detailed accounts of the work of Darwin and Wallace have been published recently, and for those looking for sources of additional information, suggestions can be found in the Chapter Notes at the end of this book.

In the early 1830s, Darwin accepted an offer to sail with the HMS *Beagle* in what would be a long (five-year) journey around the world. Darwin began the trip with reasonably standard beliefs for the time, for example, that God had created all species, and notably, the standard view that species had key essential characteristics that defined them as the species they were. Going along with this view of species having essential characteristics, Darwin also shared the related belief, extremely well entrenched at the time, that species were immutable, that is, that species were unchanging and that new species did not appear by evolution or by any other natural process.

While on the *Beagle*, Darwin made extensive observations and notes, and collected a huge number of specimens and fossils. His specimens and observations led him to recognize that organisms, even those classified as members of the same species, display an astonishing array of variation. So although Darwin began the journey with the standard view of species having a core of essential characteristics, during the voyage, he began to question this standard view. That is, we can see him beginning to recognize ingredient (1).

Upon returning to England, and for about the next five years (roughly the second half of the 1830s), Darwin began a series of notebooks in which he started to explore the idea that species might undergo "transmutation." In these notebooks, we can see Darwin becoming convinced that new species do come into existence. But this left him with the problem of explaining *how* new species might emerge, and he began thinking hard about this question.

We have already noted how Darwin, from his work on the *Beagle*, had come to recognize ingredient (1). During the late 1830s he would read a well-known work by Thomas Malthus (1766–1834), *Essay on the Principle of Population*, and this would help him recognize ingredient (2). A key part of Malthus's view involved the observation that plants and animals, including humans, tend to reproduce beyond what the environment can support. Malthus used this fact primarily to argue for certain types of social policies, but Darwin recognized how it would it would help address the problem on which he was focused. The fact that organisms breed beyond what the environment can support would result in a struggle for existence. Coupled with the range of variations in organisms he had come to appreciate while on the *Beagle*, the result, he realized, would be differences in the success of individuals with different variations. In short, he now had both the key ingredients (1) and (2). This provided him with the explanation he was looking for, that is, for how populations of organisms can undergo change over time. And from this it is not a huge step to realize that, given sufficient time, the slow accumulation of changes resulting from (1) and (2) could result in substantial changes in organisms, substantial enough to result in a population of organisms we would classify as a new species.

This process is what Darwin would later term "natural selection." The idea behind this terminology is straightforward. In a way analogous to how breeders artificially produce desirable traits in domesticated animals by selecting organisms with those traits and breeding them, so too nature "selects" for certain traits, traits that are advantageous relative to others for survival and reproduction. And whereas the sort of selection performed by breeders can appropriately be termed "artificial selection," the selection produced by the natural process summarized in (1) and (2) can appropriately be termed "natural selection."

In short, by about 1840 Darwin had found a natural, nondivine mechanism that explained how populations of organisms can undergo substantial change, and also how new species can appear. And the key ingredients in his explanation were essentially ingredients (1) and (2) discussed in the first section above. He had, he was sure, hit upon an extremely important idea.

But ... Darwin did not publish. For that matter, he shared his important idea with only a small handful of trusted friends. He did, in 1844, complete a short (by his standards) work of just under 200 pages, explaining his key ideas and providing arguments and evidence in support of those views. But Darwin did not intend the manuscript to be published, at least not while he was alive. Instead, Darwin stashed the manuscript away with a note to his wife, asking her to see to its publication should he die unexpectedly.

It would be 20 years from the time he first recognized ingredients (1) and (2) to the time he finally published his ideas. During this period, through the late 1840s and much of the 1850s, Darwin continued to work incessantly (Darwin was always, up to the day he died, at work on one project or another). Much of this would eventually make significant contributions to his later work, most notably by providing him with a large store of empirical data to support his eventual writings on evolution.

In short, by the time Darwin did get around to publishing his big idea, he was able to bring an enormous amount of data to bear on the subject. And it is exactly this wealth of data in support of his ideas that sets Darwin off from all others. Whereas others might have had the key ideas summarized in (1) and (2), Darwin had both the ideas and the data to support it.

The development of Wallace's views   In the late 1840s, while Darwin was engaged in various other projects, Alfred Russel Wallace was beginning the first of multiple voyages that would be, at least in some respects, reminiscent of Darwin's voyage on the *Beagle*. The differences between Darwin and Wallace, however, are notable. Unlike Darwin, Wallace did not come from a wealthy, established family. He lacked the connections needed for a university education, and for that matter, could not have afforded such an education. Unlike Darwin, Wallace had to pay his own way, largely by collecting specimens to send back to England for sale to wealthy collectors.

In some ways, Wallace was also just plain unlucky. For example, after four years of traveling, filling notebooks on his observations, collecting specimens and the like, the ship on which Wallace was returning to England caught fire and sank, along with most of his specimens (save those he had shipped back earlier) and most of his notebooks.

It seems clear that, very much like Darwin, during this voyage Wallace was struck by the extraordinary range of variation he observed among organisms, notably among organisms that, according to the received view, should have been unified by a core of essential characteristics. It also seems clear that Wallace was, by this time, beginning to question the received view, and that like Darwin, he began to think that populations of organisms do undergo substantial changes over time, and that new species might emerge.

Like Darwin, though, Wallace did not at this time have an account of how such changes might occur, or of how new species might emerge. He was, though, convinced enough that new species do emerge that he published a short paper in 1855, saying as much, though the paper provides no explanation of how they might do so. In other words, like Darwin at this stage of his career, Wallace had only the first piece of the puzzle, that is, ingredient (1), that organisms display a wide range of variations.

According to Wallace's account, ingredient (2) came to him in early 1858 while on a subsequent voyage (given his initial experience with the burning and sinking ship and rescue after over a week spent drifting in a lifeboat, you have to give Wallace credit for perseverance). At any rate, laid up in bed for days struggling through a malarial fever, Wallace says he hit upon ingredient (2) and recognized that this, coupled with the variation he had already noted, would provide an explanation for changes in populations of organisms. And, like Darwin, he recognized that this mechanism would provide an explanation for how new species can appear.

After recovering from the fever, Wallace quickly wrote out a short (roughly 20-page) explanation of the key ideas. And, in an intriguing twist of history, Wallace chose to mail the paper to Darwin. I say this is an intriguing twist because Wallace could have no way of knowing that Darwin shared similar ideas, or even that Darwin was at all sympathetic to such ideas, much less that Darwin had been convinced of these ideas for 20 years.

Wallace seems to have sent the paper to Darwin because of Darwin's connections. Recall that Wallace and Darwin moved in very different social circles. Darwin was on close terms with the most prominent names in British science. Wallace had no such connections. So Wallace, in the cover letter with his paper, asked Darwin if he could pass the paper on to these prominent figures.

When Darwin received Wallace's paper, he was, to say the least, disturbed. The title of Wallace paper itself says a lot: "On the Tendencies of Varieties to Depart Indefinitely from the Original Type." In this short paper, Wallace does an excellent job of characterizing the key ingredient (1) concerning variation. In particular, Wallace's point (Darwin has the same idea) is that populations can exhibit variations that are indefinitely different from ancestors of those populations. Notice how this runs exactly counter to the prevailing view that species have a definite essential core of characteristics. On the Wallace/Darwin view, the standard view of species is entirely wrong. There is no set of core characteristics for a species. In contrast, members of a population can differ, as the title suggests, indefinitely.

So with respect to the key ingredient (1), about the centrality of variations, what appears in Wallace's paper is along exactly the same lines as what Darwin had been thinking and writing (although not publishing) for 20 years. And with respect to ingredient (2), the struggle for existence resulting in differences in success of individuals with different variations, what Wallace writes is again exactly in line with what Darwin had been thinking and writing for 20 years.

In fact, in this paper Wallace even uses the same phrase that Darwin typically used and that I have been using in this chapter, namely "struggle for existence." In short, Darwin's and Wallace's key ideas, as articulated in Wallace's paper and in Darwin's earlier, unpublished writings, are so close as to be virtually indistinguishable.

It was a tricky situation. Darwin clearly had the key ideas earlier than Wallace, but he had not prepared anything for publication. Wallace, in contrast, clearly intended his paper to be published. To make a long story short, the tricky situation was more or less resolved, more or less to everyone's eventual satisfaction, when some of Darwin's friends arranged to have Wallace's paper, along with Darwin's 1844 manuscript and a new summary prepared by Darwin, presented together at an upcoming meeting, in late 1858, of a London scientific society. This 1858 presentation was the first public presentation of the key ideas of evolutionary theory.

However, this presentation of Wallace and Darwin's key ideas made little impact and led to little discussion. Shortly after this, Darwin set to work on an extended presentation and defense of these key ideas, in a work he would title *On the Origin of Species by Means of Natural Selection*.

Darwin's On the Origin of Species   As mentioned, following the presentation of his and Wallace's theory, Darwin went to work on a manuscript suitable for publication. This would turn out to be an extraordinarily influential work, and the main goal of this subsection is to provide a brief overview of this work.

Darwin had, during various periods over the past 10 years, worked on a detailed, scholarly, extremely thorough presentation of his key ideas, together with a detailed presentation, drawing on decades of his research, of empirical data supporting his views. This manuscript, which Darwin sometimes referred to as his "Big Book," had grown to hundreds of dense pages, and was far from complete. Darwin wisely took a new approach and started on a new, less dense presentation of his ideas, and one aimed at a wider audience. This work would be completed and published toward the end of 1859, and would be titled *On the Origin of Species by Means of Natural Selection* (now standardly referred to as the *Origin of Species*, or sometimes just the *Origin*).

It is one thing to articulate a theory clearly, as Wallace did in his 20-page 1858 paper, and another thing to defend and provide compelling arguments for the theory, which is what Darwin did. The *Origin of Species* is, to my mind, as pivotal a work as Newton's *Principia*, discussed in earlier chapters. Much as Newton provided a slow, accumulating amount of support for his key ideas, such that by the end of the *Principia* you have been treated to an impressive treatment of the explanatory power of the new ideas, so also is the case with Darwin's *Origin of Species*. Darwin presents the key ideas carefully, with a cumulative effect such that, by the end of the book, you have been treated, as with the *Principia*, to an impressive display of the explanatory power of his key new ideas. (In an oft-quoted phrase from the last chapter, Darwin refers to his book as "one long argument." The description is fitting.)

Of the 14 chapters comprising the *Origin of Species*, it is the first four that contain the key points focused on above, and so a few words about these early chapters are in order. Darwin focuses the first chapter, "Variation under domestication," on the almost entirely uncontroversial subject of artificial selection, that is, of deliberate cultivation of traits in domestic animals by means of selective breeding. In this chapter he manages to emphasize, using well-known examples, the surprising range of variations found within domestic animals, and how almost endless variations can be produced by artificial selection.

The next chapter, "Variation under Nature," focused on the key ingredient (1), that is, of establishing that there is in fact an astonishing array of variation among wild populations of plants and animals. Here as elsewhere, Darwin is able to draw upon decades of his extensive observations and note-taking to establish how much variation there is in the wild.

He then focuses on the key ingredient (2) in the third chapter titled "Struggle for Existence." Here too the reasoning and evidence are almost undeniable. So by the third chapter, Darwin has provided compelling reasoning and evidence supporting the two key ingredients (1) and (2).

As noted in the first section of this chapter, any time you have a situation in which these two ingredients are present, change over time is bound to occur. And in the fourth chapter, "Natural Selection," Darwin makes this explicit. He does so largely by making explicit the comparison between natural selection and the artificial selection discussed in the first chapter. That is, in just the same way that artificial selection produces an astonishing array of changes within populations of domestic animals, so too we should expect natural selection, that is, ingredients (1) and (2), to result in a wide range of changes in populations among wild organisms. And, again drawing on his experiences and data, he argues that this account explains better than any other the relationships we see among organisms in the wild.

In short, by the end of the fourth chapter, Darwin has presented a compelling case that natural selection must occur, and that its effects will be similar to the effects of artificial selection, namely, the production of organisms that can differ to any degree from their earlier ancestors. The remainder of the book deals with a variety of topics, for example, objections to the theory, the issue of the age of the earth and whether there has been sufficient geological time for small changes to accumulate sufficiently so as to result in the range of organisms we see today, the issue of the incompleteness of the fossil record, and the like.

As stressed earlier, the key ideas of Wallace and Darwin ran counter to some deeply entrenched beliefs of the time. The scientific community needed a work such as the *Origin of Species* to provide a compelling case that these long-held, deeply entrenched views were mistaken. And only Darwin, with his astonishing collection of data and his extraordinary breadth of facts, could have written such a book.

Overview of evolutionary theory, 1850–1900

The reception of the Origin of Species   The *Origin of Species* sold exceptionally well, and Darwin oversaw six editions of the work during his life, each with multiple printings. In addition, the work was translated into a number of languages other than English. The book quickly became very well known.

However, throughout the remainder of the nineteenth century, and to some extent into the first decades of the twentieth century as well, Darwin's key ideas were only partly accepted. Darwin's views that evolution occurs, that is, that populations of organisms undergo change over time, and that new species emerge, were widely accepted. This itself is a substantial accomplishment, given the standard belief prior to and during Darwin's time that species are immutable and that new species do not come into existence.

What may be surprising is that Darwin and Wallace's view that natural selection is the principal mechanism by which evolution occurs was widely rejected until the early decades of the twentieth century. As it turns out, Darwin and Wallace were entirely correct with respect to natural selection being the main driving force behind evolution. Modern evolutionary theory recognizes several additional means by which evolution occurs (these are outlined in a subsection below), but even today natural selection is seen as by far the most important factor behind evolution.

To understand the reluctance of those in the latter part of the 1800s to accept natural selection it helps to remind ourselves that, during that period, the means by which traits are passed on from one generation to another was almost entirely unknown. Contrast that with today, where even if you have not taken a biology class and are not familiar with the details of inheritance, it is very likely that you are at least familiar with the idea that the passing on of traits from one generation to the next has something to do with some sort of "units" of inheritance involving genes and DNA.

But for the most part, nothing like this was known or even suspected until the twentieth century. In contrast, the more popular ideas on how inheritance might happen tended, by and large, not to fit well with the idea that evolution occurs primarily by natural selection. There were a number of views of inheritance at the time, but two in particular bear mentioning. The first is what might be called a *blending* view of inheritance, and the second is what is commonly termed a *Lamarckian* view of inheritance.

The blending view of inheritance is basically what the name implies. The key idea was that passing on traits from one generation to the next involved some sort of blending of the parents' traits. For example, if one parent is tall and the other short, a common view was that the offspring would receive a blend of the tall and short traits, and thus tend to be of middle height. This view turns out to be mostly misguided, but without the knowledge that we have from recent discoveries, it would seem a reasonable account of inheritance.

Notably, this blending view of inheritance does not fit well with the idea of natural selection. Suppose an organism, by whatever means, has acquired an unusual trait that enhances its survival. And suppose, perhaps partly owing to its having that trait, that organism survives and reproduces. If that organism has offspring with a mate that does not have the advantageous trait (which is highly likely if the trait in question is a new trait), then on the blending model of inheritance, the offspring will receive only a diluted version of the trait, that is, a trait that has been blended with another, presumably less advantageous, trait. And in each subsequent generation that trait will be diluted even further, until it quickly gets to the point where it is difficult to see how it could any longer provide any advantage to the organism.

In short, natural selection does not fit well with a blending model of inheritance. Darwin recognized this potential problem as early as the late 1830s, long before writing the *Origin of Species*, but he was never able to formulate a completely convincing response to it. The right answer to this problem, as it turned out, lies in rejecting the blending model, but that was not understood until well into the twentieth century.

Another notable and common view of inheritance during this period (which in fact dates back much earlier) was the Lamarckian view, named after Jean Baptiste Lamarck (1744–1829). Lamarck was an influential French biologist who, around 1800, was one of the rare early defenders of the view that new species can come into existence. Part of Larmarck's view involved the inheritance of acquired characteristics. The key idea is that if during an organism's lifetime that organism acquired certain characteristics – large muscles from hard work, say – then those acquired characteristics could be passed on to the next generation.

If this Lamarckian view of inheritance is correct, then it leaves little room for natural selection to play a key role, though many Lamarckians thought natural selection might play some small role. But on this view, the inheritance of acquired characteristics provides the main mechanism by which organisms can acquire new and advantageous traits, and natural selection is left with little or no work to do.

Incidentally, it should be noted that Lamarck was a very important figure of the time, and made major contributions over a long and distinguished scientific career. He turned out to be wrong about the inheritance of acquired characteristics, and unfortunately he is now often remembered and written about in a negative way, largely because of this incorrect view. But that is quite unfair to Lamarck. Even his view of acquired characteristics was, at the time, quite reasonable (for example, Darwin accepted the idea, though he thought it did not play much of a role compared to natural selection).

The blending and Lamarckian views of inheritance were not the only views of inheritance at the time, but they were two of the most common, and serve to illustrate the ways in which the existing views of inheritance did not, in general, point toward natural selection as the main mechanism behind evolution.

In summary, during the second half of the nineteenth century, and largely as a result of the publication of the *Origin of Species*, it was common to accept that evolution did occur, and that new species did arise, through some mechanism. The mechanism, however, was widely viewed as being something other than natural selection. This left open the door for investigations into what that mechanism might be. Scientists in the twentieth century would eventually realize that Darwin and Wallace were right, and that natural selection was the key. We will look at these twentieth-century developments following a brief overview of one other bit of key work from the second half of the nineteenth century, that being Mendel's work on inheritance.

Mendelian genetics   At about the same time as Darwin was publishing the first edition of the *Origin of Species*, Gregor Mendel (1822–84) was performing a series of breeding studies that focused on a particular variety of pea plants. Mendel published his conclusions in the mid-1860s, shortly after the publication of the *Origin of Species*, and these studies include what are now seen as important insights into how inheritance works. Mendel's studies are now quite famous, but it was not until almost the beginning of the twentieth century that the importance of his work was fully recognized. What follows is a brief summary of Mendel's main insights into inheritance.

First, Mendel demonstrated that for at least some traits, and directly counter to the prevailing view, inheritance did not proceed by a blending of traits. Instead, at least some traits had to be passed on by some sort of unit of inheritance, and a unit that was passed on unchanged from the parent to the offspring. (These units would later come to be called genes.)

Second, Mendel's studies showed that, for at least some traits, an offspring inherited, for each such trait, one unit from each parent. And third, Mendel showed that, and again for at least some traits, even though the offspring did not express the trait the parent had, that trait could show up in later generations. In other words, organisms could have units for traits that they did not express, and could pass these on to later generations which might well express that trait. (The distinction Mendel recognized here would later be described as the difference between an organism's *genotype*, which is basically the collection of all the units of inheritance it has acquired from its parents, and an organism's *phenotype*, which is basically all those characteristics that are expressed by the organism. For example, part of a pea plant's phenotype might be shortness, that is, it might in fact be short, yet it might have, as part of its genotype, an unexpressed unit of inheritance for tallness.)

In the discussion above, I have repeatedly used the phrase "for at least some traits." Mendel's work was quite rigorous, and showed that at least for the limited number of traits involved in his studies, there was no question that the insights noted above held. But that still left countless numbers of traits that were not part of his investigations, and for those traits it still seemed reasonable to maintain the blending theory of inheritance. Partly because of this, Mendel's work, while not entirely unknown during his lifetime, was generally not viewed as having far-reaching consequences and did not have a major impact during his lifetime.

However, in 1900 Mendel's work was rediscovered, and in the context of new investigations into inheritance, his work was recognized, and still is, as being a major contribution to our understanding of inheritance.

Overview of evolutionary theory, 1900–1950

In this section we look at a rough sketch of developments from the first half of the twentieth century. By the beginning of this century Mendel's work had been rediscovered, and there was a vigorous debate over how to interpret that work, and also over the role (if any appreciable one) played by natural selection. We'll begin with an overview of this debate.

Early 1900s: The gradualists versus the saltationists   As noted earlier, that evolution occurred was widely accepted following the publication of the *Origin of Species*, although the means by which evolution proceeds was much debated. By the late nineteenth century, and continuing into the early decades of the twentieth century, two main camps had formed concerning the main means by which evolution proceeds. Following common practice, in this section I will refer to these as the "gradualist" and the "saltationist" camps.

The debate between these two groups was a vigorous one, and beyond their arguments over the main mechanism driving evolution, the two groups also disagreed over philosophical attitudes toward science, for example, whether it was legitimate to appeal to unobservable theoretical entities in one's theories, or whether such entities should be avoided in a properly rigorous science. The groups also disagreed over whether evolutionary changes occurred gradually in small steps, as Darwin thought and the gradualists agreed, or discontinuously and in larger jumps, as the saltationists believed ("saltation" basically means a large movement or jump, hence the name for this group). The two groups also took different basic approaches to research, with the saltationists focusing on experimental work. This group produced some important early experimental work on inheritance, pioneering the now common use of fruit flies as a model organism. They also coined now common terms such as "genetics," "genes," and "mutations." The gradualists, in contrast, tended to be skilled mathematicians, and were especially skilled in statistical analysis. As such, their approach tended to be more theoretical than experimental, and this camp produced a sizable number of important mathematical results about how evolution could and could not work.

As noted, and as happens often in science, these two camps engaged in a vigorous debate. When Mendel's work was rediscovered at the beginning of the twentieth century, it was recognized as important by both camps, but interpreted differently. To shorten a long story, the saltationists accepted Mendel's results as a model for how all inheritance worked. But they thought Mendel's model of inheritance was incompatible with the view of evolution proceeding by the slow accumulation of small changes selected for by natural selection. Thus they thought evolution must proceed intermittently, in large discontinuous jumps. In short, their acceptance of a Mendelian model of inheritance as representative of how inheritance in general worked led them to reject natural selection as playing a major role in evolution.

In contrast, the gradualists accepted natural selection as the main driving force behind evolution, and accepted the implication that evolution proceeds by the slow accumulation of small changes. But they also accepted the saltationist argument that Mendelian genetics was incompatible with natural selection as the main means by which evolution occurs, and thus thought the Mendelian model must apply to only a limited number of traits. In short, their acceptance of natural selection as the main mechanism behind evolution led them to reject the Mendelian account as a general model for inheritance.

By way of a general summary, one camp (the saltationists) accepted the Mendelian model as a model for all inheritance, and from this concluded (wrongly, as it turns out) that natural selection could not play a major role. The other camp (the gradualists) accepted natural selection as the major mechanism behind evolution, and concluded (wrongly, as it turns out) that the Mendelian model could not be a model for how inheritance in general worked. In short, it would turn out that each camp got part of the big picture right, and part of it wrong.

In the early decades of the twentieth century, the standard view was that the positions of the saltationists and gradualists could not be reconciled. The next major development came with the surprising discovery that some of the central components of each of these camps were in fact compatible with one another. The recognition of this compatibility led to what is often termed the "new synthesis" or the "modern synthesis." A brief overview of the modern synthesis is the focus of the next subsection.

The modern synthesis Shortly before 1920, one of the key early figures in what would later be termed the "modern synthesis," R. A. Fisher (1890–1962) began investigating whether the seeming impasse between the saltationists and gradualists was as difficult as it appeared to be. Fisher was not, incidentally, the first to try this route, but he had the greatest success with it. Once again, to cut a long story short, Fisher was able to show that one view generally accepted by both saltationist and gradualist camps, namely that the view of evolution proceeding primarily by means of natural selection was incompatible with Mendelian genetics as a general model of inheritance, was mistaken.

Over approximately the next 10 years, Fisher was key in developing a mathematically based approach, largely consistent with the earlier results of the gradualist's camp but that also fit in with the empirical findings of the saltationist camp. Importantly, in his key work in this area, *The Genetical Theory of Natural Selection* published in 1930, Fisher showed, contrary to the widely accepted view, that Mendelian genetics was fully compatible with natural selection being the primary mechanism behind evolutionary change.

Fisher's work, along with the work of a number of other key figures of the time, provided foundational work for an important area of evolutionary research referred to as *population genetics*. In a nutshell, population genetics focuses on the genetic makeup of populations of breeding organisms, and especially on how the distribution of genes within that population can change over time, that is, how the population evolves.

The early workers in population genetics recognized and studied four factors contributing to evolution, namely, natural selection, genetic drift, gene flow, and genetic mutation. These four factors are still widely recognized and investigated today as the four mechanisms by which evolution occurs. We have already discussed natural selection, and there is general agreement that natural selection is the primary mechanism by which evolution occurs. By way of a brief description of the other three factors in evolution, *genetic drift* refers to changes in the genetic makeup of a population due to chance events, for example, effects on the genetic makeup of a population as a result of a chance occurrence such as a large portion of the population being killed by a natural disaster. *Gene flow* refers to change in the genetic makeup of a population due to migration, for example, new mixtures of genes resulting from populations of Europeans moving to the North American continent roughly 500 years ago. Finally, *mutations* are changes due to alterations of DNA caused by factors such as radiation, chemicals that affect DNA, and the like.

The work of the population geneticists and others would provide a synthesis of an extraordinarily wide range of results and data, from fieldwork involving studies of populations in their natural environment, to controlled laboratory experiments involving fruit flies and other model organisms, to mathematical results showing what can result from evolutionary factors, and many other areas of research as well. As noted, this synthesis is now commonly referred to as the *modern synthesis*, and this synthesis brought evolutionary theory into a unified whole encompassing almost the entire field of biological research. In an oft-quoted phrase that comes from the title of a 1973 article by a key figure in population genetics, Theodosius Dobzhansky (1900–75), "nothing in biology makes sense except in the light of evolution."

Brief notes on early twentieth-century work on the physical basis of genetics   Earlier I mentioned that almost everyone, even if they have never had a biology course or only have a minimal background in biology, has at least a general sense that the physical basis of inheritance involves some sort of unit of inheritance, and that chromosomes, genes, and DNA are involved. Especially in the past 100 years, our understanding of this story has developed most impressively. This brief subsection outlines some of the early work from the first part of the twentieth century.

The existence of chromosomes, and the way they are distributed during cell division, was first recognized in the late 1800s, and by the beginning of the twentieth century researchers had speculated that chromosomes might be involved in inheritance. It did not take long to demonstrate that this was correct. At about the same time, the term "gene" was coined as a way of referring to a unit of inheritance, and it was soon shown that genes, whatever their physical structure might be, must reside on chromosomes.

Two other discoveries, crucial to the beginnings of what would eventually become the field of molecular genetics, are worth noting. By the 1930s, it was shown that chromosomes contain DNA. The discovery of the structure of DNA would eventually be the key in figuring out the molecular processes involved in heredity. In addition, proteins were recognized as central to the differences in structure and function of different organisms, and in the 1940s it was shown that genes must somehow provide a way of coding for proteins.

So by about 1950, a general outline of key structures involved in heredity had been identified. The discovery of the structure of DNA in the early 1950s, and the subsequent working out of how DNA codes for proteins, would open up entirely new and productive avenues of research in the latter half of the twentieth century.

Overview of evolutionary theory, 1950 to the present

In the early 1950s the molecular structure of DNA was discovered, and from the structure of DNA it seemed likely that it played a key role in inheritance. It would quickly be recognized that DNA did indeed provide the underlying genetic code for the proteins that make life possible, and that largely account for the differences between different organisms and different species. A key research project was uncovering how DNA coded for proteins, and while there was still a great deal of detail to be filled in, by about the mid-1960s the general picture of the molecular basis of inheritance was well understood.

Another huge change came in the late 1960s and early 1970s, with the discovery of what are termed *restriction enzymes*. These are enzymes that cut DNA at predictable points, and they provided a new and extraordinarily useful tool in research. Restriction enzymes enable researchers to cut and recombine DNA in controlled ways, turn genes on and off so as to better understand their function, enable DNA and genes to be sequenced (that is, they enable one to get a detailed, molecule-by-molecule picture of the structure of genes), and a variety of other avenues of research. This in turn led to the ability to sequence the entire genome (that is, the entire set of genes) of organisms. Notably, by the early 1990s the Human Genome Project had identified essentially the entire human genetic structure, with a high degree of accuracy, and similar studies have identified the complete genetic structure of a number of other complex organisms. It is difficult to overstate the incredible quantity of information we now have, with huge amounts of new information added daily, concerning the detailed genetic makeup of an extraordinarily wide variety of organisms.

As noted earlier, in the first half of the twentieth century the modern synthesis showed how evolutionary theory united work from a wide range of biological areas, from field biologists studying natural populations of organisms, laboratory biologists conducting controlled experiments, mathematically based investigators, studies of now extinct organisms, studies of human origins, and others. The emergence of molecular genetics in the latter half of the twentieth century has proved an enormously fruitful addition to this synthesis.

Concluding Remarks

I have tried, in the first major section above, to give an account of the key aspects of evolution, to do so in a way that is widely accessible, and also to clarify what evolution is and what it is not. The second major section traced the development of evolutionary theory from the work of Darwin and Wallace to the present time, with an eye toward indicating how evolutionary theory provides a key to a great deal of biology.

It is difficult to overstate how much we have learned about our origins, and about the origins of life in general, over the past 150 years. These are remarkable times. In this chapter the focus has been primarily on the basic facts, and historical development, of evolutionary theory. As mentioned at the outset, however, evolutionary theory raises some difficult and controversial issues, largely issues of a more philosophical and conceptual nature. In the next chapter, we explore some of these difficult issues.

Chapter Twenty-Eight

Philosophical and Conceptual Implications of Evolution

In the previous chapter we focused on relatively uncontroversial topics. We explored the basics of evolutionary theory, including what the theory of evolution is and what it is not, and we saw an overview of the historical development of evolutionary theory. In this chapter we explore some implications of what we have discovered about evolution. The implications are complex and controversial, and as we will see, evolutionary theory forces us to confront issues, some aspects of which many find uncomfortable. The main goal of this chapter is to get a sense of some of these issues.

Not surprisingly, evolutionary theory raises more questions than we could possibly discuss in a single chapter. In what follows, I've opted to focus on what I take to be the two biggest ticket items, namely, the implications of evolutionary theory for religious beliefs, and the implications for morality.

Implications for Religion

The main goal of this section is to explore some implications evolution seems to have for religious beliefs, and also some differing views on those implications. Let's begin with some background information.

Background

Recall that during the Aristotelian period God, or at least something like a god, was needed to explain the continual motion of the heavenly bodies. In this way, God, or at least something roughly like God, played an important role in a key scientific theory about the workings of the universe.

Worldviews: An Introduction to the History and Philosophy of Science. Richard DeWitt
© 2010 Richard DeWitt

But as we saw, with the new science of the 1600s the movement of the heavenly bodies was explained in entirely natural terms, largely as a result of the principle of inertia coupled with an understanding of universal gravitation. In short, in the 1600s we discovered that nothing like a God or gods was needed to explain what it had explained for most of the previous 2,000 years.

In this way, the discoveries of the 1600s had nontrivial religious implications, in that those discoveries removed the need for a God or gods in the everyday workings of the universe. But still, the apparent organization of living organisms seemed to require an explanation, and it was difficult to imagine a purely natural explanation. As various authors argued, for example William Paley (1743–1805), the apparent organization and design of living organisms suggested a designer. In his best-remembered argument, Paley appealed to the analogy of a watch requiring a watchmaker – if we came upon a watch, with its intricate design and parts working together to achieve goals, we would immediately conclude that the watch was a product of an intelligent designer. We should draw the same conclusion, Paley argued, in observing living organisms. The apparent design of such organisms, with parts working together to achieve goals, likewise suggests that living things have an intelligent designer.

One can certainly criticize these sorts of arguments, and David Hume has the best-known such criticism in his *Dialogues Concerning Natural Religion* (he was not, incidentally, addressing Paley's specific version of the design argument, but nonetheless his analysis applies to Paley's argument). Hume's conclusion is that such arguments, at best, get you a vague sort of designer, but nothing remotely like, say, the Judaic, Christian, or Islamic God of western traditions. Still, in spite of Hume's criticism, design arguments such as Paley's continued to have appeal.

But it is exactly here that Darwin and Wallace's account seems, at least on the face of it, to have substantial implications for religious views, at least for religious views that include a God who is in some way responsible for what we find in the universe, in particular responsible for the forms of life we see on Earth. Evolutionary considerations also call into question the notion that humans are special, and that the universe has some sort of overall purpose to it. The evolutionary picture seems, again at least on the face of it, to make belief in such a God, and in a purposeful universe with humans occupying some sort of privileged place, at least superfluous if not outright contrary to the empirical evidence.

Evolutionary theory is supported by overwhelming empirical evidence, such that if one is going to respect the empirical evidence, one has to accept the general picture provided by evolutionary theory. So the questions arise, can one fully accept the general account of evolution and still believe, in a consistent and intellectually honest manner, in a God that is in some meaningful way involved in what goes on in the universe? And is the Darwinian evolutionary account compatible with a belief, common and perhaps central to most versions of western religions, in a universe that has some sort of overall purpose to it, and in which humans are in some interesting way special?

Debates surrounding such questions have heated up considerably in recent years. On the one hand are a number of scholars who, in a nutshell, maintain that any traditional notion of a creator God, who is involved in and influences the everyday workings of the universe, and any traditional notion of humans occupying a special place in a universe with any sort of overall purpose to it, have been shown by the evolutionary picture to be at least superfluous if not almost certainly false. In contrast, a number of scholars have argued that one can fully accept Darwinian evolution and natural science in general, and still hold to a belief in a God who is involved in meaningful ways in what goes on in the universe, and in a universe with a purpose and in which humans, are in some way special. My goal in the remainder of this section is to provide a sense of some of the considerations and arguments presented on both sides of this issue.

Some problems for religious views

The scholars whose views I want to outline in the remainder of this section all agree on a particular point, and it might be helpful to begin with this point of agreement. It is not uncommon to hear claims that evolution and a traditional account of God can be made compatible simply by allowing God to play a role in the evolutionary process. The idea is of a God who is involved in the evolutionary process, mixing ingredients into the evolutionary stew as needed, stirring when necessary, and generally guiding the evolutionary process so that it comes out in accordance with some sort of divine recipe or divine plan. This involvement in the evolutionary stew would include, on most such accounts, assuring that humans are one of the products of evolution.

But the scholars we discuss below agree that taking evolutionary theory (and natural science in general) seriously does not allow for such an option. An example might help to clarify why. As mentioned above and discussed in Chapter 20, within Newtonian theory, planetary orbits received a natural explanation by way of the principle of inertia and the principle of universal gravitation, rather than receiving an explanation in terms of a God or gods or some other sort of unmoved mover. This sort of emphasis on explaining natural phenomena naturally, rather than by appeal to supernatural beings and forces, is a core part of modern science. So being intellectually honest about taking natural science seriously, and taking evolutionary theory seriously, requires accepting this aspect of modern science.

Recall that natural selection plays a central role in the Darwinian account. And all the scholars we discuss below would agree that the "natural" part of natural selection is core. So if one adds a supernatural involvement into the account of evolution by natural selection, say by allowing a God to meddle in the evolutionary process, then it is no longer natural selection. One is no longer taking natural science, and evolutionary theory, seriously. In short, taking natural science seriously means that an account of evolutionary development that is importantly influenced by a supernatural being is not an intellectually honest option.

Again, the scholars whose views we consider below all agree on this point. And from just this initial point of agreement, we can see several immediate and substantial consequences. First, taking evolutionary theory seriously requires giving up the idea that the specific life we see around us, the specific organisms and species, are here because they are part of a divine blueprint. The species that currently inhabit the Earth, including humans, are in part the result of countless chance occurrences over the course of billions of years. Such random occurrences would include chance events that affected the environment and thus the survivability of organisms within that environment, random mutations that became established in a population, and an endless number of other chance events. Given this, the idea that the specific species and organisms we see around us are here because they were included long ago in a divine blueprint is not compatible with the evolutionary account.

As a corollary to the above, humans cannot be the (or even an) intended product of evolution. Again, taking evolution seriously means accepting that the species that developed came about because of a good number of chance events. To give just one well-known example, there is good reason to think that a massive asteroid impact about 65 million years ago substantially affected the environment and likely contributed in major ways to the extinction of the dinosaurs. The extinction of the dinosaurs in turn opened up space for the development of larger mammals, eventually including humans. But if the asteroid had missed, then evolutionary history would have played out differently, and it is extremely unlikely that humans would have appeared. The asteroid impact is just one of countless chance events that played an important role in our evolutionary history. In short, the appearance of humans, as with all other species, is in large part due to such chance events. So taking the evolutionary account seriously requires giving up the idea that humans are the intended product of evolution.

One more point of broad agreement is worth noting. If one is going to take natural science in general seriously, including the evolutionary account, then one must allow that events proceed according to natural causes, not supernatural causes. Again, the "natural" part of natural science does not leave room for supernatural influences on how specific events unfold. From this it follows that one must reject the view, common in the western religious tradition, of prayer as a means of influencing natural events. Such a belief in the effectiveness of prayer would be a belief that supernatural causes influence how natural events unfold, which again would be incompatible with taking natural science seriously.

In short, taking evolutionary theory, and natural science in general, seriously has substantial implications for religious beliefs, and in particular for the sorts of beliefs that have been a part of the general western religious view for quite some time. Does taking evolutionary theory seriously leave *any* room for anything like the traditional western view of God?

Dennett, Dawkins, Weinberg, and others: "No"

A number of prominent scholars, including physicists, biologists, philosophers of science, and many others, have argued that the answer to the above question is a clear "no." Taking evolutionary theory, and natural science in general, seriously does not leave room for anything like the traditional western view of God. Such scholars would include Daniel Dennett, Richard Dawkins, E. O. Wilson, Stephen Weinberg, and many others.

The specific views vary from person to person, but there is general consensus that the sorts of questions at stake here – questions involving the origins of the universe, the development of life, the day-to-day workings of the universe including how particular sequences of events unfold, and so on – are empirical issues. And to deal with empirical issues, we should look to our best empirical theories in forming our beliefs about such matters.

For such scholars, largely for the reasons sketched above, our best empirical theories (including evolutionary theory) leave no room for anything like the view of God common in the western world. There is no room for a God who planned the universe with any sort of detailed blueprint. Such a detailed plan for the universe and for life in the universe runs directly counter to what we have discovered about the evolutionary origins of life. So such a belief in a blueprint for the universe and for the development of life is simply incompatible with modern scientific discoveries, especially evolutionary theory.

Likewise for the common conception of a God who intervenes and affects the course of everyday events. Science since the 1600s has amply demonstrated that the universe unfolds according to natural principles, and evolutionary developments over the past 150 years show that this applies to the way life has unfolded as well. In short, the account provided by modern science, and as regards life the account provided by modern evolutionary theory, leaves no room for a God who intervenes and affects the course of everyday events. Such a God is again incompatible with modern scientific developments, especially evolutionary theory.

And to repeat a point made in the subsection immediately above, the evolutionary account is incompatible with the view that humans are special in any interesting sense of the word. In contrast, our understanding of the evolutionary development of life forces us to accept that all life currently existing is here because of a large number of chance events, which leaves no room for believing that humans are in some sense special, in some sense the intended product of the evolutionary development of life.

In short, these scholars argue that evolutionary theory provides an important final piece of a picture that has been developing for some time. In particular, evolutionary theory provides a natural explanation for the last remaining phenomenon that seemed previously to call for a supernatural explanation, that is, evolutionary theory provides a natural explanation for the complexities we find in living things. At this point, then, in a scientifically informed and intellectually honest worldview there is no longer any room left for a belief in a God having the key characteristics typically associated with the western view of God, or for belief in a universe with a large-scale grand purpose.

Again, these scholars would agree that the sorts of questions we are concerned with here – questions about the origin of the universe, about the way events in the universe unfold, about the development of life – are empirical questions. As they are empirical questions, we should look to the empirical evidence to decide on the most reasonable views about such matters. And if the empirical evidence suggests, as these scholars argue it does, that there cannot be a God of the sort common in western religions, then so be it. We have to accept this, and move on.

Haught, process philosophy, and process theology

The reasoning above seems compelling. But some scholars have argued that one can, in an intellectually honest way, fully accept natural science in general and evolutionary theory in particular, and still believe in a God that has, at least in some sense, the important characteristics envisioned by most western religions. The main goal of this section is to outline the views of one such scholar. He is not the only scholar taking the position that one can fully accept the evolutionary account and still believe in an interesting concept of God, but in his recent writings he has articulated this position more thoroughly than most.

John Haught is a modern theologian, well acquainted with evolutionary theory and willing to fully accept the general correctness of the evolutionary account. Haught would agree with a good deal of what has been said in the previous two sections. For example, he would agree that the "natural" part of natural science and natural selection is central. And if one is to be intellectually honest, one must accept that modern science, especially evolutionary theory, leaves no room for a God that interferes with the evolutionary process or directly interferes with the natural workings of the universe.

Likewise, Haught agrees that modern science, and evolutionary theory in particular, leaves no room for belief in a universe that develops according to a detailed blueprint. And he also accepts that humans cannot be viewed as special in the way western religions have traditionally viewed humans as special. In particular, humans cannot be thought of as the intended product of the evolutionary process.

In short, Haught certainly thinks that modern science, especially evolutionary theory, has substantial implications for religious views. And he thinks such considerations call for substantial changes in the way God has been conceived in past centuries. But he sees the changes that would need to be made if we are to take evolution seriously as good changes. Haught argues that careful consideration of the implications of evolution leads to a conception of God that is better than any the western world has had previously. Thus, Haught speaks of "Darwin's gift to theology." In the next few paragraphs, I will attempt to outline some features of what Haught sees as Darwin's contribution to theology. Haught's views are subtle and nuanced, and I should note that what follows is but an overview of some aspects of those views.

First, Haught argues that giving up the typical notion of a creator God who, working from a detailed blueprint, produced a finished (or nearly finished) universe ages ago, is a good thing to give up. To get an idea of his reasoning here, consider the following analogy. Suppose I decide to build something, for the sake of the discussion suppose it is a small pool for our back deck, built from natural stone and maybe with a nice fountain in the center. I acquire some blueprints, assemble the materials, and over the course of some days or weeks build it. Once I am finished, what is left to do? My wife and I would presumably enjoy it, probably a lot at first and less so as the years went on and the novelty wore off. We could perhaps stock the pool with some goldfish or similar species, and occasionally restock the pool as needed. But in general, with this sort of creation, one that is essentially completed shortly after it is begun, there is not much to do once the creation is finished.

This sort of creation seems to be a common view among adherents of typical western religions. For example, recent polls indicate that in the United States close to half the population believes the creation of the universe was begun and finished by God within the last 10,000 years, with the creation process being essentially completed at the beginning. And if one asks what goes on in the world following this initial act of creation, a common view is of the world as a sort of testing place where one proves, for example, one's worthiness for salvation.

Haught does not describe it exactly this way, but one gets the sense that he views this notion of a God, who produced a finished universe shortly after the beginning of creation, with the world subsequently serving as a testing place, as just not a very interesting conception of God or of creation. God creates and finishes the universe at the outset, then century after century after century the same thing happens. People are born, are tested, pass the test or not, and then the same process is repeated over and over and over. It just seems not to be a very interesting conception of God, or of creation, or of the overall purpose of the world. And this is, I think, part of what Haught has in mind when he speaks of Darwin providing a gift to theology. Part of the gift he sees was forcing theologians to rethink these big questions about God, creation, and the purpose of the universe.

In place of the blueprint-creator God, Haught envisions a very different sort of God and a different sort of creative process. He argues that this God, and this creative process, are not only compatible with modern science but also more interesting and more in keeping with the key principles underlying western religion. His theological views have ties to the work of the philosopher Alfred North Whitehead (1861–1947), and also to the scientist/theologian Pierre Teilhard de Chardin (1881–1955). A few brief words on Whitehead and de Chardin are in order.

In addition to some important foundational work in logic and mathematics early in the twentieth century, Whitehead is most closely associated with what is termed "process philosophy." In a nutshell, process philosophy views processes as more fundamental than objects. That is, instead of viewing objects as the fundamental constituents of reality, and then viewing events, change, and other processes as arising out of the interactions of those basic objects, process philosophy reverses this order. Processes are now viewed as fundamental, and objects are viewed as arising out of processes and events.

This is, of course, nothing more than the most basic sketch of one aspect of process philosophy. But it is enough to see how from such a perspective, the world, including objects in the world, can be understood only in terms of the constantly ongoing events and changes and processes and relationships out of which objects in the world arise. The world, then, and things in the world, are not static entities, but rather are constantly evolving processes.

Whereas Whitehead is closely associated with process philosophy, de Chardin is closely associated with *process theology*. Different process theologians differ on the details, but in general, they tend to agree that the old conception of God as an independent agent who created the world and now stands separate from that creation should be replaced with a notion more in keeping with process philosophy. In such a conception, God is seen not as a thing existing separately from the world, but instead God is viewed as part of (or perhaps the sum, or future, of) the ongoing, constantly changing, constantly evolving processes that underlie the world. In this way God is viewed as being a constant participant in the world. Not a participant in the sense of a being who coercively interferes with how events unfold in the world, but rather, by being involved in the processes that are constantly unfolding, and unfolding according to natural principles.

Again, this is nothing more than the most cursory sketch of one aspect of process theology, but it is enough to see how process theologians view God quite differently from the way God has traditionally been viewed in western religions. Haught is working within this broad tradition of process philosophy and process theology. And, he argues, evolution is an ongoing creative process in which new types of organisms, and at times even entirely new species, are coming into being. As such, evolutionary theory not only fits in with this sort of theology, but enhances it.

Moreover, Haught argues that evolution could take place only in a universe that has the right sort of balance between order and randomness. Too much order would not allow the sort of chance events required for the evolutionary development of life on Earth. On the other hand, a universe with too much chance, too much randomness, would not allow for the regularity needed for life to develop as it has in our evolutionary past. And given the sort of balance between order and randomness we see in the universe (which presumably can be traced back to the conditions present at the beginning of the universe), the universe could be expected to unfold in a certain broad way. Not unfold in a precise, predetermined direction, but rather unfold at least roughly in a certain direction, a direction that includes the eventual development of some sort of intelligent life. Not necessarily (or even probably) human life, or any of the particular species that happen to be alive today. But, given the initial conditions of the universe, Haught thinks life was likely to develop somewhere in the universe, and among the varieties of life that could be expected to appear, some sort of intelligent life, capable of understanding and appreciating the universe, was likely to appear.

So in Haught's theology, the universe was not created by God such that it was, like the pool and fountain on my deck, almost fully completed shortly after the creation. Nor was it created so that it would unfold according to a specific blue-print. Nor are the day-to-day workings of the universe planned by God, nor does this God interfere in the day-to-day workings of the universe. God is not some sort of "thing" existing separately and independently from the universe. Rather, the ongoing unpredictable processes, including evolutionary processes, that make up the universe mean that each moment the universe is different, each moment there is an ongoing creative process. And it is a universe that is always being drawn, so to speak, toward the future. For Haught there is a sense in which God can be viewed as this future toward which, somewhat metaphorically, the universe is being drawn. In this way God is intimately involved in the universe and in the ongoing creative processes that arise from the balance between chaos and order. So on this view, the universe is not engaged in just some sort of "mean-ingless rambling." Rather, the universe has a purpose to it, tied to the way it is unfolding roughly in a certain direction, a direction that has its roots in the balance between the deterministic and chance processes that allow the universe to unfold in accordance with natural principles, including the principles at the core of evo-lutionary theory, yet such that there is genuine indeterminacy and unpredictability in the processes.

The above is but a brief sketch of parts of Haught's views. It is a view that clearly goes beyond science (more on this below). But it is a view, Haught would maintain, that fully accepts evolutionary theory, and fully accepts the implications of modern natural science in general. Yet within this framework, he argues, there is still room for a God who participates in the universe. Not by interfering with what takes place, but rather by being part of the processes that underlie the universe. And humans are in a sense special. Not in the sense that the universe was made for humans, or that humans were the intended outcome of the evolutionary process. But rather, given the initial conditions at the beginning of the universe, and the appropriate balance between randomness and order, some sort of intel-ligent beings were likely to result. And at least one intelligent being that emerged happens to be us.

Discussion

A brief review: We began this section with the questions of whether one could, in an intellectually honest way, fully accept the implications of evolutionary theory, and more generally of natural science, and still believe in some sort of God who is involved in the workings of the universe in an interesting way, and in a view of humans as in some way special, and in a universe that has some sort of overall purpose to it. Above we looked at the views of a number of prominent scholars who have argued, in a nutshell, that the answer is "no." In contrast, Haught argues that we can fully accept evolutionary theory, and natural science in general, and still believe in an interesting sort of God, with humans (or at least some sort of intelligent life) as being in some sense special, and in a universe with some sort of overall purpose to it.

What are we to make of this disagreement? I think much of it stems from differing views on what emphasis one thinks should be put on empirical evidence. The scholars in the first camp would argue that beliefs about the universe are empirical beliefs, and that the only, or at least the primary evidence for such beliefs must be empirical. And, they argue that the empirical evidence does not leave much, if any, room for a traditional sort of God.

Haught agrees that empirical evidence is very important. However, he is quite clear that he thinks some beliefs about the universe – for example, providing what he calls an "ultimate explanation" for certain features of the universe – is outside the realm of natural science and is a legitimate function of theology, rather than being solely a matter of straightforward empirical evidence.

Note that the two camps are not arguing about the empirical evidence. By and large they agree on that. Rather, the disagreement stems from differences in fundamental beliefs about how much emphasis to put on empirical evidence. Those in the first camp would say that it comes to beliefs about the nature of the universe, empirical evidence is the only evidence to consult, or at least it should be the overwhelmingly favored evidence. Beyond empirical evidence, there is no further to go. But Haught, on the other hand, would not accept that inquiry must stop where the empirical evidence stops.

At this point it might help to remind ourselves of the disagreement, discussed in Chapter 17, between Galileo and Bellarmine. We can see a somewhat similar situation (though there are important differences) with respect to these two camps. If one has at the core of their beliefs, as Galileo did, that when forming one's beliefs about the universe empirical evidence should overwhelmingly have priority, then I think one *cannot* consistently accept both a Darwinian evolutionary picture and the sort of God that Haught envisions. That is, one cannot consistently and in an intellectually honest way answer "yes" to the central questions of this section.

But if one has a different set of core beliefs, then one *can* consistently accept both the evolutionary picture and the sort of God that Haught envisions. That is, if one has the sort of core jigsaw puzzle pieces similar to what are almost certainly some of Haught's core pieces, then one can fully accept the Darwinian evolutionary account, and natural science in general, and still provide affirmative answers to our central questions.

At bottom, I think we are here seeing what we have seen earlier about differences in individual jigsaw puzzles of belief, with Steve and his beliefs about the moon (from Chapter 7), and differences between the core beliefs of Galileo and Bellarmine (from chapter 17). Such disputes raise quite difficult questions about the reasonableness of different jigsaw puzzles of belief. As with the earlier examples, neither side in this dispute can simply assert, dogmatically, that their preferred system of beliefs is better. As we discussed in Chapter 7, to do so would be to take an unfalsifiable attitude toward one's basic belief system. There can be reasoned debate over the reasonableness of various systems of beliefs, but that debate has to consider difficult and broad issues, broader than is often appreciated.

Before closing this section, I want to pause for a moment, to be careful so as not to be misunderstood. Importantly, I am *not* arguing or in any way suggesting that the views of both the above camps are equally reasonable. Nor am I arguing the opposite, that one is reasonable and the other not. Rather, I am being (or trying to be) entirely agnostic on this question. What I *am* trying to convey is that the disagreement between these two camps stems, importantly, from key pieces of their respective (individual) jigsaw puzzles of beliefs. Any ongoing debate about the respective reasonableness of the camps' positions has to consider very difficult questions about the reasonableness of those overall sets of beliefs.

Morality and Ethics

There has been an enormous amount written over the past two centuries about what implications, if any, evolutionary theory has for our ethical views. For example, 50 years before Darwin's *Origin of Species* Lamarck (discussed briefly in the previous chapter) wrote on the implications of evolution on ethics (here, "evolution" has to be understood in a pre-Darwinian sense, that is, as referring generally to changes in populations of organisms, but not involving natural selection). Darwin's grandfather Erasmus Darwin likewise wrote on evolutionary implications for ethics. Darwin did too, as have numerous others before, during, and since Darwin's time.

In short, far more has been written on this topic than we could possibly survey in this section. But we can get at least a sense of some of the positions that have been articulated concerning this issue, and also look at some recent work concerning evolution and ethics. In particular, we will first clarify some background material, especially concerning a fundamental distinction between metaethics and normative ethics. We will then look at some recent empirical work involving evolutionary theory that sheds some light on metaethical issues concerning the origins of cooperative and altruistic behavior. Finally, we will turn to issues concerned with how, if at all, evolutionary considerations influence normative ethical issues. Let's begin with the distinction between normative ethics and metaethics.

Background material: Normative ethics and metaethics

The study of ethics is generally divided into two broad areas, *normative ethics* on the one hand, and *metaethics* on the other. Let's begin by clarifying these two areas.

Normative ethics is the branch of ethics concerned primarily with ethical norms, that is, this branch of ethics has to do with issues involving how one ought to act. For example, suppose you are on a medical review board whose task it is to decide who should receive organ transplants. Cases will often arise where there is only one organ available but many patients who need it, and these sorts of situations raise difficult ethical questions. In making your decision do you follow some sort of agreed-upon ethical rule, for example, a rule to favor the one who is in the worst shape and most likely to die if he or she does not receive a transplant immediately? Or do you perhaps base your decision on the likely consequences of your choice, that is, a consideration of how much good will result from the various options, perhaps favoring the candidate who can be expected to derive the most years of additional healthy life from the transplant? Questions such as these, which involve how you ought to act (in this case, what recommendation you ought to make), are questions involving normative ethics. Again, in general, normative ethics is the branch of ethics concerned with issues and theories about how we ought to act.

In contrast, metaethics is concerned with broader questions about ethics. Again an example might help. Consider that within our language, we have different types of sentences that play different sorts of roles. The sentence "The sun is about 93 million miles from Earth" is a straightforward declarative sentence, seemingly expressing an objective fact about the world, and typically viewed as being either true or false. Sentences such as "Pass the ketchup" or "Close the door" play a different linguistic role, namely, directing someone to do something. And utterances such as "Soft-shelled crabs for dinner … yuck" play yet another role, expressing preferences, in this case, a food choice.

But what kind of expressions are ethical expressions? When someone says "Abortion is wrong," is this supposed to be expressing what is believed to be an objective fact about the world? Or is it more like a command, along the lines of "Don't have or support abortions?" Or is it an expression of an individual preference, in this case an expression that the individual uttering the sentence does not personally like the idea of abortion? Or some other option?

Debates over what kind of statements moral expressions are is one topic within metaethics. Metaethics is also concerned with questions surrounding the origins of our ethical inclinations, the nature of ethical judgments, and the like. In general, metaethics is not concerned, as is normative ethics, with questions about how we ought to act, but rather with broader questions *about* ethics.

Concerning evolution, we can distinguish different types of implications evolutionary theory might have on ethics. First, there are questions within normative ethics. For example, and most straightforwardly, does our understanding of evolution shed any light on how we ought to behave? Do evolutionary considerations lead us to any new and interesting conclusions about what sort of behavior is morally proper? These sorts of questions involve implications evolution might have for normative ethical issues.

Quite separate from these sorts of questions are questions arising within the framework of metaethics. To give one example, a good amount of research has been conducted geared toward shedding light on the evolutionary origins of our ethical behavior. Such research would be within the domain of metaethics, and as noted, we will explore some of this research below.

This brief distinction between normative ethical concerns and metaethical concerns will help as we explore possible implications evolutionary considerations have for ethics. In particular, we will divide our discussion into two broad categories. First, we will look at some recent empirical work concerning the evolution of behavior we often consider morally praiseworthy, notably cooperative and altruistic behavior. Such research is beginning to shed light on metaethical issues involving the origins of our ethical inclinations. Then we will turn to some controversial topics concerning evolution and normative ethics.

Metaethical considerations: The evolution of cooperative and altruistic behavior

An enormous amount has been written, especially in recent decades, on possible evolutionary origins of some of our common behaviors, including differences between men and women regarding courtship behavior, monogamy, divorce tendencies, and a wide range of other behaviors. As interesting as some of this work is, in this section I want to consider much more narrowly focused work. The work we will explore is relevant to broad questions, but the empirical work itself is narrowly focused. Since this work tends to be better grounded empirically than the accounts mentioned immediately above, these examples provide better illustrations of how empirical work can shed light on metaethical issues. In this section we will explore, in particular, a sampling of recent work involving the evolution of cooperative and altruistic behavior. Let's begin with a look at why cooperative, and especially altruistic, behavior seems problematic from an evolutionary perspective.

Why is altruistic behavior problematic from an evolutionary perspective? Altruistic behavior is behavior that is detrimental (or potentially detrimental) to the agent performing the behavior while it is beneficial to others. Such behavior, say risking oneself to pull a drowning child from a river, is a prime example of behavior we tend to think of as morally praiseworthy. A quick example might help illustrate why altruistic behavior is puzzling from an evolutionary perspective. Shortly after graduating from college I lived and worked in France for a period of time. One afternoon, while I was bicycling on a remote country road on a day off, the driver of a tractor-trailer lost control just after passing me. The truck rolled down a steep embankment, coming to rest upside down with the engine running, diesel fuel gushing everywhere, and with the driver and passenger trapped in the cab. Without much if any thought, I scrambled down the embankment and helped free the driver and passenger, even though I had been raised on American television and movies and thus believed that such scenarios often end badly, say with an explosion. (I later found this was not true – diesel fuel does not erupt the way gasoline can, so we really were in no substantial danger.)

This is an example of an altruistic act, that is, an act that is detrimental (or potentially detrimental) to the one performing the act, while being beneficial to someone else. Some altruistic acts make perfectly good evolutionary sense. Consider a bird that puts itself in harm's way to try to draw a predator away from its offspring in a nearby nest. Such acts fall under the heading of *kin altruism*. It is not surprising that natural selection would have resulted in these sorts of altruistic acts, as they are directly related to the reproductive success of the organism. Another form of altruism, referred to as *reciprocal altruism*, likewise need not be puzzling from an evolutionary perspective. Reciprocal altruism involves situations in which the one performing the action can reasonably think that he or she will receive something beneficial in return. This is sort of a "you scratch my back, I'll scratch yours" sort of altruism.

But the truck example does not easily fit either the kin or reciprocal altruism scenario. Because it took place in a country other than mine, there was no chance the person I was helping was related to me in any way that would make sense in terms of kin altruism. And I lacked the paperwork needed to be in the country legally for as long as I had been, not to mention that I was working illegally. Because of this, I had to leave as soon as I thought the driver and passenger were safe, certainly before the police came, and without identifying who I was. So there was no real chance of my gaining anything in return for the action.

Look at this from an evolutionary perspective. I was in my early twenties, with my entire reproductive future ahead of me, and there I was risking that reproductive future for strangers genetically unrelated to me and in a situation where I had no hope of getting anything in return. These sorts of altruistic acts are not at all uncommon, yet how could an evolutionary process geared toward the survival and reproductive success of individual organisms have resulted in a tendency to act in ways such as this, ways that seem to make no sense in terms of the organism's survival and reproductive success? Such behavior puzzled Darwin, and such behavior has been the focus of much discussion and, especially in recent years, a good deal of empirical research. In the remainder of this section, I want to illustrate a few bits of recent empirical research relevant to the evolution of cooperative and altruistic behavior.

The iterated prisoner's dilemma and the evolution of cooperation   Some important and ongoing empirical work concerning the evolution of cooperation involves what is called the iterated prisoner's dilemma. Of this work, some of the best known involves the computer simulations begun by Robert Axelrod in the late 1970s, originally published in a series of articles in 1980 and more extensively in Axelrod's 1984 book *The Evolution of Cooperation*.

In a nutshell, Axelrod was investigating questions such as how cooperative behavior could arise from a situation (such as evolution by natural selection) geared toward promoting individual self-interest. As noted, this work involved the iterated prisoner's dilemma. To understand the iterated prisoner's dilemma, it helps to begin with what is now commonly referred to as the "classical" or "one shot" prisoner's dilemma. The classical prisoner's dilemma dates back to Thomas Hobbes (1588–1679). The key ingredients of the classical prisoner's dilemma are as follows.

Suppose that two agents, A and B, have the opportunity to interact; that it will be a one-shot interaction in that neither agent will interact with the other again; that each is acting so as to maximize individual self-interest; and that each, without knowing what the other agent intends to do, has to decide whether to cooperate or not cooperate in the interaction. Suppose (a,b) represents the payoff for A and B respectively (so that, for example, (13, 0) means that A received 13 points and B received 0 points as a result of the interaction). Suppose the possible payoffs are summarized in the matrix in Figure 28.1.

If A is acting to maximize self-interest without regard for B, then A will reason as follows. I don't know whether B is going to cooperate or not cooperate. If B cooperates, then if I cooperate I receive 10 and if I do not cooperate I receive 13. Since 13 is better than 10, in this scenario I am better off not cooperating. On the other hand, if B does not cooperate, then if I cooperate I'll receive 0 and if I do not cooperate I'll receive 3. Since 3 is better than 0, in this scenario I am also better off if I do not cooperate. So either way – whether B cooperates or not – I am better off not cooperating.

B, of course, will reason in an exactly parallel manner. The result will be that both A and B, if acting to maximize self-interest without regard for the other, will not cooperate, leading to the (3,3) payoff. It is an interesting situation in

		B's Options	
		Cooperate	**Not Cooperate**
A's Options	**Cooperate**	**(10 , 10)**	**(0 , 13)**
	Not Cooperate	**(13 , 0)**	**(3 , 3)**

Figure 28.1 Prisoner's dilemma payoff matrix

which the individuals, acting in a rational manner so as to maximize individual self-interest, are driven to the worst-case outcome from the perspective of joint interest.

The iterated prisoner's dilemma is similar to the classical prisoner's dilemma. In particular, the payoff for every individual interaction is the same as in the payoff matrix above. However, in a typical iterated prisoner's dilemma scenario there will be more than just two agents (there may well be hundreds or thousands), and each agent interacts with every other agent many times, each time with knowledge of how previous interactions played out.

Whereas the rational strategy for maximizing self-interest in the classical prisoner's dilemma is quite easy to work out – both agents maximize individual self-interest by not cooperating – the best strategy for an iterated prisoner's dilemma scenario generally cannot be worked out in advance. In most such scenarios there will be too many unknowns involved, most notably, what strategies other agents will be using.

To get an insight into what strategies might be best, in the late 1970s Axelrod solicited from researchers around the world strategies, encapsulated in computer programs, to compete against one another in an iterated prisoner's dilemma tournament. There were no restrictions on the strategies, and they could be as simple or as complex as one wished. The only goal of each strategy was to try to accumulate, over the course of hundreds of interactions with the other programs, the highest number of points, with the winning program being the one who finished with the highest total points. (Axelrod was not, by the way, the first to use the iterated prisoner's dilemma as a research tool. It had been used extensively for some time, although Axelrod's approach was interestingly different.)

The results of the competition were surprising. Many thought the adage "Nice guys finish last" would apply, and that "friendly," cooperative programs would fare poorly against more devious, uncooperative programs. But instead, the simplest program submitted, and a cooperative one at that, won the tournament. Even more interesting, the same program won the tournament when it was held again a few months later, this time against a much wider field of competing programs. And in both tournaments, most people knew this program would be included in the competition and so could program strategies specifically designed to exploit that program.

The program that won these first two competitions, and that continues to fare extremely well in regular competitions of this sort that have been held since Axelrod's initial tournament, was a program called Tit for Tat (TfT). As suggested, TfT is a "cooperative" program. That is, this program will always cooperate the first time it interacts with another program, and TfT will never be the first not to cooperate with a program. TfT will retaliate, however – if a program fails to cooperate with TfT, then during the next interaction with that program, TfT will not cooperate with it. TfT's entire strategy can be summed up quite simply: TfT cooperates the first time it interacts with another program, and thereafter TfT does whatever the other program did the last time they interacted.

TfT is what is called a "nice" program, which by definition is a program that will always cooperate with another program so long as that other program has always cooperated. A nice program may retaliate if another program fails to cooperate with it, but a nice program will never be the first to not cooperate. In short, nice programs are highly cooperative programs.

Axelrod's original work, and the work conducted since then, provides good empirical data which strongly suggests that cooperative behavior can be an evolutionarily advantageous form of behavior. Further analysis of these tournaments also provides some suggestive empirical data concerning behavior.

For example, not just TfT, but cooperative, nice programs in general do overwhelmingly better than non-nice programs. For example, the second tournament Axelrod conducted included over 60 programs encapsulating different strategies, some nice, some not. Among the top 15 finishers (that is, the 15 with the highest overall points), all but one (which came in eighth) were nice programs. And even that program (more on this below) was generally a cooperative program.

This research has also shed light on other factors important to cooperative behavior, such as the role of retaliation. For example, the one non-nice program just discussed, the one that came in eighth in the second tournament, generally cooperated with programs that had previously cooperated with it. But unlike a purely nice program, this program would also at some point not cooperate even with programs that had always cooperated with it. The program was basically testing to see what it could get away with. If the other program immediately retaliated, this program went back to the cooperative Tit for Tat strategy. But if it did not receive immediate retaliation, then it would increase its frequency of noncooperation, basically exploiting programs that were super-nice, that is, ones that hesitated to retaliate when another program failed to cooperate.

Another interesting and general suggestion coming from these studies involves the notion of forgiveness. In analyzing the interactions of individual programs, it became clear that some programs became locked in a cycle of retaliation. One strategy that can help break out of such a cycle is to include a sort of "forgiveness" policy in a program. For example, a program might try "forgiving" another program that has recently been not cooperating. The idea, roughly, is to try forgiving the other program, and seeing what happens. If the other program goes back to cooperating, the two programs can then break out of the cycle of retaliation and go back to a mutually beneficial cycle of cooperation.

But equally clear from the analysis of the interactions is that, generally, it does not pay to be immediately forgiving. Programs such as TfT, which immediately retaliate for noncooperative behavior, can be seen to do better generally than those that do not immediately retaliate. For example, consider the program Tit for Two Tats. This program in essence immediately forgives noncooperative behavior by cooperating on the next interaction rather than retaliating, and will retaliate only if the other program fails to cooperate twice in a row. Such programs do well with other nice programs, but tend to be badly exploited by certain non-nice programs. In contrast, programs that retaliate immediately, ones that are, so to speak, easily provoked, tend to do better than programs which are not easily provoked.

As a brief side note, you might note the use of terms such as "nice," "forgiving," "easily provoked," and the like, that is, terms we usually use to describe people's behavior, and some of which have ethical overtones. It is interesting how apt such terms seem, and how much easier it is to describe the programs' behavior using such terms than it is to describe that behavior without such terms. And in fact, researchers in this area tend to use such terms, seemingly because they are so apt.

From the time of Darwin, there have been speculative proposals about how cooperative and altruistic behavior might have been evolutionarily advantageous. But speculation is one thing, and empirical data another. A key aspect of Axelrod's work is that it provides firm empirical data relevant to questions such as how cooperative behavior (and as we will see below, altruistic behavior as well) can be evolutionarily advantageous, and thus how such behavior could have arisen from an evolutionary process that is, at bottom, a selfish process.

The ultimatum game　　In addition to studies involving iterated prisoner dilemma scenarios, much additional work has been done in recent decades that provides additional data relevant to our cooperative and altruistic tendencies. The main goal of the next three subsections is to provide an overview of a few such studies.

In recent years, what is generally termed the "ultimatum game" has been a widely used scenario designed to gather data about cooperative and altruistic behavior. In what follows I provide a brief description of a typical ultimatum game scenario, and a summary of some of the basic results coming from such studies.

Suppose you and I are part of a study in which the two of us are to play the ultimatum game. The setup is quite simple, and goes as follows. You are provided with a sum of money, let's say $10. We'll call you the "proposer," as your job is to propose how that $10 is to be split between the two of us. Your options are to offer me a part of the $10, from $1 to the entire $10, in $1 increments. I will be called the "responder," as my job is to respond to your proposal. In particular, I can either accept or reject your proposal. If I accept it, we split the money as you have proposed. If I reject your offer, we each get nothing.

The possible payoff scenarios for the ultimatum game can be summarized in the payoff matrix shown in Figure 28.2. As with the classical prisoner's dilemma scenario, here too it is easy to see what the rational course of action is for two agents who are acting only to maximize their self-interest. Notice that in every possible payoff scenario, I derive a greater benefit by accepting your offer than I do by rejecting it. So if I am acting in a purely self-interested manner, I should accept whatever you propose, no matter how little. Again, no matter how little you offer, I am better off taking it than rejecting it. You, on the other hand, know that the self-interested and rational thing for me to do is to accept your offer no matter how low, and so if you are acting selfishly and believe I will act selfishly also, you should make the minimum offer allowed, that is, $1, since this is the scenario where you derive the greatest benefit.

**Responder's
Options**

Proposer's Options

Propose giving one of these amounts to responder:

	$1	$2	$3	$4	$5	$6	$7	$8	$9	$10
Accept Proposal	(1,9)	(2,8)	(3,7)	(4,6)	(5,5)	(6,4)	(7,3)	(8,2)	(9,1)	(10,0)
Reject Propoaal	(0,0)	(0,0)	(0,0)	(0,0)	(0,0)	(0,0)	(0,0)	(0,0)	(0,0)	(0,0)

Figure 28.2   Ultimatum game payoff matrix

Note that given the setup, if you make what I consider too low an offer, I have the option of punishing you by rejecting the offer and therefore depriving you of any money. But importantly, notice that I can punish you only by sacrificing something of benefit for myself, that is, I can punish you only by depriving myself of any payoff.

Notably, in studies involving this game interactions are generally done anonymously, that is, if you and I are interacting we do not do so face to face, nor do you know who I am or vice versa. An important consequence of this setup is that if I punish you, I cannot do so with any expectation that it will benefit me. I might expect that punishing you might benefit others, since my punishing you might make you more likely to give higher offers when you interact with others in the study. But I cannot expect it to benefit me.

In short, in this scenario my punishing you would be a form of altruistic behavior, as it can at best benefit others but only at a cost to myself. It is a modest form of altruistic behavior, not as grand as risking one's life to pull a drowning child from a raging river, but it is a form of altruistic behavior nonetheless. Note also that this would be a form of altruistic behavior not easily explainable by kin altruism or reciprocal altruism.

In such studies, under a wide variety of conditions (for example, varying the amount of money involved) and a wide variety of subjects (for example, from different countries and cultures – more on this below), selfish behavior is never the common outcome. The studies reveal that there are always a percentage of subjects (usually around 25 percent) who do act selfishly, but they are always a minority. The most common offer from proposers is roughly 50 percent, that is, most proposers offer to split the money roughly evenly, even though they are under no obligation to do so. And it is not at all uncommon to see proposers offering more than 50 percent.

Also, altruistic behavior in the form of punishing proposers who are seen as making too low an offer is standard behavior, even though such behavior is detrimental to the one doing the punishing. In particular, offers below about 30 percent are routinely rejected.

In short, unselfish and altruistic behavior is the normal outcome. This outcome may not be surprising, as one might speculate that something along these lines would be the outcome. But again, speculation is one thing, empirical data another. These and similar studies, though focused on relatively modest forms of cooperative and altruistic behavior, provide such empirical data.

Questions analogous to those discussed above involving cooperative behavior might well arise here with respect to the observed altruistic behavior. For example, can we do studies that will shed light on why the sort of altruistic punishment found in the ultimatum game might have been, and might continue to be, evolutionarily advantageous? Can we gather data that will tell us something about the conditions under which such behavior would be advantageous, and conditions under which it would not be? And if we can identify conditions under which such behavior would be evolutionarily advantageous, are those conditions likely to have been present in the evolutionary past of humans?

The design of the ultimatum game is too simple to shed light on these sorts of questions. But in a minute we will turn to other studies that do provide data relevant to such questions. Before turning to such studies, it is worth making a couple of additional observations about the ultimatum game.

Most studies similar to the ultimatum game are conducted almost exclusively on college students, and for studies published in English-language journals, conducted almost exclusively on students from the United States and Britain. (Many such studies are done by academics associated with universities, so college students provide a convenient source of subjects.) Notably, studies involving the ultimatum game and variations on it have been conducted on a vastly larger variety of subjects than is typical. Subjects studied in the ultimatum game range from a large number of college students from across the world (not just the US and Britain), and also include a wide range of cultures. But subjects also include a large number of noncollege students, including members of small whale-hunting societies in the east Indonesian islands, hunting societies of Tanzania, nomadic tribes, indigenous cultures of southern Chile and Argentina, and many others.

The results of these studies are robust – no matter the country or the type of culture, one rarely observes the outcomes predicted by the view that agents act in primarily a selfish manner. And altruistic behavior, in the form of punishing proposers with offers considered too low, is standard behavior across cultures and across the world.

Behavior of this sort, found across cultures and around the world, suggests that the behavior is not simply the result of one's cultural heritage. Rather, behavior that is this consistent across cultures suggests that the behavior is stemming from more deeply ingrained tendencies. And tendencies that are deeply ingrained in this way are almost certainly a result of our evolutionary past.

Additional studies on cooperation and altruism Studies such the ultimatum game show that certain types of cooperative and altruistic behavior are common. But they also show that a percentage of subjects do tend to act selfishly. Additional studies have probed whether the behavior of such selfish subjects is influenced by the behavior of others, and if so, what kind of behavior influences them. The results show that the answer is clearly yes, and that altruistic behavior plays a key role in affecting the behavior of selfish agents. Such research has also probed the conditions under which such altruistic influences will work and not work, with an eye toward seeing if the conditions under which altruistic behavior is effective are conditions likely to have been present in our evolutionary past. The material below outlines a few such studies.

The setup of these studies tends to be a bit more complex than that of the ultimatum game, and so I will provide an overview of a typical such study without going into the details. In a typical such study, the experimental setup is somewhat analogous to a prisoner's dilemma setup. However, groups of four or more are typically involved in the interactions, rather than just two agents interacting with one another. And the more the agents cooperate, the more the group benefits. However, the study is set up so that the best outcome for any individual agent is to act selfishly while having others in the group act cooperatively. In a situation where most of the group is acting cooperatively, the benefit to the group increases, but the selfish agent gets the largest share of the reward gotten by the cooperative behavior of the others.

Further, such studies are typically set up so that agents can punish those who are acting selfishly, but only in such a way that is substantially detrimental to the one doing the punishing. This would again be a form of altruistic behavior, in that those doing such punishing hurt themselves, but with the likely outcome that the group will benefit.

In such studies, a percentage of people reliably show a tendency to engage in such altruistic behavior. And notably, the altruistic behavior tends to have an effect on the selfish agents. In particular, selfish agents tend to stop acting so selfishly, with the result being additional benefits to the group. In short, the agent acting altruistically individually ends up worse off than if he or she had not acted altruistically, but the group as a whole ends up better off.

It is worth noting that such altruistic behavior will not necessarily be an evolutionarily stable strategy. That is, one can show (this is generally done through computer modeling) that in some contexts, agents acting altruistically in these sorts of ways will not be reproductively successful enough to ensure that the altruistic tendencies remain in the population. In other contexts, however, such altruistic tendencies can be an evolutionarily stable strategy.

Additional studies have investigated the conditions under which these sorts of cooperative and altruistic behaviors are and are not successful. Such studies suggest that the conditions under which these behaviors would be successful are the sorts of conditions that were likely present when modern humans first appeared about 100,000 to 200,000 years ago. Such conditions include small group size (early humans were almost certainly members of relatively small communities), conditions in which migration in and out of the group is relatively low (again a likely condition present with early humans), and conditions where there is likely to be substantial competition with other groups (it is far less clear what the level of competition and conflict would have been among groups of early humans, but it would not be surprising if such competition and conflict was common). In short, there is reason to think that the conditions under which these sorts of altruistic behaviors can be successful were conditions that were likely present when modern humans branched off from our last common ancestor with other (now extinct) species of humans about 100,000 to 200,000 years ago.

Again, these results are preliminary. But results such as these nicely illustrate some of the ways issues involved in our ethical behavior – cooperation, altruism, punishment, trust (more on this below), and the like – can be studied empirically. In particular, it appears increasingly clear that our evolutionary past had a heavy hand in shaping our ethical leanings and ethical inclinations. An enormous amount remains to be done, but it is clear that empirical studies such as the ones noted above, which explore the evolutionary advantages and disadvantages of behavior we tend to consider ethical behavior, will continue to shed light on the origins of our ethical inclinations.

The trust game   Before closing this section, I want to provide a brief overview of related studies relevant to aspects of our ethical inclinations. These studies illustrate the way in which such tendencies can be investigated at another level, in this case the biochemical level.

Trust is a crucial element of a wide range of human interactions, including our ethical interactions. For example, you and I will behave in entirely different ways toward each other depending on the level of trust between us. Such behavior includes behavior involving moral elements, such as whether we share resources such as food and other goods with each other. Again, if we trust each other, our behavior toward each other will be quite different than if we do not trust each other.

Recently, researchers have begun studying some of the biochemical influences on trusting behavior. Consider the following study involving what the authors call the trust game. This game, as with the ultimatum game, involves money to be divided up between two players. Suppose again you and I are playing this game. At the start of the game, we are both given $12. Let's call my role the role of the "giver," and your role the role of the "sharer." My role is to give you a portion (maybe none) of my money. In particular, I can give you $0, $4, $8, or $12 of my money.

Whichever of these amounts I give you, the people conducting the experiment will give you an additional amount equal to twice what I gave. For example, if I give you $4 of my money, the experimenters additionally give you double that, that is, they give you an additional $8. So you now have your original $12, plus $4 of my money, plus an additional $8 provided by the experimenters, giving you a total of $24. If I give you my entire $12, the experimenters double it and give you an additional $24, which, added to your original $12, gives you $48 in total.

Giver's Options

Give one of these amounts to sharer:

		$0	$4	$8	$12
Sharer's Options	**Share the money in one of these ways:**	(12,12)	(24,8)	(36,4)	(48,0)
		(11,13)	(23,9)	(35,5)	(47,1)
		(10,14)	(22,10)	(34,6)	(46,2)
		⋮	⋮	⋮	⋮
		(2,22)	(2,30)	(2, 38)	(2,46)
		(1,23)	(1,31)	(1,39)	(1,47)
		(0,24)	(0,32)	(0,40)	(0,48)

Figure 28.3   Trust game payoff matrix

So depending on whether I give you $0, $4, $8, or $12 of my money, you will end up with a total pool of $12, $24, $36, or $48. At this point it is entirely up to you as to how, if at all, you will share your pool with me. You can give me any amount from $0 to the entire pool of money you have. Importantly, unlike the ultimatum game I do not have the option of rejecting your offer. Again, once I have initially given you a portion of my money, I have no more influence on how things proceed – how much money of the total pool we each get is entirely up to you.

The payoff scenario for this game can be summarized in the payoff matrix in Figure 28.3. Here, there is one large box for each of the four amounts of my money I might turn over to you, and in each large box is a summary of the possible payoffs for each of those options.

This is a trickier game, in a sense, than the ultimatum game. If I am purely self-interested and think you are as well, I will give you none of my money. The reason is simply that if I believe you will act in a purely self-interested way, then I will believe that you will keep any money I give you. So I am better off giving you nothing and keeping my money for myself. On the other hand, if I trust that you will not behave in a purely self-interested way, then I will give you some of my money, trusting that you will share with me some of the additional proceeds gotten from the people conducting the study. In short, the results of this game depend a good deal on whether people act in self-interested or in cooperative ways, on how people believe others will act, and, importantly, on how much trust the giver has and thus how much (if any) money the giver hopes will be shared with him or her.

Given the results of the experiments discussed in the sections above, it may not be surprising to learn that givers rarely acted in an untrusting manner by giving no money to the sharers, and likewise very few of the sharers acted in a purely selfish manner by not returning any (or even only a very small amount of) money.

These results alone provide additional data about our behavior, not unlike the data discussed earlier. But in this study, the experimenters were mainly interested in factors involved in the biological basis of trust. Studies with nonhuman subjects had suggested that the neurotransmitter oxytocin (a molecule involved in communication between neurons) affects social behavior. The researchers were curious as to whether oxytocin would influence the level of trust among players of the trust game. (By the way, oxytocin is not to be confused with oxycontin, the painkiller that has been in the news recently.)

In the study, the results were striking. Oxytocin substantially increased trust. Before participating in the game, about half the givers were given oxytocin before engaging in the game, and the other givers were given a placebo (as with good experiments, this was done in a double blind way, that is, neither the subjects nor the researchers knew who was receiving the oxytocin and who was receiving the placebo). The level of trust among those receiving oxytocin skyrocketed, as reflected in a dramatic increase in the amount of money these subjects were willing to give over to the other member of the pair.

The researchers went on to investigate whether the effects could be attributed to factors other than trust. In a nutshell, the additional studies strongly indicate that the results above are due to oxytocin's influence on the subject's level of trust, and not to any other factors.

As noted, trust plays a huge role in a large number of our behaviors, including friendship, politics, economics, and almost all social interactions. Trust plays a central role in much of our moral behavior as well. As mentioned earlier, how you and I treat each other, including behavior we would consider morally praiseworthy – sharing food and other resources with each other, assisting each other in times of need, and so on – depends crucially on whether you and I trust each other. Studies such as the one outlined above are preliminary, but they are beginning to shed interesting light on biochemical influences underlying moral behavior.

Evolution and normative ethics

The research outlined above applies to metaethical considerations, such as those involving the origins of our ethical inclinations, what advantage such inclinations might have provided in our past, and the like. But can evolutionary considerations shed any light on how we *ought* to behave? In short, can evolutionary considerations shed any light on normative ethics?

There is a common argument concluding that the answer to this question is "no." But on the other hand, recent scholars have argued the opposite. In the three subsections below, we first look at a traditional argument concluding that evolutionary considerations cannot shed light on normative ethics. We then look at two camps who disagree, although they draw very different conclusions on what implications evolutionary considerations have for normative ethics.

The naturalistic fallacy   The main focus of this subsection is to outline a standard argument as to why evolutionary considerations cannot shed light, at least not in any substantial way, on normative ethics. The basic point behind this argument was most famously articulated in the 1700s by Hume (Hume was first discussed in Chapter 6). Hume noted a tendency for authors to move, without additional justification, from claims that such and such *is* the case, to claims that such and such *ought* to be the case. And, he noted, *ought* claims do not follow logically from *is* claims, at least, not unless some sort of further justification is provided.

It is important to note that, contrary to how Hume's point is often presented, he was not saying that there is no way to derive *ought* claims from *is* claims. Rather, he was pointing out that such reasoning requires additional premises beyond merely noting that such and such is the case. And, he noted, authors making *ought* claims based on *is* claims rarely, if ever, provide these additional needed premises.

This problem of deriving *ought* from *is* has, in the past century, come to be referred to as the "naturalistic fallacy." More precisely, the "ought from is" issue is one version of the naturalistic fallacy. (The other version is discussed briefly in the Chapter Notes, but need not concern us here.) Following convention, I too will adopt the term "naturalistic fallacy" as a convenient label for this problem, that is, the problem that one cannot derive, at least not without further justification, an *ought* claim from an *is* claim.

The relevance of the naturalistic fallacy for any simple version of an ethical theory based on evolutionary observations is straightforward. For example, consider altruistic actions. As discussed above, these are actions that are detrimental to the one performing the action while being beneficial to someone else (for example, risking one's life to pull someone from a burning building). As noted in the sections above, there has been a good deal of interesting work on how altruistic behavior might have arisen through a process of natural selection. Suppose this research continues and that it fills in a fairly complete and convincing account of how our altruistic tendencies provided an evolutionary advantage.

This would be an example of an *is* claim, that is, it *is* the case that humans have altruistic tendencies because such tendencies provided an evolutionary advantage in the past. And thus, as a result of natural selection, such tendencies have continued to be represented in current populations.

But even if this *is* claim is correct, it does not follow that one *ought* to act altruistically. Factual claims, such as *is* claims, at most tell us how things are, not how things ought to be. And in general, *ought* claims, by themselves, do not follow logically from *is* claims.

Empirical investigations into our evolutionary origins, including those discussed in earlier sections of this chapter, presumably will be purely descriptive, that is, they will at best provide evidence for what *is* the case with respect to evolutionary origins of our ethical tendencies. But insofar as normative ethics is viewed as a field geared toward telling us what *ought* to be the case, and if one accepts that one cannot legitimately derive an *ought* claim from a mere *is* claim, the implication is that evolutionary considerations cannot, at least not in any straightforward sense, provide us with any sort of substantial insight into normative ethical issues.

This, then, is an outline of a common position with respect to evolutionary theory and normative ethics. Not everyone agrees with this point, but nonetheless it is quite a common view. Let's next explore two views maintaining that evolutionary considerations do lead us to important insights concerning normative ethics, albeit in quite different ways.

Haught: Evolution as providing a grounding for normative ethics   The first view I want to outline is that of John Haught. As these views are largely an extension of the views we discussed in the section above on evolution and religion, I will briefly outline them here.

Haught accepts that our normative ethical inclinations are a product of the evolutionary process, but rather than thinking that this in some way undermines morality, he views evolutionary theory as providing us with a much better understanding of what grounds our ethical behavior. Haught argues that evolutionary considerations provide us with a deeper understanding of why we ought to behave in a moral way, and in this way, evolutionary considerations are directly relevant to normative ethics.

Recall that for Haught, the balance in the universe between too much order and too much chaos is crucial. Such balance allows for the universe to unfold according to natural principles, to be "drawn toward the future," so to speak, yet still unfold in ways that are genuinely new and unpredictable. And this unfolding of the universe, not in a designed or deterministic way, but in a way that has a rough sort of direction to it, allows there to be an interesting sort of purpose to the universe.

As discussed above, Haught argues that evolution not merely fits into this picture, but enhances it. Our evolutionary origins are part of the way the universe is unfolding. And morality, in particular, is a key part of the process that resulted in modern humans, and a key part of how things will continue to unfold. Evolutionary considerations help us to better understand our moral sentiments, and to better understand how our ethical actions contribute to, and fit in with, this constantly changing and constantly unfolding universe.

Suppose one accepts Haught's account. Then in following the moral inclinations that we now understand resulted from our evolutionary past – in short, in behaving ethically – we are contributing to the ongoing development, the ongoing processes, that make up the universe and that contribute to the purpose Haught sees in the universe. On this view, our ethical actions are part of a much bigger picture and, to borrow a phrase Haught uses on several occasions, in this way our morality has the backing of the universe. In this way, understanding the evolutionary origins of our moral behavior, and understanding the way our moral behavior fits into the broader picture, helps us understand why we ought to act morally. In this way evolutionary considerations contribute to our understanding of normative ethics, not because we deduce specific moral actions from evolutionary considerations, but rather because evolutionary considerations provide a deeper understanding of why we ought to behave morally.

Ruse and Wilson: Normative ethics as an illusion of our evolutionary heritage   Writers such as Michael Ruse, E. O. Wilson, Richard Dawkins, and others have argued, like Haught, that evolutionary considerations can provide interesting insights into normative ethics. But the insights they see are rather different. The main task of this subsection is to sketch their alternative view of the evolutionary implications for normative ethics.

To begin, it is worth making a fairly uncontroversial observation, namely, that new discoveries very often change our understanding of key concepts involved. To take one example, consider a key concept from Newtonian science, that of mass. In the earlier chapter on general relativity (Chapter 24), we discussed how two manifestations of mass were recognized in Newtonian physics, generally termed "gravitational mass" (effects of gravitational mass would include, for example, the weight you feel when in a gravitational field such as that of the Earth), and "inertial mass" (effects of inertial mass would include the heavy feeling you experience when accelerating). Then in Chapter 24, we got a glimpse of how our understanding of mass changed with the development of Einstein's theory of general relativity. Most notably, in general relativity the distinction between gravitational and inertial mass disappears, so that there is no longer any distinction between the two. In short, the acceptance of general relativity led to a change in our concept of mass. This situation is not unusual. It often happens that new discoveries require that we change our understanding of certain concepts.

To take one other example: During Aristotle's time and for much of the time of the Aristotelian worldview, the concept of weight was tied to the effect an object would have on a balance scale. In particular, heavier objects can be seen to be heavier because they displace a balance scale more than lighter objects, and the relative amounts by which two objects displace a balance scale tells us the relative weights of those objects. Later, following the work of Newton, the concept of weight came to be understood in terms of the way a body's mass manifests itself in the presence of a gravitational field.

These are two very different concepts of weight. For example, on the earlier concept of weight, an object's weight will vary depending on how fast it is falling. And if we have two objects that are more or less identical except that one is falling twice as fast as the other, then the one falling twice as fast will weigh twice as much (because it will displace a balance scale twice as much). This is the sort of statement that is trivially true on the earlier concept of weight, but sounds nonsensical on the later concept of weight (and the failure to appreciate the difference in concepts has likely led to some substantial misreadings of early statements relating weight to the speed of falling objects).

These examples of mass and weight illustrate a key point, namely, that empirical discoveries often change our understanding of basic concepts. And the writers mentioned at the outset of this section would agree that there is no reason why empirical discoveries involving evolution should be any different. In particular, if, as seems likely, there comes to be consensus that our ethical feelings, feelings of right and wrong, morally correct and incorrect behavior, morally praiseworthy and morally reprehensible behavior, and the like, stem from our evolutionary heritage, and we come to understand better the biological and evolutionary bases for these ethical inclinations, then it is almost inevitable that this will change, and change dramatically, our understanding of these same key ethical concepts.

In particular, many writers, including the ones mentioned at the outset of this section, argue that evolutionary considerations concerning our ethical inclinations force us to revise our understanding of key aspects of our normative ethical inclinations. Take, for example, the sense of objectivity that accompanies many of our moral judgments. Our understanding of the evolutionary origins of morality have made it overwhelmingly likely that although we often have the distinct sense that our moral judgments are objective, they really are not. Morality stems from human nature, and our nature is as it is because of our evolutionary past. In short, these writers argue that our moral sentiments are as they are because they provided an evolutionary advantage, and not because our moral sentiments reflect an objective feature of the world.

However, these writers argue, the sense that morality is objective is crucial to morality performing the evolutionary task that it does. That is, the apparent objectivity of moral judgments is a crucial component of morality. The sense of moral outrage we feel when we hear of cases of murder, rape, child abuse, and the like, the sense that the action really is wrong – in short, the sense we have that expressions of moral outrage are not just expressions of preferences but rather are expressions of fact – is crucial to the evolutionary role morality plays. Morality could not have played the evolutionary role it played, such writers argue, without this sense of objectivity.

But we can now see, from understanding the evolutionary origins of our moral sentiments, that this sense of objectivity is an illusion. It is an important illusion, granted, and not one that disappears once it is pointed out, but an illusion nonetheless.

Where does this view leave us with respect to normative ethics, that is, with respect to questions as to how we ought to behave? Importantly, this new understanding of our ethical inclinations does *not*, these writers argue, lead us to behave in morally different ways. In particular, it will not lead us to behave in ways that would be considered immoral.

An analogy might help clarify this. There is broad agreement that phenomenal colors, for example, our experience of red, the redness of red, so to speak, is not an objective feature of objects that appear to us to be red. Rather, objects appear red as a result of the way our particular visual systems evolved and the way our visual systems respond to characteristics of the light striking our retinas. Had organisms such as us, with the types of visual systems we have, never appeared, there would be no redness. That redness is a subjective feature of the way our visual systems (and some other organisms' visual systems as well) respond to certain features of light, but not an objective feature of the world.

But even if we understand all this about the lack of objectivity of a color such as red, we will continue, and must continue, seeing certain types of objects as being red. We are simply built that way. Likewise, scholars such as Ruse argue, our morality is part of how we're built. We can no more decide to no longer view certain acts as morally wrong any more than we can decide to no longer see a typical ripe apple as red.

Importantly – and this is a distinction that is easy to miss – such scholars are not arguing that morality is not real. It is real. Likewise, redness is real. What redness is not, and what morality is not, is objective. Redness and morality are subjective features resulting from how we are built and that resulted in large part because they were evolutionarily advantageous. Had humans (or similar organisms) not existed, neither redness nor morality would have existed. But humans do exist, and we did have the evolutionary past we had, and we do have the visual systems and moral sense that we have. Redness and morality are real, but they are not objective, independently existing features of the world. Again, that sense of objectivity is an illusion.

The next key point can be a rather subtle one, and here I will largely focus on these views as articulated by Ruse. Let's go back to the *ought/is* issue discussed in the section above on the naturalistic fallacy. Recall the basic point was that one cannot deduce an *ought* claim from a mere *is* claim, at least without providing additional premises. Ruse would agree with this. And he would agree that evolutionary considerations are descriptive, that is, they are *is* claims. So how can evolutionary considerations tell us anything about normative ethics?

Notice that the naturalistic fallacy allegedly occurs when one deduces an *ought* claim from an *is* claim. But, Ruse argues, evolutionary considerations are not showing us how to *deduce* normative ethical claims; rather, evolutionary considerations are *explaining* normative ethical claims. For example, evolutionary considerations can explain why we have the normative ethical leanings we do, how such ethical leanings were and probably still are advantageous, why they have the feel of being objective, and the like. But having explained this, *there is nothing more to be explained*, nothing more to be done, with respect to normative ethical issues.

To clarify this, it might help to contrast this view with the way normative ethics has traditionally been viewed. A standard approach to normative ethics is to argue for a particular normative ethical theory. Consider, for example, the normative ethical theory termed utilitarianism. This is basically the normative ethical theory that takes, as its basic principle, the "greatest good for the greatest number" view. That is, on this view one ought to do whatever action will maximize the greatest good for the greatest number. (What I am describing here is actually a simplified version of a particular type of utilitarianism, but it will suffice to illustrate the point I want to make.) So, if you find yourself in a unique situation, say having to decide who should get a much needed organ for an organ transplant, and you are a good utilitarian, you calculate which choice would result in the greatest good for the greatest number, and that option is the option you ought to do.

Notice the way, in such a traditional approach, that the normative ethical theory is used to deduce which actions one ought to do. But, writers such as Ruse and others argue, our understanding of the evolutionary origins of our ethical inclinations force us to give up this conception of normative ethics. Ruse and others argue that evolutionary considerations show us that there are no ultimate reasons behind moral actions, no fundamental normative ethical principles from which to deduce morally correct actions. Our understanding of the evolutionary origins of our moral sentiments means that we can no longer view our moral sentiments as being justified by any ultimate, objective principles. In this way, developments in evolutionary theory have substantial implications for normative ethics – they force us to dramatically change our traditional concept of the way normative ethics works.

In short, as with the way our earlier concepts of weight and mass changed as a result of new scientific discoveries, so too discoveries about evolution force us to change our concept of the workings of normative ethics. The traditional concept of normative ethics, in which one deduces correct actions from moral principles, is a concept evolutionary considerations force us to give up. Beyond the explanations of our normative ethical leanings provided by evolutionary considerations (which no doubt will fill out considerably as more work is done), there is nothing left to be explained, and no additional work left for normative ethics to do.

Concluding Remarks

As we saw in earlier chapters, the discoveries of the seventeenth century required our predecessors to rethink core beliefs that had long been taken as established empirical facts. But the challenges to our core beliefs from more recent discoveries are, to my mind at least, substantially more dramatic. In the preceding chapters we looked at discoveries such as relativity and quantum theory, and in the past two chapters we have been focusing on evolutionary theory. Suffice it to say that evolution, like relativity and quantum theory, forces us to substantially rethink basic views we have held for a long, long time. It seems clear that we can never return to our earlier views. Rather, like our predecessors in the early seventeenth century, we are at a point where we can see that major changes in our overall worldview are required, but it is still too early to discern just what sorts of new views will emerge. We live in interesting times.

Before closing, let me make one final observation related to the issues above. There seems to be a common belief that the evolutionary account forces on us some sort of dismal, less interesting view of the universe and our place in it. But the evolutionary account need not be taken in any sort of negative light. Evolution forces us to view our place in the big scheme of things in a very different way. But not, I think, in a worse way.

To take one example, and admittedly speaking personally, I *like* the idea that we are but one of an estimated 10 million species currently on Earth, and I like that we are related to every single organism currently alive as well as all those that have gone extinct before us. I have had the good fortune to live in a variety of places around the world and to travel to more, and I like it that wherever I go, the new flora and fauna I see are all part of a large family. It is an amazing idea: *every* organism on Earth, every plant and every animal, is our relative. There is no reason for this to be viewed in a negative light.

Darwin, too, seemed to feel the same way. Darwin ended his unpublished 1844 summary of his views, and the 1859 *On the Origin of Species*, with this oft-quoted and beautifully phrased passage:

> There is grandeur in this view of life, with its several powers, having originally been breathed into a few forms or into one; and that, whilst this planet has gone cycling on according to the fixed law of gravity, from so simple a beginning forms most beautiful and most wonderful have been, and are being, evolved. (Darwin 1964, p. 490)

The changes required by our discoveries about evolution are substantial. But as Darwin notes, there is grandeur in this view.

Chapter Twenty-Nine

Worldviews: Concluding Thoughts

In this final chapter, we take a broad look at what we have discussed so far, consider some implications of the discoveries explored in the previous chapters, and speculate on some of the changes that will likely be required of our worldview.

Overview

Early in the text, we explored the Aristotelian worldview. Again, this worldview was a jigsaw puzzle-like, interlocking system of beliefs. The pieces fit together well. The universe made sense. Within that jigsaw puzzle, we had a sense of having a good understanding of all the important questions – the structure of the universe, our place in the universe, how and why things behaved as they did, and so on.

Not only did we have answers to individual questions about the world, but we also had a sense of what the universe was like, that is, of what sort of universe we inhabited. The universe was teleological and essentialistic, a purposeful universe full of objects working toward essential, inherent, natural goals. The picture seemed so complete, and so clearly correct, that we find Aristotle himself suggesting that the understanding of the world was pretty much complete. All that remained was to fill in some small gaps, to cross some ts and dot some is. In doing so, Aristotle was voicing an opinion that would be repeated in every time period, up to and including the present.

Within the Aristotelian worldview, the usual metaphor, that is, the usual way of thinking broadly about the sort of universe we inhabited, was to think of the universe as being like an organism. In something like the way an organism has parts that function so as to accomplish goals – the heart has the function of pumping blood, the digestive system of processing food, and so on – so also

Worldviews: An Introduction to the History and Philosophy of Science.   Richard DeWitt
© 2010 Richard DeWitt

the universe is conceived of as having parts with natural functions and goals. We understood – or thought we understood – what kind of universe we inhabited.

In Chapters 9 through 22, we explored the transition from the Aristotelian worldview into the Newtonian worldview. As we saw, the jigsaw puzzle of beliefs that formed the Aristotelian worldview could not continue in light of the new discoveries of the 1600s. Some of the mistaken beliefs in this worldview, such as the perfect circle and uniform motion facts, had looked like reasonably straightforward empirical facts but turned out to be mistaken, philosophical/conceptual facts. Other beliefs within this worldview, such as the belief that the Earth was the center of the universe, were well supported with empirical observations and solid reasoning, but turned out to be mistaken nonetheless.

We saw that it was not simply that individual, peripheral pieces of the Aristotelian jigsaw puzzle had to be abandoned. Rather, core pieces of the puzzle had to be abandoned, and with those core pieces went the entire jigsaw puzzle. Notably, the Aristotelian worldview turned out to be wrong in a broad sort of way. In other words, it was not merely that individual beliefs of the Aristotelian jigsaw puzzle turned out to be wrong; rather, the Aristotelian jigsaw puzzle turned out to be the wrong *sort* of jigsaw puzzle. The universe turned out not to be at all like the way it was conceived of on the Aristotelian worldview. The universe was not like an organism after all.

As we saw, the Aristotelian jigsaw puzzle was replaced with a new jigsaw puzzle, one compatible with the new discoveries. The new jigsaw puzzle, the Newtonian worldview, seemed to work well. The pieces fit together well. The universe made sense. Once again we had answers to what we took to be the major individual questions – the structure of the universe, how things behave, and so on.

And once again, not only did we have answers to individual questions about the universe, but we also had a good sense of the sort of universe we live in. We lived in a mechanistic universe, one in which objects behave as they do largely because of external forces acting on those objects. And we could understand these forces, and characterize them with precise, mathematical laws.

And once again we had a good metaphor to sum up the sort of universe we inhabited. The universe, we came to believe, is like a machine. The universe, we thought, consisted of objects that interact as the parts of a machine interact. As parts of machines interact by pushing and pulling against other parts, so too we viewed objects in the universe as interacting in this sort of mechanical, mechanistic way. And implicit in this machine-like view was the notion that interactions are local interactions, with one object influencing only objects with which it has some sort of connection. The parts worked together in ways we thought we understood, and, like Aristotle, we thought we were close to a complete understanding of the world.

And we had a general sense of our place in the grand scheme of things. We were no longer located at the physical center of the universe, but in another sense we were, we thought, still the center of creation. Life, the general view went, was the product of a divine influence. How else could the apparent design found in living organisms be explained? And along with this view, it was natural to view humans as the apex of life, as special.

All this worked well for quite some time. But as a result of more recent discoveries, we see that relativity theory and quantum theory have nontrivial implications regarding the sort of universe we live in, and evolutionary theory has equally substantial implications for our place within that universe. Do these new discoveries require changes merely in some of the peripheral beliefs of the older, Newtonian jigsaw puzzle? Or, as was the case with the new discoveries in the 1600s, are we being forced to abandon core pieces of that jigsaw puzzle? Let's explore these questions next.

Reflections on Relativity Theory

At first glance, the implications of relativity theory seem quite substantial. The implications of relativity – for example, that space and time can differ for different observers – run counter to strong intuitions we have about the nature of space and time. Again, we tend to believe, usually strongly, that space and time are absolute, that, roughly, space and time are the same for everyone everywhere.

These views on space and time – that is, that space and time are absolute – can be found explicitly stated in Newton's *Principia*. But in fact they go back long before Newton. The belief that space and time are absolute is a belief that is implicitly found at least as far back as the ancient Greeks. In short, absolute space and time have been around, at least implicitly, for a long time, and they are explicitly found within the Newtonian framework.

But suppose we consider the question of whether, within the Newtonian jigsaw puzzle, absolute space and absolute time are more like core beliefs or peripheral beliefs. There is little question that Newton himself, and most of us as well, tend to have strong beliefs in absolute space and time. But recall our discussion from Chapter 1, that the distinction between core and peripheral beliefs is not a matter of the depth of conviction in the beliefs. Rather, the distinction hinges on whether the belief, the piece of the jigsaw puzzle, can be replaced without substantially altering the overall jigsaw puzzle.

Viewed this way, the beliefs in absolute space and absolute time, although strongly held, are not core beliefs. Within the Newtonian framework, these beliefs can be replaced without substantially altering the overall Newtonian jigsaw puzzle. Replacing these beliefs does require changing some other beliefs, of course. But replacing the belief in absolute space and time with a belief in relativistic space and time does not require replacing the overall mechanistic, Newtonian view described above. One can still view the universe as consisting of objects that interact with one another in a mechanistic way, and a way describable by precise laws. What must change is our understanding of issues such as the location and timing of events, but otherwise, the overall Newtonian jigsaw puzzle can remain more or less intact. In short, although it is surprising – quite surprising, I think – to discover that space and time are not absolute, such facts are compatible with the overall mechanistic, Newtonian jigsaw puzzle.

A similar story goes for implications such as the curvature of spacetime, as well as relativity's quite different account (that is, different from the account usually associated with the Newtonian worldview) of gravity. It is surprising to find out that spacetime itself can be influenced and curved by the presence of matter. And it is likewise surprising to discover that the account of gravity most of us have taken for granted, that is, a realistic notion of gravity as an attractive force, can at best be maintained with an instrumentalist attitude in light of the relativistic account of gravity.

But these implications also, although quite surprising, do not require rejecting the core pieces of the Newtonian jigsaw puzzle. That is, we can accept the curvature of spacetime and the relativistic account of gravity without drastically changing the overall mechanistic, Newtonian jigsaw puzzle.

This is not to say that relativity theory has no substantial implications. Even if the implications discussed above do not require rejecting core pieces of the overall Newtonian jigsaw puzzle, neither are they completely trivial implications (they are not, for example, purely peripheral in the sense that ice cream preferences are peripheral). But putting aside for the moment the issue of what particular beliefs relativity theory forces us to change, the more important implication of relativity theory, the real moral, I think, is that it dramatically shows how wrong we can be about issues that seem so obvious. Or in other words, how easy it can be for philosophical/conceptual facts to masquerade as obvious empirical facts. For example, everyone I know, before being introduced to relativity theory, took it as an obvious empirical fact that space and time are the same for everyone. Everyone just *knew* – it was just obvious – that time doesn't pass at different rates, that people do not miraculously age more slowly, just because someone or something happens to be moving. And everyone just knew that space doesn't shrink, like some balloon left out in the cold, just because someone or something is moving. These seemed like such obvious, empirical facts. But not only do these turn out not to be obvious; they turn out to be wrong.

Consider again our predecessors' beliefs in the perfect circle and uniform motion facts, and compare those beliefs with our beliefs in absolute space and time. Our predecessors took it as an obvious empirical fact – everyone just *knew*, it was just obvious to anyone with even the smallest bit of common sense – that heavenly bodies move in perfect circles with uniform speed. These seemed like obvious empirical facts. From our perspective, from within a quite different worldview, it seems odd that anyone could have been so committed to beliefs such as the perfect circle and uniform motion facts. Most people react, when first learning about the perfect circle and uniform motion facts, with sentiments along the lines of "Why would anyone ever have believed *that*?"

But now think about our descendants. At some point they will look back at our beliefs in the same way. Our grandchildren and great grandchildren will look back at us and wonder why we would ever have believed something as odd as that space and time were the same for everyone.

In short, our mistaking absolute space and time for empirical facts is much like our predecessors' mistaking uniform motion and perfectly circular motion for empirical facts. In both cases, what appeared to be obvious empirical facts turned out to be mistaken, philosophical/conceptual facts. And this, I think, is the most important implication of relativity theory. It vividly illustrates how a belief that seems so commonsensical, so obviously correct, can turn out to be so wrong. And this should make us a bit more cautious about how sure we are of other facts that seem to us to be obvious, undeniable facts. So it is not so much that relativity theory forces us to change major aspects of our worldview; rather, relativity theory should lead us to rethink the degree of confidence we have in our picture of the world.

Reflections on Quantum Theory

In contrast to relativity theory, the implications of new discoveries involving quantum theory, especially Bell's theorem and the Aspect experiments, are likely to require substantial changes in the general Newtonian picture. Again, on the Newtonian worldview, the universe is viewed as proceeding as a machine-like series of events. Central to our concept of a machine is a sort of push–pull type of interaction of parts. Gears push against other gears, pulleys move other pulleys but always through some sort of connection such as belts, and, generally, one part of a machine influences only those parts with which it has some sort of contact. And likewise for the universe. We were convinced we live in a universe with the same sort of push–pull interactions. Objects and events influence other objects and events in the same sort of machine-like way, and interactions are local interactions, with influence only among objects and events that have some sort of connection between them.

But it seems that just this core feature of the Newtonian view of the universe cannot be maintained in light of the discovery of the new quantum facts revealed by the Aspect experiments. We may not understand how it is possible, but we live in a universe that allows for instantaneous, nonlocal influences between events, even events separated by substantial distances and for which there is apparently no possibility of any sort of communication or connection between them. No one knows *how* the universe can be like this, only that the universe *is* like this.

For the sake of having a convenient term, I will refer to the sort of instantaneous influences between distant events, of the sort demonstrated by the Aspect experiments, as *Bell-like influences*. It is worth noting that the Bell-like influences demonstrated thus far – for example, those demonstrated in the Aspect and similar experiments – have involved what we might consider micro-level entities, rather than the sort of ordinary objects that get more of our attention in a typical day. That is, although instantaneous influences have been fairly well established for entities such as photons, electrons, and the like, thus far no such experiments have shown Bell-like influences between ordinary-sized objects such as tables, trees, rocks, and the like. Is it possible, then, that the instantaneous, nonmechanistic influences might be confined to micro-level entities? If so, might we be able to retain our views that macro-level objects behave in a Newtonian, mechanistic way, even though we have to abandon our views that micro-level entities behave in this way?

It is too early to give a definitive answer to this question. My sense, though, is that the answer will be "no." Since the original Aspect experiments, physicists have successfully demonstrated Bell-like influences between ever larger entities and over ever larger distances. For example, such influences have been demonstrated between two separated, roughly golf-ball-sized collections of atoms. And in other experiments, such influences have been demonstrated for objects separated by miles rather than merely the distance across a lab room. The fact that Bell-like influences are being demonstrated for more and more, and larger, and more separated entities is one of the reasons for thinking that we will not be able to confine nonmechanistic, Bell-like influences to only a small portion of the world.

Another reason comes from looking at past history. One lesson from history is that we should not underestimate the ingenuity of scientists to find new and novel ways to exploit new discoveries. In the past, fundamental new discoveries have led to changes – including theoretical, technological, and conceptual changes – that could not even be imagined when the discoveries were first made. The discovery that we live in a universe that, deep down, allows for Bell-like influences strikes me as just such a fundamental, important new discovery. It is the sort of discovery that is likely to snowball. Right now the snowball is still fairly small, but I suspect it will become ever larger, leading to changes – again theoretical, technological, and conceptual changes – whose outlines we can barely discern at the moment.

If I am right about this, then we are living in a period that is in many ways like that of the early 1600s. At that time, new discoveries, such as those involving Galileo and the telescope, eventually led to an entirely new way of thinking about the sort of universe we live in. Today, at the very least, the discovery of Bell-like influences forces us to give up the Newtonian view that the universe is an entirely mechanistic universe. And I suspect this is only the tip of the iceberg, and that this discovery, like those in the 1600s, will lead to quite a different view of the sort of universe we live in.

Reflections on Evolutionary Theory

Whereas relativity theory and quantum theory have implications for the sort of universe we live in, evolutionary theory has implications mainly for our place in that universe. If we are to accept the empirical evidence for what it is – and I think we must – then discoveries in evolutionary theory require us to give up the long-held view that humans are special. We have to accept that we are the result of a natural, not supernatural, process, and that rather than being the apex of life, we are instead one type of organism among roughly 10 million currently existing species that, from an evolutionary perspective, all have equal status.

As our predecessors in the 1600s dealt with the discovery that we were no longer the physical center of the universe, so too we will have to deal with the discovery that we are not in any way the center of the universe. This realization requires, among other things, a rethinking of religious views. But this is not the first time empirical discoveries have forced such a rethinking. In Chapter 20 we discussed how the Newtonian account of the motion of heavenly bodies removed the need for a supernatural explanation of such motion. This in turn required a rethinking of the earlier conceptions of God's role (in particular, the role in explaining the motion of heavenly bodies). But as noted in that chapter, religious beliefs tend to be well entrenched. The discoveries in the 1600s forced a rethinking of the conception of God, but it certainly did not lead to an abandonment of religious belief. And I suspect the same will be true in the coming years. An increasing appreciation of the implications of evolution will, I hope, at least lead to a substantial rethinking of traditional religious beliefs, but as in the 1600s, I suspect this is unlikely to lead to a complete abandonment of such beliefs.

Likewise for some of our basic ethical concepts. As discussed in the previous chapter, it is likely that, as we better understand the evolutionary origins of our ethical tendencies, this understanding will lead to a rethinking of key ethical concepts. In short, our understanding of our evolutionary origins requires us to rethink our place in the universe, and almost certainly will require that we rethink traditional views on religion and ethics. It is too early to predict exactly how such changes will play out, but as happened in the 1600s, such changes are almost surely forthcoming. As I've mentioned at several points, we live in exciting times.

But there is no reason a new outlook need be a gloomy one. As Darwin put it, and as I tried to say in Chapter 28, there is grandeur in this view of life. Our predecessors worked out respectable new philosophical and conceptual views compatible with the empirical discoveries of their time, and I suspect we will as well.

Metaphors

I will close with a final observation. As noted above, worldviews have tended to be accompanied by a prevailing metaphor or analogy. Again, within the Aristotelian worldview, the universe was viewed as like an organism, with parts functioning together to achieve natural goals and purposes. Within the Newtonian worldview, the universe was viewed as like a machine, with parts pushing and pulling and interacting with other parts, much like the way the parts of a machine interact.

It is easy to understand the appeal and usefulness of such metaphors, in that they provide a convenient and simple way of summarizing the overall view of what the universe is like. But one intriguing feature of recent discoveries is this: the universe they suggest is not like anything we have ever experienced. That is, the nonlocal influences demonstrated by the Aspect experiments suggest a universe that is not like anything with which we are familiar. A universe that allows for instantaneous influences between events that have no connection what-soever between them is not a universe that is like anything familiar to us.

Notably, because of this, the universe suggested by recent developments may be a universe that does not lend itself to being summarized by any convenient metaphor. We may live in a universe that is like – well, that is not like anything with which we are familiar. For the first time in (at least recorded) history, we may be metaphorless. And we may have reached the point where we will never again be able to summarize the world we inhabit by appealing to a convenient metaphor.

Nonetheless, even though the emerging view may not lend itself to being sum-marized by a convenient metaphor, some general view of the universe will prob-ably emerge. And although it is difficult to predict what this view will be exactly, it seems likely that our children and grandchildren will develop a view of the universe that is substantially different from our own. This view will likely be shaped not only by the discoveries we discussed in the final part of this book, but also by developments occurring now and in the near future. Again, we live in interesting times. Stay tuned.

Chapter Notes and Suggested Reading

In a moment, I will turn to notes and suggested reading for the various sections and chapters in this book. But before turning to specific notes and suggested readings for the individual chapters, let me make a few general suggestions. The works mentioned here are primarily suggested with an eye toward works that would be good starting points for further investigating topics you might find of interest.

History of Science

With respect to general histories of science, Mason's *A History of the Sciences* (1962) is an excellent, single-volume source. Mason provides an overview of the sciences from the ancient Babylonians and Egyptians through the twentieth century, and although it is a single-volume work, Mason's book contains a surprising amount of detail. Lindberg's *The Beginnings of Western Science: The European Scientific Tradition in Philosophical, Religious, and Institutional Context, 600 BC to AD 1450* (1992) provides a more detailed account of ancient and medieval science, while Kuhn's *The Copernican Revolution: Planetary Astronomy in the Development of Western Thought* (1957) is a classic work exploring the changes in the 1500s and 1600s. Cohen's *The Birth of a New Physics* (1985) is a somewhat more general, and quite accessible, account of these changes. For more recent developments, Kragh's *Quantum Generations: A History of Physics in the Twentieth Century* (1999) is an excellent and thorough history of physics from the late 1800s to the present day. Pyenson's and Sheets-Pyenson's *Servants of Nature: A History of Scientific Institutions, Enterprises, and Sensibilities* (1999) provides a history of somewhat different, but important, aspects of the scientific enterprise.

Women in the History of Science

You may have noticed that, aside from a brief mention of Marie Curie in Chapter 21, there is almost no mention of the role of women in the present book. It is certainly not the case that women played no role in the history of science. But there is little question that, during most of our history, society's attitudes discouraged women from playing prominent roles in the sciences focused on in this book, especially physics and astronomy. But again, this is not to say that women played no important roles in these sciences. To take just one example, the astronomical work that began in the 1600s required copious amounts of careful (not to mention tedious) observation and calculation, and much of the observations and calculations were performed by women (for example, Sophia Brahe, Tycho Brahe's younger sister, provided important assistance in his observations). The role of women in the history and philosophy of science is another general area of interest you might want to explore. If so, I would recommend Margaret Alic's *Hypatia's Heritage: A History of Women in Science from Antiquity through the Nineteenth Century* (1986) as a good starting point. Also, an extensive list of biographies of women in science, from the ancients to the present, can be found at www.astr.ua.edu/4000WS/4000WS.html, and this too provides a good starting point.

Philosophical Issues in Physics and Astronomy

For those interested in further exploring issues along the lines of those discussed in this book, especially historical examples and issues involving astronomy and physics, Cushing's *Philosophical Concepts in Physics: The Historical Relation between Philosophy and Scientific Theories* (1998) is a good starting point. Cushing (1937–2002) was a physicist, albeit one with a long-held interest in philosophical issues. His book provides a detailed look at numerous discoveries in physics, with an eye toward illuminating the philosophical issues involved in those discoveries. Kosso's *Appearance and Reality: An Introduction to the Philosophy of Physics* (1998) is another interesting and accessible exploration of philosophical issues in physics. Likewise, Lange's *An Introduction to the Philosophy of Physics: Locality, Fields, Energy, and Mass* (2002a) is an accessible, but more detailed, exploration of some key philosophical questions arising in the context of modern physics. For those interested in further exploring issues related to astronomy, a book mentioned earlier, Kuhn's *The Copernican Revolution* (1957), is a good starting point.

Areas Other than Physics and Astronomy

Although the historical examples in the present text (with the exception of the discussion in Chapter 27 of the historical development of evolutionary theory), and most of those in the texts mentioned above, primarily involve physics and astronomy, these fields are of course not the whole of science. Nor, of course, does philosophy of science involve only these fields. An interesting general introduction to the philosophy of science, but one focusing on biology (specifically immunology) rather than astronomy and physics, is Klee's *Introduction to the Philosophy of Science* (1997). Hull and Ruse's *The Philosophy of Biology* (1998) is a good anthology, organized by central topics in the philosophy of biology, and would provide a good starting point for investigating issues in the philosophy of biology. Likewise, Part IV of Brody and Grandy's *Readings in the Philosophy of Science* (1971) provides a range of introductory readings in the philosophy of biology. And for philosophical issues more closely tied to evolutionary theory, Ruse's *Taking Darwin Seriously* (1998) is a good starting point.

The history and philosophy of chemistry has, in recent years, become a well-established area within the broader umbrella of the history and philosophy of science. The main journal in this area is *Hyle: International Journal for Philosophy of Chemistry*, and this journal is probably the best starting point for getting a sense of the topics investigated within this area. The journal can be found online at www.hyle.org.

Another area that has become well established in recent decades involves feminist issues in the philosophy of science. Feminist approaches to the philosophy of science cover a broad area, including (but not limited to) general methodological and epistemological issues, as well as more specific issues involving particular disciplines (for example, such as is found in writings concerning feminist archaeology). Section 5 of Klee's *Scientific Inquiry: Readings in the Philosophy of Science* (1999) provides introductory readings involving feminist issues in the philosophy of science. Harding's *The Science Question in Feminism* (1986) would provide another good starting point. The issues discussed by Harding range from some that are not at all contentious, to others that are highly contentious. Given this, her book would provide a good idea of the range of issues involved in feminist approaches to science and the philosophy of science.

Miscellaneous Works

Gale's *Theory of Science: An Introduction to the History, Logic, and Philosophy of Science* (1979) is a good general introduction to the philosophy of science, making good use of many examples from the history of science. And another good introductory book, also drawing heavily on historical examples, is Losse's *A Historical Introduction* to the Philosophy of Science (1972). A book I've always rather liked, though one somewhat broader in scope and somewhat difficult to classify, is Pine's *Science and the Human Prospect* (1989). This book is unfortunately now out of print, but it is available on the web at http://home.honolulu.hawaii.edu/~pine/book1-2.html. Gingerich's *The Eye of Heaven: Ptolemy, Copernicus, Kepler* (1993) provides good examples of much more specific and detailed studies in the history and philosophy of science. Lindberg's *Science in the Middle Ages* (1978), and Clagett's *Critical Problems in the History of Science* (1969), likewise provide collections of papers that deal with much more specific issues, while at the same time remaining accessible to nonspecialists.

   With these general recommendations in place, I'll move on to more specific notes and suggestions for further readings for the individual chapters.

Part I: Fundamental Issues

In terms of the general issues covered in Part I, most of these issues are discussed in a number of introductory books and anthologies (such anthologies generally consist of collections of papers, usually by philosophers of science, organized by theme and typically preceded by introductory comments by the editors). Introductory books include Losse (1972), Gale (1979), and Klee (1997), and anthologies include such works as Brody and Grandy (1971), Klemke, Hollinger, and Kline (1988), Curd and Cover (1998), and Klee (1999).

Chapter 1: Worldviews

It is worth noting that the notion of a worldview is related to a variety of ideas presented by Thomas Kuhn (1922–96) in *The Structure of Scientific Revolutions*, first published in 1962. One of the key concepts in this work is the notion of a "paradigm," which is, very roughly, a shared collection (shared by the relevant group of scientists) of beliefs and shared approaches to problems (so in a sense, a paradigm is a subset of a shared worldview). On Kuhn's view, "paradigm shifts" occasionally occur, when an existing scientific paradigm is replaced by a new paradigm, and the existing worldview is replaced by another. The changes in the 1600s from Aristotelian to Newtonian science, explored in Part Two of the present book, is one example of such a paradigm shift. It is worth noting that according to Kuhn paradigm shifts occur very rarely, and he cautioned against using the term too broadly. In spite of his warning, his notion of a paradigm shift has become one of the most popular and overworked notions in recent years. Kuhn's work, especially *The Structure of Scientific Revolutions*, has been one of the most influential works of recent decades in the history and philosophy of science. If you are interested in further exploring issues in this area, Kuhn's *Structure* would certainly be recommended reading.

The notion of a worldview, and especially the jigsaw puzzle analogy, also bears some resemblance to Willard Van Orman Quine's notion of a web of belief as found in, for example, Quine (1964). Quine's preferred analogy was that of a web, in which core beliefs are represented by the inner parts of the web. The idea is that changes in an inner part of a web require changes throughout the web, in a similar way to the fact that changes in core beliefs require changes throughout one's collection of beliefs. In contrast, changes in the outer regions of a web can be made without substantially altering the more inner regions of a web, similar to the way changes in peripheral beliefs can be made without substantially altering the overall collection of beliefs. Most of the anthologies mentioned in the notes above will contain selected papers written by Quine, as well as discussions of Quine's views. Also, some of Quine's views are discussed more fully in Chapter 5 of this book.

As noted in the chapter, Aristotle's own views are complex and, moreover, his writing itself is difficult. The translations closest to the original works are probably those of Apostle, as found in Aristotle (1966, 1969, 1991), but these are also the most difficult to follow. McKeon's translations, as found in Aristotle (1973) are more accessible and probably the more widely read translations. Also, many of Aristotle's works are available on the web, for example, at classics.mit.edu/aristotle. For a quick overview of Aristotle, see Robinson (1995). For brief but good discussions of Greek science before and up to Aristotle, and Greek science after Aristotle, see Lloyd (1970, 1973).

A discussion of, and references for, Newton's work and the Newtonian worldview can be found in the Chapter Notes for Chapter 20.

Chapter 2: Truth

Generally, working scientists, especially physicists, are hesitant to discuss issues involving truth, often characterizing truth (with some justification) as a philosophical rather than scientific issue. One notable exception to this is Steven Weinberg (one of the leading modern physicists), who unabashedly says things like "scientists like myself … think the task of science is to bring us closer and closer to objective truth" (*New York Times Review of Books*, 45(15), 1998). Some of Weinberg's further reflections, including reflections on broader issues surrounding physics, can be found in Weinberg (1992), which is also a reasonably accessible overview of the state of the art in modern physics.

With respect to philosophical treatments of theories of truth, Kirkham (1992) is the most recent and comprehensive discussion of issues surrounding theories of truth, and would be the most thorough source for those interested in investigating theories of truth. With respect to the discussion of Descartes in this chapter, a relatively recent translation of Descartes' *Meditations* is Descartes (1960). An interesting introduction to the *Meditations*, in which the work is used to investigate a variety of philosophical issues, is Feldman (1986).

Chapter 3: Empirical Facts and Philosophical/Conceptual Facts

Issues discussed in this chapter have close ties to what is commonly referred to as the "theory-ladenness of observation." The idea is, roughly, that even seemingly straightforward empirical observations are generally intertwined with various theories. For example, if we use a voltmeter to measure the voltage of the current in the light socket next to our desk, all we will literally observe is the position of a needle on a scale. To infer from this that we have a voltage of, say, 110 volts requires accepting certain theories about the nature of electricity, the way electric current interacts with measuring devices such as voltmeters, the way voltmeters work, and so on. Many of the more contentious issues presented by Kuhn (1962) involve such interplays between observation and theory. Laymon (1984) is another interesting look at the various ways theory and observation are intertwined in one of the more famous experiments in the twentieth century (this experiment involved the bending of starlight, and is discussed more in the next chapter).

Chapter 4: Confirming and Disconfirming Evidence and Reasoning

Further discussion of the range of issues involved in reasoning, especially confirmation and disconfirmation reasoning, can be found in most introductory books in the philosophy of science, as well as in most introductory anthologies of papers in the philosophy of science. Such books and anthologies would include Brody and Grandy (1971), Gale (1979), Klemke, Hollinger, and Kline (1988), Klee (1997), Curd and Cover (1998), and Klee (1999).

Also, as mentioned above, Laymon (1984) is an interesting account of the complexities involved in the observation of the bending of starlight during the 1919 eclipse. Laymon's paper provides a detailed analysis nicely illustrating how intertwined observations and theories are, and how difficult it can be to determine if an observation predicted by a theory has actually been observed.

Chapter 5: The Quine–Duhem Thesis and Implications for Scientific Method

With respect to issues involved in the Quine–Duhem thesis, the main sources from the principal players are Duhem (1954, originally published in 1906) and Quine (1964, 1969, 1980). A good discussion of many of the issues involved can be found in Klee (1997). Papers related to these issues can be found in the anthologies of Curd and Cover (1998) and Klee (1999).

A good overview of Aristotle's approach to science can be found in Robinson (1995). The best original source for the material concerning Descartes is Descartes (1960), and a more thorough collection of his works can be found in Descartes (1931). As noted earlier, Feldman (1986) provides an introductory and very accessible discussion of Descartes' approach. Popper (1992) is the classic source for his views on science, while further discussion of Popper, and of approaches to science in general, can be found in the anthologies mentioned earlier, notably Klemke, Hollinger, and Kline (1988), Curd and Cover (1998), and Klee (1999).

Chapter 6: Philosophical Interlude: Problems and Puzzles of Induction

The original source of what is now called Hume's problem of induction is Hume (1992, first published 1739), especially Book I, Part III (although issues involving induction are a common theme in the work). Hempel's raven paradox can be found in his "Studies in the Logic of Confirmation," originally published in 1945 and reprinted in Hempel (1965). Goodman's views on the "new" riddle of induction (that is, the one involving the predicate "grue") can be found in Goodman (1972, 1983). The example drawn from Heinlein's novel *Job* can be found in Heinlein (1990). Discussion of a wider variety of issues involving induction, including discussion of Hempel and Goodman, can be found in Brody and Grandy (1971) and Curd and Cover (1998).

Chapter 7: Falsifiability

As noted at the outset of this chapter, issues surrounding falsifiability are surprisingly complex, and probably the best way to explore the topic further is to see the way such issues play out in examples from the history of science. The conflict between Galileo and the church, discussed more fully in Chapter 17, is one such case. Good discussions of the Galileo case can be found (among others) in Santillana (1955), Biagioli (1993), Machamer (1998), and Sobel (2000). Another case that involves many of the issues discussed in this chapter is the creation science trials from the 1980s, and the discussion of this case in Part 1 of Curd and Cover (1998) is a good starting point. The cold fusion controversy is another ongoing case illustrating many of the issues discussed in this chapter (and especially of the way in which both sides claim it is the other side that is treating their theories as unfalsifiable). Park (2001) is a good source illustrating the ongoing problems with cold fusion. See www.lenr-canr.org for the point of view of remaining defenders of cold fusion. Finally, the current defenders of an Earth-centered view of the universe provide another example illustrating many of the topics discussed in this chapter. See www.geocentricity.com for their point of view.

Chapter 8: Instrumentalism and Realism

Many of the standard anthologies mentioned in regard to earlier chapters, including Brody and Grandy (1971), Klemke, Hollinger, and Kline (1988), and Curd and Cover (1998), provide further discussion of issues central to this chapter, especially issues involving explanation. See Salmon (1998) for a more thorough look at such issues.

An issue currently much discussed in certain areas of the philosophy of science, and one related to the instrumentalism/realism distinction, is the realism/anti-realism debate. The realism/anti-realism distinction and debate is similar, but not identical, to the realist/instrumentalist issue. The exact focus of the realism/anti-realism debate has changed a number of times over the years, but, roughly, realists maintain that descriptions given by our scientific theories (or at least, by our mature theories) reflect the way things really are, and the entities that are at the heart of such theories actually exist. Anti-realists maintain otherwise: even in the case of our best theories, although they may be convenient and useful, there are no good reasons to think that such theories reflect the way things really are, or to think that the entities suggested by such theories really exist. Jones (1991) provides a good introduction to many of the issues involved in this debate. (Jones's article is not intended as an introductory article, but I think it nonetheless provides an accessible and interesting introduction.) Klee (1997) provides a good introductory discussion of many of the key issues. And Leplin (1984) and French, Uehling, and Wettstein (1988) are good collections of papers centering on the realism/anti-realism issue.

Part II: The Transition from the Aristotelitan Worldview to the Newtonian Worldview

In terms of the general issues covered in Part II, good overviews of the scientific developments can be found in Mason (1962), Cohen (1985), and Lindberg (1992). Overviews of the interplay between scientific developments and philosophical developments can be found in Burtt (1954), Kuhn (1957), Dijksterhuis (1961), Toulmin and Goodfield (1961, 1962), and Matthews (1989).

Chapter 9: The Structure of the Universe on the Aristotelian Worldview

Cohen (1985) provides an introductory and very readable overview of the Aristotelian view of the universe, especially the views on the physical structure of the universe. Dreyer (1953), Kuhn (1957), Dijksterhuis (1961), Toulmin and Goodfield (1961), and Lindberg (1992) provide substantially more detailed discussion both of views on the physical structure of the universe and of conceptual beliefs about the universe. For those interested in exploring more specific issues in the history of western science, especially during the medieval period, Lindberg (1978) is a good starting point.

Chapter 10: The Preface to Ptolemy's *Almagest*: The Earth as Spherical, Stationary, and at the Center of the Universe

Ptolemy (1998) provides a recent translation of the *Almagest*. The preface to the *Almagest* can be found in Munitz (1957), which is the source for the excerpt included in this chapter. Munitz's book also includes Aristotle's arguments, from *On the Heavens*, for the earth as spherical, stationary, and at the center of the universe. In generally, Munitz's book provides a nice collection of excerpts concerning views of the universe, ranging from early Babylonian writings to twentieth-century views.

Chapter 11: Astromomical Data: The Empirical Facts; Chapter 12: Astronomical Data: The Philosophical/Conceptual Facts

For further discussion of the material discussed in Chapters 11 and 12, Dreyer (1953) and Kuhn (1957) are good starting points. More general discussions of these topics can be found in Cohen (1985) and Pine (1989).

Chapter 13: The Ptolemaic System; Chapter 14: The Copernican System; Chapter 15: The Tychonic System; Chapter 16: Kepler's System

Many of the original sources for the material discussed in Chapters 13 through 16 are readily attainable in new translations. See Ptolemy (1998) for a new translation, with commentary by the translator, of Ptolemy's *Almagest*. Likewise, Copernicus

(1995) is a new translation, again with commentary, of Copernicus' major work *On the Revolution of Heavenly Spheres*. And Kepler (1995) is a new translation of some of his key works.

Dreyer (1953) and Kuhn (1957) are the best secondary sources for general descriptions of these systems. Gingerich (1993) is a collection of much more specific and detailed papers by a leading scholar in this area, and provides a good indication of the sort of more detailed research done in the history of science.

As mentioned at the end of Chapter 15, the Tychonic system is the preferred system for those still maintaining (generally for religious reasons) that the Earth is the center of the universe. Further information on modern-day defenders of the Tychonic system can be found at www.geocentricity.com. The writings of the defenders of geocentricity provide interesting illustrations of many of the issues discussed in Part One, especially issues involving competing worldviews, falsifiability, evidence, and confirmation and disconfirmation reasoning.

Chapter 17: Galileo and the Evidence from the Telescope

Galileo's own writing is quite accessible, and translations of Galileo's main writings involving his work with the telescope, as well as his views on the Earth-centered versus sun-centered issue, can be found in Galileo (1957, 2001). Fantoli (1996) is a detailed and well-referenced work on Galileo, especially on the issues involving the church, and highly recommended for additional reading on Galileo. Santillana (1955) is a more general account of Galileo's work, also with an emphasis on issues involving Galileo and the church. Machamer (1998) is a collection of more narrowly focused papers on aspects of Galileo's work, and provides a good sense of the sort of more detailed scholarship involving Galileo. Biagioli (1993) and Sobel (2000) are somewhat different, but quite enjoyable, approaches to Galileo's life and work. The former focuses on the role of court politics on Galileo's work (as noted in the chapter, Galileo was a member of the Medici court), while the latter emphasizes his relationship with his daughter, using the surviving letters from his daughter to provide a somewhat different look at the life and work of Galileo and his daughter.

Chapter 18: A Summary of Problems Facing the Aristotelian Worldview

Kuhn (1957) and Cohen (1985) provide good general accounts of the sorts of problems facing the Aristotelian worldview in the early 1600s. A more detailed account of developments at this time can be found in Dijksterhuis (1961) and Mason (1962).

Chapter 19: Philosophical and Conceptual Connections in the Development of the New Science

Kuhn (1957) provides further discussion on many of the topics discussed in this chapter. Mason (1962), although focusing primarily on the history of science, provides good detailed discussion of some of the broader issues discussed in this chapter. Dijksterhuis (1961) and Toulmin and Goodfield (1961) likewise provide somewhat more detailed accounts of many of the topics touched on in this chapter, and are good sources for further investigating these topics.

Chapter 20: Overview of the New Science and the Newtonian Worldview

Newton's *Principia* is highly recommended, and Newton (1999) is a new translation, with extensive commentary provided by the translators. Cohen (1985) again provides a general overview of these developments, while Dijksterhuis (1961) and Mason (1962) provide more detailed accounts. Also, I want to thank Charles Ess for the suggestion to include the discussion of instrumentalist and realist attitudes toward Newton's notion of gravity.

Chapter 21: Philosophical Interlude: What is a Scientific Law?

An early and still classic work on issues involving laws is Hempel and Oppenheim (1948). Armstrong (1983) and Carroll (1994) provide thorough treatments of these issues, and Lange (2000) provides a good summary of standard views and offers an alternative account. Cartwright (1983) provides an interesting and somewhat different perspective on some of the common attitudes toward laws, as does Gierre (1999). For early discussions of problems involving counterfactuals, Quine (1964) and Goodman (1983) are good sources, as is Lewis (1973). For more on *ceteris paribus* clauses, see Lange (2002b) and Earman, Glymour, and Mitchell (2003).

Chapter 22: The Development of the Newtonian Worldview, 1700–1900

Mason (1962) provides a good account of the scientific developments in the period covered in this chapter, and Kragh (1999) is a good source for the situation in physics at the end of the 1800s. Cushing (1998) likewise provides a good investigation of many of the topics covered in this chapter, with more of an eye toward the interactions between scientific questions and philosophical issues. See Everitt (1975) for a detailed account of Maxwell's contributions.

Part III: Recent Developments in Science and Worldviews

For general sources dealing with the material in Part III, Kragh (1999) is a relatively new and quite detailed history of physics in the twentieth century, ranging from the situation in physics from the late 1800s to the end of the 1900s. Mason (1962) likewise covers developments in science, both biology and physics, albeit somewhat more briefly. The interplay between philosophy and physics in recent years is nicely illustrated by many of the case studies in Cushing (1998).

Chapter 23: The Special Theory of Relativity

The original source for the special theory of relativity is Einstein (1905), and Einstein (1920) provides an accessible account of the theory. Mermin (1968) is an excellent presentation of special relativity, which is thorough and precise while presupposing nothing beyond basic algebra. A recent revision of this book, Mermin (2005) is likewise a good source. D'Abro (1950) is another good presentation of special relativity, while Kosso (1998) provides a general overview of special relativity with an emphasis on certain philosophical consequences.

It is worth noting that "absolute space" and "absolute time" are often used in a different way than I use them in this chapter. From the time of Newton and Leibniz, and continuing today, there has been debate over whether space is an entity existing separately from the objects that (presumably) exist within that space. That is, is space a substance existing independently of objects, or does space consist of nothing over and beyond the relations between objects? The former view is generally considered a *substantivalist* view of space, with the latter view a *relational* view of space. To use a common analogy, the substantivalist view maintains that space is like a container, and objects exist within that container. Importantly, on this view the container – that is, space – exists independently of, and has properties independent of, the objects within the container. The relational view of space rejects this "container" view of space, maintaining instead that space consists of nothing more than the relations between objects. The term "absolute space" is sometimes used to refer to the substantivalist view of space, and again, this is a somewhat different use of the term from the way I use it in this chapter. Similar questions arise for time as well – is time something that exists independently of objects and events, or is time nothing over and above the relations between objects and events?

Finally, in the body of the chapter I make reference to the Lorentz transformations, without specifying what those transformations are. For those who are curious, suppose we let x, y, z, and t represent spatial and time dimensions in a stationary coordinate system, and x', y', z', and t' represent spatial and time dimensions in a coordinate system moving in the x direction with velocity v (that is, moving relative to the first coordinate system, and as usual, moving in a straight line with uniform speed). Define γ as follows:

$$\gamma = \frac{1}{\sqrt{1-\left(\dfrac{v}{c}\right)^2}}$$

Then the Lorentz transformations are

$$t' = \gamma\left(t - \frac{vx}{c^2}\right)$$

$$x' = \gamma(x - vt)$$

$$y' = y$$

$$z' = z$$

These are the transformations used, in the discussion of Joe's spacetime coordinate system and Sara's spacetime coordinate system, to convert the coordinates from one coordinate system to the other.

Chapter 24: The General Theory of Relativity

The original source of the general theory of relativity is Einstein (1916), and Einstein (1920) again is an accessible discussion of the theory. D'Abro (1950) is likewise a good source for discussion of the general theory.

As mentioned in the body of the chapter, figures such as Figure 24.2 are typically two-dimensional "slices" through the four-dimensional spacetime. It is not crucial for the discussion in the chapter, but you might notice that, in the diagram, this two-dimensional slice is "embedded" in a three-dimensional space, and for this reason such diagrams are typically referred to as *embedding diagrams*. Also, I want to thank an anonymous reviewer for catching a glaring mistake in my description of geodesics in an earlier draft of this chapter.

Chapter 25: Overview of the Empirical Facts, Mathematics, and Interpretations of Quantum Theory

The emphasis on distinguishing quantum facts, quantum theory itself, and the interpretation of quantum theory is drawn largely from Herbert (1985). The distinction is, I think, a very useful one to keep in mind when reading the vast literature on quantum theory.

The quantum facts described in this chapter are fairly standard examples widely used to illustrate some of the oddities associated with quantum facts. Pine (1989) discusses similar experiments.

A general descriptive overview of the mathematics of quantum theory, similar to that in this chapter, can be found in Herbert (1985). For more detailed accounts of the mathematics, the best approach probably depends largely on one's mathematical background. My favorites are Hughes (1989) and Baggott (1992, 2004).

There are an enormous number of books, of quite uneven quality, generally revolving around the interpretation of quantum theory. Here I will note a few of the books I think are of better quality. Herbert (1985) is written by a physicist but for a general audience, and although he has his preferred interpretation, he provides an even-handed treatment of the options. Baggott (1992) likewise is a good discussion of quantum theory and issues involved in the interpretation of quantum theory. Baggott's book, incidentally, is subtitled *A Guide for Students of Chemistry and Physics*. I would ignore the subtitle, in that the book is a good guide to quantum theory and the interpretation of quantum theory, regardless of whether you are a student of physics or chemistry. Baggott (2004) is a substantial revision and expansion of this work, and also recommended. The last chapter of Lange (2002a) likewise provides a good discussion of these issues. Finally, I would like to thank Marc Lange for catching a substantial mistake in my discussion of Bohm's interpretation in an earlier version of this chapter.

Finally, in the body of the chapter I mentioned that I would provide a more detailed summary of the mathematics of quantum theory. In summary: (a) a (pure) state of a quantum system is represented by a vector in a Hilbert space; (b) each type of measurement one might perform on a quantum system is associated with a particular operator in the Hilbert space; and (c) predictions about the outcome of a measurement on the quantum system are arrived at by finding the eigenvalues associated with that operator (that is, the operator associated with that measurement).

To understand (a), consider the two-dimensional Cartesian coordinate system we studied in our earlier mathematical training, and suppose we have a line drawn from the point (0,0) to some other arbitrary point, say the point (11,7). Such a line would be an example of a *vector*, and a collection of such vectors is a *vector space over the two-dimensional space of the real numbers*. Vector spaces can involve three, four, or any (including infinitely many) dimensions. Vector spaces can likewise involve numbers other than the real numbers, and in the mathematics of quantum theory, some particularly important vector spaces involve complex numbers (that is, a number of the form $a + bi$, where a and b are real numbers, and i is the imaginary number equal to the square root of -1).

Let's use an analogy to get a rough idea of a *Hilbert space*. Consider the set of vector spaces over the two-dimensional space of the real numbers. Note that some of these vector spaces will meet various criteria. Some of these vector spaces consist only of vectors that are specified by pairs of even numbers, some might consist only of vectors specified by positive numbers, and others might consist only of vectors for which certain mathematical operations can be applied. A Hilbert space is a vector space that meets certain well-defined, well-understood criteria, and admits of certain types of mathematical operations. The specific criteria are beyond the scope of our discussion, but this should suffice to give at least a general idea of a Hilbert space.

An *operator* in a Hilbert space is a function that operates on vectors, transforming one vector into another. To understand the idea behind an eigenvalue, consider again the set of all vectors over the space of the two-dimensional space of the real numbers. Consider a particular operator O and a particular vector v. Suppose Ov results in a vector twice the length of v, which we will write as $2v$. Note that O may not simply be a doubling operator, that is, it may not double the length of all vectors. But there may be certain vectors, such as v, for which O results in a vector twice the length of the original. That is, for this particular vector, O$v = 2v$. In this case, v is called an *eigenvector* of O, and 2 is the corresponding *eigenvalue*. Likewise, if O$v = 3v$, then v is an eigenvector of O and 3 is the corresponding eigenvalue. Also, some operators have no eigenvectors and hence no corresponding eigenvalues. Hilbert spaces are more complicated vector spaces, and the notion of eigenvectors and eigenvalues for such spaces are more difficult to picture. But the analogy with the simpler vector space over the space of the two-dimensional Cartesian coordinate system should give a flavor of what these are like.

Recall that an operator is a function that transforms one vector into another, and, from (b) above, we know that possible measurements on a quantum system are associated with particular operators on the Hilbert space. With respect to (c) above: most of the time (though not always) the operators associated with measurements will have eigenvectors and corresponding eigenvalues. The eigenvalues represent the possible results of the measurement associated with the operator. In particular, from the eigenvalue and the vector representing the state of the system, together with a special operator called a projection operator, one can calculate a probability between 0 and 1. And that probability represents the probability of observing the particular measurement outcome associated with that eigenvalue.

Chapter 26: Quantum Theory and Locality: EPR, Bell's Theorem, and the Aspect Experiments

Herbert (1985) provides a good general discussion of the topics in this chapter, and as noted, my explanation of Bell's theorem draws substantially from Herbert's presentation.

Baggott (1992, 2004) also provide excellent, and somewhat more detailed, accounts of these issues. For issues involving locality, Maudlin (1994) is a thorough and careful analysis of the quite complex issues involved, and is highly recommended for anyone interested in further pursuing topics involving locality and nonlocality.

Finally, Bell's writings are well worth reading. His key papers on this subject are collected in Bell (1988).

Chapter 27: Overview of the Theory of Evolution

Desmond and Moore (1991) is a relatively recent, thorough, and highly recommended biography of Darwin and his work. Quammen (2006) is a much shorter account, though informative and accessible, and also recommended. For the first edition of Darwin's key work, Darwin (1964) is a good facsimile of the first edition of the *Origin of Species*.

Mayr (1982) is an extensive and detailed account of developments in biology in recent centuries, including evolutionary theory. Provine (1971) is likewise a good source. For briefer overviews, Mason (1962) provides a quick overview of key developments, as does Silver (1998). Wilson (1969) and Greene (1969) provide a more detailed discussion within the context of broader developments in biology during the time surrounding Darwin and Wallace's key work.

Fisher (1999) is a key work in the eventual development of population genetics, and Williams (1966) and Hartl (1981) provide thorough accounts of this field. The relevant chapters in Mayr (1982) also provide good additional information on this subject.

Watson and Crick (1953) is a classic paper announcing the discovery of the structure of DNA, and well worth reading. Olby (1974) is a good thorough account of this discovery. This time is also a rich period for exploring broader questions in the conduct of science, especially the reluctance to acknowledge important contributions of women working in this field. Sayre (1975) and Fox Keller (1983) provide good starting points for further exploration of this.

As noted in the body of the chapter, it is difficult to overemphasize the areas of research that recent discoveries have made possible, especially the discovery of restriction enzymes and subsequent tools for manipulating DNA. Bergman and Siegal (2003), Abzhanov et al. (2006), and Mecklenburg (2010) are good examples.

Finally, I'd like to thank Jim Long for assistance and helpful discussions of possible translations of Latin phrases from Darwin's notebooks mentioned in the chapter.

Chapter 28: Philosophical and Conceptual Implications of Evolution

Good sources for the views of those discussed in the chapter making the case that evolution, and modern science in general, leaves little or no room for anything like a traditional God are Dennett (1995, 2006), Dawkins (2006), and to a lesser extent but still foreshadowing writings to come later, Dawkins (1976) and Weinberg (1992). Two prominent writers taking a somewhat different line of argument, though not mentioned in the text are Harris (2004, 2007) and Hitchens (2007). For Haught's views, see primarily Haught (2008a, 2008b), but also Haught (2001). For some of the earlier writings on process philosophy, see Whitehead (1978). For other related approaches to reconciling religion and evolution, see Mooney (1996) and Miller (1999). A work somewhat further afield of the material in this chapter, but still an interesting take on evolution and religion, is Miller (2008).

For early work on the iterated prisoner's dilemma, the best sources are Axelrod (1980a, 1980b, 1984). Studies involving the ultimatum game, trust game, and others can be found in Gintis et al. (2002), Bohnet and Zeckhauser (2004), and Kosfeld et al. (2005). A detailed and thorough treatment of a range of issues involving altruism and unselfish behavior can be found in Sober and Wilson (1998).

The classic source for the naturalistic fallacy is Moore (1962), which was first published in 1903, though the version discussed in this chapter dates back to Hume (1992), first published in 1739. The other version of what is often termed the naturalistic fallacy, and the version of primary interest to Moore, is an argument that any attempt to provide a naturalistic foundation for normative ethical claims is misguided. In support of this, Moore notes what is now commonly referred to as the "open question" argument: given any alleged identification of something morally good with some natural property, it will still make sense to ask of something having that natural property whether it is good. In other words, given that something has the property is question, it is still an open question as to whether it is good, and if it is still an open question, good must not consist in having that natural property.

For Haught's views as outlined in this part of the chapter, Haught (2008a) is again the best source. Ruse's and Wilson's views are best found in Ruse (1998) and Wilson (1978). Ruse (2009) is an excellent collection of articles on these and related matters.

Chapter 29: Worldviews: Concluding Thoughts

For those interested in the role of metaphors and analogies in science, Hesse (1966) would be a good starting point. Other than this, the chapter is an overview of topics we have discussed, and of what the future might bring, so there is not a great deal in the way of notes, sources, and suggested reading. The best idea is, I think, the one suggested at the end of the chapter: with apologies to Timothy Leary, turn off (the TV) and tune in!

References

Abzhanov, A., Kuo, W., Hartmann, C., Grant, B., Grant, P., and Tabin, C. (2006) "The Calmodulin Pathway and Evolution of Elongated Beak Morphology in Darwin's Finches," *Nature* 442, 563–567.

Alic, M. (1986) *Hypatia's Heritage: A History of Women in Science from Antiquity through the Nineteenth Century*, Beacon Press, Boston.

Aristotle (1966) *Aristotle's Metaphysics*, translated by H. Apostle, Indiana University Press, Bloomington.

Aristotle (1969) *Aristotle's Physics*, translated by H. Apostle, Indiana University Press, Bloomington.

Aristotle (1973) *Introduction to Aristotle*, second edition, translated by R. McKeon, University of Chicago Press, Chicago.

Aristotle (1991) *Aristotle: Selected Works*, third edition, translated by H. Apostle and L. Gerson, Peripatetic Press, Grinnell, IA.

Armstrong, D. (1983) *What is a Law of Nature?* Cambridge University Press, Cambridge.

Axelrod, R. (1980a) "Effective Choice in the Prisoner's Dilemma," *Journal of Conflict Resolution* 24(1), 3–25.

Axelrod, R. (1980b) "More Effective Choice in the Prisoner's Dilemma," *Journal of Conflict Resolution* 24(3), 379–403.

Axelrod, R. (1984) *The Evolution of Cooperation*, Basic Books, New York.

Baggott, J. (1992) *The Meaning of Quantum Theory*, Oxford University Press, Oxford.

Baggott, J. (2004) *Beyond Measure: Modern Physics, Philosophy and the Meaning of Quantum Theory*, Oxford University Press, Oxford.

Bell, J. S. (1964) "On the Einstein Podolsky Rosen Paradox," *Physics* 1, 195–200.

Bell, J. (1988) *Speakable and Unspeakable in Quantum Mechanics: Collected Papers on Quantum Philosophy*, Cambridge University Press, Cambridge.

Bergman, A., and Siegal, M. L. (2003) "Evolutionary Capacitance as a General Feature of Complex Gene Networks," *Nature* 424, 549–552.

Biagioli, M. (1993) *Galileo, Courtier*, University of Chicago Press, Chicago.

Bohnet, I., and Zeckhauser, R. (2004) "Trust, Risk and Betrayal," *Journal of Economic Behavior and Organization* 55, 467–484.

Brody, B., and Grandy, R. (eds.) (1971) *Readings in the Philosophy of Science*, Prentice Hall, Englewood Cliffs, NJ.

Burtt, E. (1954) *The Metaphysical Foundations of Modern Science*, Doubleday, New York.

Carroll, J. (1994) *Laws of Nature*, Cambridge University Press, Cambridge.

Cartwright, N. (1983) *How the Laws of Physics Lie*, Oxford University Press, Oxford.

Clagett, M. (ed.) (1969) *Critical Problems in the History of Science*, University of Wisconsin Press, Madison.

Cohen, I. (1985) *The Birth of a New Physics*, W. W. Norton, New York.

Copernicus, N. (1995) *On the Revolution of Heavenly Spheres*, translated by C. Wallis, Prometheus Books, Buffalo, NY.

Curd, M., and Cover, J. (eds.) (1998) *Philosophy of Science: The Central Issues*, W. W. Norton, New York.

Cushing, J. (1998) *Philosophical Concepts in Physics: The Historical Relation between Philosophy and Scientific Theories*, Cambridge University Press, Cambridge.

D'Abro, A. (1950) *The Evolution of Scientific Thought*, Dover Publications, New York.

Darwin, C. (1964) *On the Origin of Species by Means of Natural Selection*, Harvard University Press, Cambridge, MA.

Dawkins, R. (1976) *The Selfish Gene*, Oxford University Press, New York.

Dawkins, R. (2006) *The God Delusion*, Houghton Mifflin, Boston.

Dennett, D. (1995) *Darwin's Dangerous Idea: Evolution and the Meanings of Life*, Simon and Schuster, New York.

Dennett, D. (2006) *Breaking the Spell: Religion as a Natural Phenomenon*, Penguin Books, New York.

Descartes, R. (1931) *The Philosophical Works of Descartes*, volume 1, translated by E. Haldane and G. Ross, Cambridge University Press, Cambridge.

Descartes, R. (1960) *Meditations on First Philosophy*, translated by L. LaFleur, Prentice Hall, Englewood Cliffs, NJ.

Deschanel, P. (1885) *Elementary Treatise on Natural Philosophy*, eighth edition, translated by J. Everett, Blackie and Son, London.

Desmond, A., and Moore, J. (1991) *Darwin: The Life of a Tormented Evolutionist*, W. W. Norton, New York.

Dijksterhuis, E. (1961) *The Mechanization of the World Picture*, translated by C. Dikshoorn, Oxford University Press, London.

Dobzhansky, T. (1973) "Nothing in Biology Makes Sense Except in the Light of Evolution," *American Biology Teacher* 35, 125–129.

Dreyer, J. (1953) *A History of Astronomy from Thales to Kepler*, Dover Publications, New York.

Duhem, P. (1954) *The Aim and Structure of Physical Theory [1906]*, translated by P. Wiener, Princeton University Press, Princeton, NJ.

Earman, J., Glymour, C., and Mitchell, S. (eds.) (2003) *Ceteris Paribus Laws*, Springer Verlag, Berlin.

Einstein, A. (1905) "On the Electrodynamics of Moving Bodies," *Annalen der Physik* 17.

Einstein, A. (1916) "The Foundations of the General Theory of Relativity," *Annalen der Physik* 49.

Einstein, A. (1920) *Relativity: The Special and General Theory*, Henry Holt, New York.

Einstein, A., Podolsky, B., and Rosen, N. (1935) "Can Quantum-Mechanical Description of Physical Reality be Considered Complete?" *Physical Review* 47, 777–780.

Everitt, C. (1975) *James Clerk Maxwell: Physicist and Natural Philosopher*, Charles Scribner's Sons, New York.

Fantoli, A. (1996) *Galileo: For Copernicanism and for the Church*, translated by G. Coyne, University of Notre Dame Press, Notre Dame, IN.

Feldman, F. (1986) *A Cartesian Introduction to Philosophy*, McGraw-Hill, New York.

Fisher, R. A. (1999) *The Genetical Theory of Natural Selection*, Oxford University Press, Oxford.

Fox Keller, E. (1983) *A Feeling for the Organism: The Life and Work of Barbara McClintock*, Freeman Publishers, San Francisco.

French, P., Uehling, T., and Wettstein, H. (eds.) (1988) *Realism and Antirealism*. Midwest Studies in Philosophy 12. University of Minnesota Press, Minneapolis.

Gale, G. (1979) *Theory of Science: An Introduction to the History, Logic, and Philosophy of Science*, McGraw-Hill, New York.

Galileo (1957) *Discoveries and Opinions of Galileo, including The Starry Messenger*, translated by S. Drake, Anchor Books, New York.

Galileo (2001) *Dialogue concerning the Two Chief World Systems*, translated by S. Drake, Modern Library, New York.

Gierre, R. (1999) *Science without Laws*, University of Chicago Press, Chicago.

Gingerich, O. (1993) *The Eye of Heaven: Ptolemy, Copernicus, Kepler*, Springer Verlag, Heidelberg.

Gintis, H., Bowles, S., Boyd, R., and Fehr, E. (2004) "Explaining Altruistic Behavior in Humans," *Evolution and Human Behavior* 24, 153–172.

Goodman, N. (1972) *Problems and Projects*, Bobbs-Merrill, Indianapolis.

Goodman, N. (1983) *Fact, Fiction, and Forecast*, Harvard University Press, Cambridge, MA.

Greene, J. (1969) "Biology and Social Theory in the Nineteenth Century: Auguste Comte and Herbert Spencer," in M. Clagett (ed.), *Critical Problems in the History of Science*, University of Wisconsin Press, Madison.

Harding, S. (1986) *The Science Question in Feminism*, Cornell University Press, Ithaca, NY.

Harris, S. (2004) *The End of Faith: Religion, Terror, and the Future of Reason*, W. W. Norton, New York.

Harris, S. (2007) *Letter to a Christian Nation*, Knopf Publishers, New York.

Hartl, D. (1981) *A Primer of Population Genetics*, Sinauer Associates, Sunderland, MA.

Haught, J. (2001) *Responses to 101 Questions on God and Evolution*, Paulist Press, New York.

Haught, J. (2008a) *God After Darwin: A Theology of Evolution*, Westview Press, Boulder, CO.

Haught, J. (2008b) *God and the New Atheism: A Critical Response to Dawkins, Harris, and Hitchens*, Westminster John Knox Press, Louisville, KY.

Heinlein, R. (1990) *Job: A Comedy of Justice*, Ballantine Books, New York.

Hempel, C., and Oppenheim, P. (1948) "Studies in the Logic of Explanation," *Philosophy of Science* 15, 135–175.

Hempel, G. (1965) *Aspects of Scientific Explanation*, Macmillan Publishing, New York.

Herbert, N. (1985) *Quantum Reality: Beyond the New Physics*, Doubleday, New York.

Hesse, M. (1966) *Models and Analogies in Science*, University of Notre Dame Press, Notre Dame, IN.

Hitchens, C. (2007) *God is Not Great: How Religion Poisons Everything*, Hachette, New York.

Hughes, R. (1989) *The Structure and Interpretation of Quantum Mechanics*, Harvard University Press, Cambridge, MA.

Hull, D., and Ruse, M. (eds.) (1998) *The Philosophy of Biology*, Oxford University Press, Oxford.

Hume, D. (1992) *Treatise of Human Nature* [1739], Prometheus Books, Buffalo, NY.

Hume, D. (1998) *Dialogues Concerning Natural Religion* [1779], Hackett Publishing, Indianapolis.

Jones, R. (1991) "Realism about What?" *Philosophy of Science* 58, 185–202.

Kepler, J. (1995) *Epitome of Copernican Astronomy and Harmonies of the World*, translated by C. Wallis, Prometheus Books, Buffalo, NY.

Kirkham, R. (1992) *Theories of Truth*, MIT Press, Cambridge, MA.

Klee, R. (1997) *Introduction to the Philosophy of Science*, Oxford University Press, Oxford.

Klee, R. (1999) *Scientific Inquiry: Readings in the Philosophy of Science*, Oxford University Press, Oxford.

Klemke, E., Hollinger, R., and Kline, A. (eds.) (1988) *Introductory Readings in the Philosophy of Science*, Prometheus Books, Buffalo, NY.

Kosfeld, M., Heinrichs, M., Zak, P., Fischbacher, U., and Fehr, E. (2005) "Oxytocin Increases Trust in Humans," *Nature* 435, 673–676.

Kosso, P. (1998) *Appearance and Reality: An Introduction to the Philosophy of Physics*, Oxford University Press, New York.

Kragh, H. (1999) *Quantum Generations: A History of Physics in the Twentieth Century*, Princeton University Press, Princeton, NJ.

Kuhn, T. (1957) *The Copernican Revolution: Planetary Astronomy in the Development of Western Thought*, Harvard University Press, Cambridge, MA.

Kuhn, T. (1962) *The Structure of Scientific Revolutions*, University of Chicago Press, Chicago.

Lange, M. (2000) *Natural Laws in Scientific Practice*, Oxford University Press, New York.

Lange, M. (2002a) *An Introduction to the Philosophy of Physics: Locality, Fields, Energy, and Mass*, Blackwell Publishers, Oxford.

Lange, M. (2002b) "Who's Afraid of Ceteris-Paribus Laws? Or: How I Learned to Stop Worrying and Love Them," *Erkenntnis* 57, 407–423.

Laymon, R. (1984) "The Path from Data to Theory," in J. Leplin, *Scientific Realism*, University of California Press, Berkeley, pp. 108–123.

Leplin, J. (ed.) (1984) *Scientific Realism*, University of California Press, Berkeley.

Lewis, D. (1973) *Counterfactuals*, Harvard University Press, Cambridge, MA.

Lindberg, D. (ed.) (1978) *Science in the Middle Ages*, University of Chicago Press, Chicago.

Lindberg, D. (1992) *The Beginnings of Western Science: The European Scientific Tradition in Philosophical, Religious, and Institutional Context, 600 BC to AD 1450*, University of Chicago Press, Chicago.

Lloyd, G. (1970) *Early Greek Science: Thales to Aristotle*, W. W. Norton, New York.

Lloyd, G. (1973) *Greek Science After Aristotle*, W. W. Norton, New York.

Losse, J. (1972) *A Historical Introduction to the Philosophy of Science*, Oxford University Press, Oxford.

Machamer, P. (ed.) (1998) *The Cambridge Companion to Galileo*, Cambridge University Press, Cambridge.

Mason, S. (1962) *A History of the Sciences*, Macmillan Publishing, New York.

Matthews, M. (ed.) (1989) *The Scientific Background to Modern Philosophy*, Hackett Publishing, Indianapolis.

Maudlin, T. (1994) *Quantum Non-Locality and Relativity*, Blackwell Publishers, Oxford.

Mayr, E. (1982) *The Growth of Biological Thought*, Harvard University Press, Cambridge, MA.

Mecklenburg, K. (2010) "Retinophilin is a Light-Regulated Phosphoprotein Required to Suppress Photoreceptor Dark Noise in *Drosophila*," *Journal of Neuroscience* 30(4), 1238–1249.

Mermin, D. (1968) *Space and Time in Special Relativity*, McGraw-Hill, New York.

Mermin, D. (2005) *It's About Time: Understanding Einstein's Relativity*, Princeton University Press, Princeton, NJ.

Miller, K. (1999) *Finding Darwin's God: A Scientist's Search for Common Ground between God and Evolution*, Harper Press, New York.

Miller, K. (2008) *Only a Theory: Evolution and the Battle for America's Soul*, Viking Press, New York.

Mooney, C. (1996) *Theology and Scientific Knowledge*, University of Notre Dame Press, Notre Dame, IN.

Moore, G. E. (1962) *Principia Ethica*, Cambridge University Press, Cambridge.

Munitz, M. (ed.) (1957) *Theories of the Universe: From Babylonian Myth to Modern Science*, Free Press, New York.

Newton, I. (1999) *The Principia: Mathematical Principles of Natural Philosophy*, translated by B. I. Cohen and A. Whitman, University of California Press, Berkeley.

Olby, R. (1974) *The Path to the Double Helix*, University of Washington Press, Seattle.

Park, R. (2001) *Voodoo Science: The Road from Foolishness to Fraud*, Oxford University Press, Oxford.

Pine, R. (1989) *Science and the Human Prospect*, Wadsworth Publishing, Belmont, CA. At http://home.honolulu.hawaii.edu/~pine/book1-2.html (accessed Apr. 26, 2010).

Popper, K. (1992) *Conjectures and Refutations: The Growth of Scientific Knowledge*, Routledge, London.

Provine, W. (1971) *Origins of Theoretical Population Genetics*, University of Chicago Press, Chicago.

Ptolemy, C. (1998) *Ptolemy's Almagest*, translated by G. Toomer, Princeton University Press, Princeton, NJ.

Pyenson, L., and Sheets-Pyenson, S. (1999) *Servants of Nature: A History of Scientific Institutions, Enterprises, and Sensibilities*, W. W. Norton, New York.

Quammen, D. (2006) *The Reluctant Mr. Darwin: An Intimate Portrait of Charles Darwin and the Making of His Theory of Evolution*, W. W. Norton, New York.

Quine, W. (1964) *Word and Object*, MIT Press, Cambridge, MA.

Quine, W. (1969) *Ontological Relativity and Other Essays*, Columbia University Press, New York.

Quine, W. (1980) *From a Logical Point of View*, second edition, Harvard University Press, Cambridge, MA.

Robinson, T. (1995) *Aristotle in Outline*, Hackett Publishing, Indianapolis.

Ruse, M. (1998) *Taking Darwin Seriously*, Prometheus Books, Amherst, NY.

Ruse, M. (ed.) (2009) *Philosophy After Darwin*, Princeton University Press, Princeton, NJ.

Salmon, W. (1998) *Causality and Explanation*, Oxford University Press, New York.

Santillana, G. (1955) *The Crime of Galileo*, University of Chicago Press, Chicago.

Sayre, A. (1975) *Rosalind Franklin and DNA*, W. W. Norton, New York.

Silver, B. (1998) *The Ascent of Science*, Oxford University Press, New York.

Sobel, D. (2000) *Galileo's Daughter*, Penguin Books, New York.

Sober, E., and Wilson, D. S. (1998) *Unto Others: The Evolution and Psychology of Unselfish Behavior*, Harvard University Press, Cambridge, MA.

Toulmin, S., and Goodfield, J. (1961) *The Fabric of the Heavens: The Development of Astronomy and Dynamics*, Harper and Row, New York.

Toulmin, S., and Goodfield, J. (1962) *The Architecture of Matter*, Harper and Row, New York.

Watson, J. D., and Crick, F. H. (1953). "Molecular Structure of Nucleic Acids: A Structure for Deoxyribose Nucleic Acid," *Nature* 171, 737–738.

Weinberg, S. (1992) *Dreams of a Final Theory*, Pantheon Books, New York.

Whitehead, A. N. (1978) *Process and Reality*, Free Press, New York.

Williams, G. (1966) *Adaptation and Natural Selection*, Princeton University Press, Princeton, NJ.

Wilson, E. O. (1978) *On Human Nature*, Harvard University Press, Cambridge, MA.

Wilson, J. W. (1969) "Biology Attains Maturity in the Nineteenth Century," in M. Clagett (ed.), *Critical Problems in the History of Science*, University of Wisconsin Press, Madison.

Index